NICOLAS BOURBAKI

ELEMENTS OF MATHEMATICS

General Topology

Chapters 5–10

Springer-Verlag
Berlin Heidelberg New York
London Paris Tokyo

Originally published as
ÉLÉMENTS DE MATHÉMATIQUE,
TOPOLOGIE GÉNÉRALE 5–10

Mathematics Subject Classification (1991): 54XX

Distribution rights worldwide:
Springer-Verlag Berlin Heidelberg New York London Paris Tokyo

ISBN 3-540-64563-2 Springer-Verlag Berlin Heidelberg New York
ISBN 3-540-19372-3 2nd printing Springer-Verlag Berlin Heidelberg New York

Softcover edition of the 2nd printing 1989

Library of Congress Cataloging-in-Publication Data
Bourbaki, Nicolas. [Topologie générale. English] General topology / Nicolas Bourbaki. p. cm.-(Elements of mathematics / Nicolas Bourbaki) Translation of: Topologie générale. Includes index.
ISBN 0-387-19372-3 (U.S.)
1. Topology. I. Title. II. Series: Bourbaki, Nicolas. Éléments de mathématique. English.
QA611.B65913 1988 514-dc 19 88-31202

Printed in Germany
SPIN 10681874
41/3143-5 4 3 2 1 0
Printed on acid-free paper

ADVICE TO THE READER

1. Examples have been frequently inserted in the text, which refer to facts the reader may already know but which have not yet been discussed in the series. Such examples are always placed between two asterisks: * ... *. Most readers will undoubtedly find that these examples will help them to understand the text, and will prefer not to leave them out, even at a first reading. Their omission would of course have no disadvantage, from a purely logical point of view.

2. This series is divided into volumes (here called " Books "). The first six Books are numbered and, in general, every statement in the text assumes as known only those results which have already been discussed in the preceding volumes. This rule holds good within each Book, but for convenience of exposition these Books are no longer arranged in a consecutive order. At the beginning of each of these Books (or of these chapters), the reader will find a precise indication of its logical relationship to the other Books and he will thus be able to satisfy himself of the absence of any vicious circle.

3. The logical framework of each chapter consists of the *definitions*, the *axioms*, and the *theorems* of the chapter. These are the parts that have mainly to be borne in mind for subsequent use. Less important results and those which can easily be deduced from the theorems are labelled as "propositions," "lemmas", "corollaries", "remarks", etc. Those which may be omitted at a first reading are printed in small type. A commentary on a particularly important theorem appears occasionally under the name of "scholium".

To avoid tedious repetitions it is sometimes convenient to introduce notations or abbreviations which are in force only within a certain chapter or a certain section of a chapter (for example, in a chapter which is concerned only with commutative rings, the word "ring" would always signify "commutative ring"). Such conventions are always explicitly mentioned, generally at the beginning of the chapter in which they occur.

4. Some passages in the text are designed to forewarn the reader against serious errors. These passages are signposted in the margin with the sign

☡ ("dangerous bend").

5. The Exercises are designed both to enable the reader to satisfy himself that he has digested the text and to bring to his notice results which have no place in the text but which are nonetheless of interest. The most difficult exercises bear the sign ¶.

6. In general, we have adhered to the commonly accepted terminology, except where there appeared to be good reasons for deviating from it.

7. We have made a particular effort to always use rigorously correct language, without sacrificing simplicity. As far as possible we have drawn attention in the text to *abuses of language*, without which any mathematical text runs the risk of pedantry, not to say unreadability.

8. Since in principle the text consists of the dogmatic exposition of a theory, it contains in general no references to the literature. Bibliographical references are gathered together in *Historical Notes*, usually at the end of each chapter. These notes also contain indications, where appropriate, of the unsolved problems of the theory.

The bibliography which follows each historical note contains in general only those books and original memoirs which have been of the greatest importance in the evolution of the theory under discussion. It makes no sort of pretence to completeness; in particular, references which serve only to determine questions of priority are almost always omitted.

As to the exercises, we have not thought it worthwhile in general to indicate their origins, since they have been taken from many different sources (original papers, textbooks, collections of exercises).

9. References to a part of this series are given as follows:

a) If reference is made to theorems, axioms, or definitions presented *in the same section*, they are quoted by their number.

b) If they occur *in another section of the same chapter*, this section is also quoted in the reference.

c) If they occur *in another chapter in the same Book*, the chapter and section are quoted.

d) If they occur *in another Book*, this Book is first quoted by its title.

The *Summaries of Results* are quoted by the letter R: thus *Set Theory*, R signifies " *Summary of Results of the Theory of Sets* ".

CONTENTS

CHAPTER V

One-parameter groups

1. SUBGROUPS AND QUOTIENT GROUPS OF **R**

1. CLOSED SUBGROUPS OF R

PROPOSITION 1. *Every closed subgroup of the additive group* **R**, *other than* **R** *and* $\{0\}$, *is a discrete group of the form* $a.\mathbf{Z}$, *where* $a > 0$ (in other words, it consists of the integer multiples of a).

We begin by showing that every non-discrete subgroup of **R** is dense in **R**. If a subgroup G of **R** is not discrete, then for every $\varepsilon > 0$ there is a point $x \neq 0$ in G which belongs to the interval $[-\varepsilon, +\varepsilon]$; since all integer multiples of x belong to G, every interval of length $> \varepsilon$ contains an element of G, and therefore G is dense in **R**.

Every closed subgroup of **R** other than **R** itself is therefore discrete. It remains to show that every discrete subgroup G of **R** other than $\{0\}$ is of the form $a.\mathbf{Z}$, where $a > 0$. Now the relation $-G = G$ shows that the set H of elements > 0 in G is not empty; if $b \in H$, the intersection of the interval $[0, b]$ and G is *compact and discrete*, and is therefore *finite*. Let a be the smallest element of H contained in $[0, b]$, and for every $x \in G$ put $m = [x/a]$, the integer part of x/a; then we have $x - ma \in G$ and $0 \leqslant x - ma < a$. By the definition of a it follows that $x - ma = 0$ and therefore $G = a.\mathbf{Z}$.

2. QUOTIENT GROUPS OF R

Every *Hausdorff* quotient group of **R** is of the form **R**/H, where H is a *closed* subgroup of **R** (Chapter III, § 2, no. 6, Proposition 18); hence, by Proposition 1 of no. 1:

PROPOSITION 2. *The Hausdorff quotient groups of* **R**, *other than* $\{0\}$, *are the groups* $\mathbf{R}/a\mathbf{Z}$ $(a \geqslant 0)$.

If a and b are > 0, the automorphism $x \to b/a\,x$ of **R** transforms $a\mathbf{Z}$ into $b\mathbf{Z}$, and therefore (Chapter III, § 2, no. 8, Remark 3) the quotient groups $\mathbf{R}/a\mathbf{Z}$ and $\mathbf{R}/b\mathbf{Z}$ are isomorphic; in other words:

PROPOSITION 3. *Every Hausdorff quotient group of* **R**, *other than* **R** *and* $\{0\}$, *is isomorphic to the group* $\mathbf{R}/\mathbf{Z}$.

DEFINITION 1. *The topological group* $\mathbf{R}/a\mathbf{Z}$ $(a > 0)$ *is called the additive group of real numbers modulo* a. *The topological group* $\mathbf{R}/\mathbf{Z}$ *is denoted by* **T**. *As a topological space,* **T** *is called the one-dimensional torus* (by abuse of language, the topological *group* **T** is also called the one-dimensional torus).

Remarks. 1) The relation $x \equiv y \pmod{a\mathbf{Z}}$ is usually written $x \equiv y \pmod{a}$, or simply $x \equiv y\ (a)$, and is read " x and y are congruent modulo a "; it means that $x - y$ is an *integer multiple* of a. When a is an integer, the relation induced on **Z** by this relation is precisely congruence modulo a; this justifies the notation.

* 2) As we shall see in Chapter VI, § 2, no. 4, the topological space **T** is homeomorphic to the *circle* $x^2 + y^2 = 1$ in the real number plane $\mathbf{R}^2$; the product space $\mathbf{T}^2$ is homeomorphic to a *torus of revolution* in $\mathbf{R}^3$ (Chapter VII, § 1, Exercise 15). This is the origin of the name " one-dimensional torus " for **T** (in Chapter VII, § 1, we shall call $\mathbf{T}^n$ the n-dimensional torus) *.

PROPOSITION 4. *The torus* **T** *is homoeomorphic to the quotient space of any closed interval of* **R** *of the form* $[a, a+1]$ *obtained by identifying the end-points of this interval; it is compact, connected and locally connected.*

Every $x \in \mathbf{R}$ is congruent (mod 1) to some number in the interval $[a, a+1]$, namely to $x - [x - a]$; hence **T** is the image of this interval under the canonical mapping φ of **R** onto $\mathbf{R}/\mathbf{Z}$, and is therefore compact and connected (Chapter I, § 9, no. 4, Theorem 2, and § 11, no. 2, Proposition 4). On the other hand, two distinct elements of the interval $[a, a+1]$ are congruent (mod 1) only if they are the end-points of the interval; from the compactness of **T** it therefore follows that **T** is homeomorphic to the quotient space of $[a, a+1]$ obtained by identifying the end-points (Chapter I, § 9, no. 4, Theorem 2, Corollary 4, and § 10, no. 4, Proposition 8). Finally, **Z** being a discrete subgroup of **R**, $\mathbf{T} = \mathbf{R}/\mathbf{Z}$ is *locally isomorphic* to **R** (Chapter III, § 2, no. 6, Proposition 19) and in particular is locally connected (this last fact is also a consequence of Chapter I, § 11, no. 6, Proposition 12).

Remark. Note that the canonical mapping φ of $\mathbf{R}$ onto $\mathbf{T} = \mathbf{R}/\mathbf{Z}$, restricted to the half-open interval $[a, a+1[$, is a *continuous bijective* mapping of this interval onto $\mathbf{T}$; the inverse mapping is *continuous* at every point of $\mathbf{T}$ other than $\varphi(a)$, but *discontinuous* at $\varphi(a)$. We shall sometimes identify the space $\mathbf{T}$ with the interval $[a, a+1[$, endowed with the topology which is the inverse image under φ of the topology of $\mathbf{T}$ (Chapter I, § 1, no. 3); this topology is of course not the same as that induced on $[a, a+1[$ by the topology of $\mathbf{R}$.

3. CONTINUOUS HOMOMORPHISMS OF R INTO ITSELF

PROPOSITION 5. *Every continuous homomorphism f of the topological group $\mathbf{R}$ into itself is of the form $x \to ax$, where $a \in \mathbf{R}$; it is an automorphism of $\mathbf{R}$ if $a \neq 0$.*

For every $x \in \mathbf{R}$ and every integer $p \in \mathbf{Z}$, we have $f(px) = pf(x)$; replacing x by $(1/p)\,x$, it follows that

$$f\left(\frac{1}{p}x\right) = \frac{1}{p}f(x) \qquad \text{if } p \neq 0;$$

hence, for all integers p and $q \neq 0$, we have

$$f\left(\frac{p}{q}x\right) = \frac{p}{q}f(x).$$

In other words, $f(rx) = rf(x)$ for all rational numbers r. If now t is any real number, by reason of the continuity of f we have

$$f(tx) = \lim_{r \to t,\, r \in \mathbf{Q}} f(rx) = \lim_{r \to t,\, r \in \mathbf{Q}} rf(x) = \Big(\lim_{r \to t,\, r \in \mathbf{Q}} r\Big).f(x) = tf(x).$$

In particular, if $a = f(1)$ we have $f(t) = at$, and the proposition is proved.

The *group of automorphisms* of the topological group $\mathbf{R}$ is therefore isomorphic to the *multiplicative group* $\mathbf{R}^*$ of non-zero real numbers.

COROLLARY. *Let* G *be a topological group isomorphic to* $\mathbf{R}$. *For each* $a \in$ G *there is exactly one continuous homomorphism* f_a *of* $\mathbf{R}$ *into* G *such that* $f_a(1) = a$, *and this homomorphism is an isomorphism of* $\mathbf{R}$ *onto* G *if* a *is not the zero element of* G.

4. LOCAL DEFINITION OF A CONTINUOUS HOMOMORPHISM OF R INTO A TOPOLOGICAL GROUP

If we are given a group G and a subset A of G *which generates* G, it is clear that two homomorphisms f, g of G into a group G′ coincide if they take the same value at every point of A. But the values on A of a homomorphism f of G into G′ cannot in general be taken arbitrarily; if G and G′ are written multiplicatively, these values must satisfy the condition $f(xy) = f(x)f(y)$ for each pair (x, y) such that $x \in A$, $y \in A$ and $xy \in A$. Moreover, this necessary condition is not in general sufficient.

☡ In particular, a *local isomorphism* of a topological group G with a topological group G′ cannot always be extended to a homomorphism (continuous or not) of G into G′. For example, a local isomorphism f of **T** with **R** cannot be extended to a homomorphism of **T** into **R**; for if f is defined on a neighbourhood V of 0, there is an integer $p > 0$ such that the class x (mod **Z**) of $\frac{1}{p}$ belongs to V; since x has order p in **T**, its image under every homomorphism of **T** into **R** is necessarily 0 and therefore distinct from $f(x)$ by hypothesis.

In this respect the topological group **R** enjoys the following property:

PROPOSITION 6. *Let* I *be an interval of* **R** *which contains* 0 *and at least one other point; let* f *be a continuous mapping of* I *into a topological group* G (written multiplicatively), *such that* $f(x + y) = f(x)f(y)$ *for each pair of points* (x, y) *such that* $x \in I, y \in I$ *and* $x + y \in I$. *Then there is a unique continuous homomorphism of* **R** *into* G *which extends* f.

The uniqueness of the extension (if it exists) follows from the preceding remarks, because I generates the group **R**. We have to establish the existence of such an extension.

If n is an integer > 0 and if $x \in I$ and $nx \in I$, we have $f(nx) = (f(x))^n$, by induction on n (since $mx \in I$ for all integers m such that $1 \leqslant m \leqslant n$). Put $J = \bigcup_{n \in N} nI$; J is either the whole line **R**, or one of the intervals $[0, +\infty[$ or $]-\infty, 0]$, according as 0 is or is not an interior point of I. If $x \in J$ we have $x/n \in I$ whenever n is a sufficiently large integer > 0. Let $x \in J$, and let m, n be two integers > 0 such that $x/n \in I$ and $x/m \in I$; then $x/mn \in I$, and therefore

$$f\left(\frac{x}{m}\right) = \left(f\left(\frac{x}{mn}\right)\right)^n \quad \text{and} \quad f\left(\frac{x}{n}\right) = \left(f\left(\frac{x}{mn}\right)\right)^m;$$

in other words, the element $(f(x/n))^n$ of G is the same for all integers $n > 0$ such that $x/n \in I$. Let us denote this element by $f_1(x)$; f_1 is a mapping of J into G, which agrees with f on I and is therefore continuous at the point 0 (with respect to J). Let x, y be two elements of J and let n be a sufficiently large integer > 0 such that $\frac{x}{n} \in \mathrm{I}$, $\frac{y}{n} \in \mathrm{I}$ and $\frac{x+y}{n} \in \mathrm{I}$; then

$$f\left(\frac{x+y}{n}\right) = f\left(\frac{x}{n}\right)f\left(\frac{y}{n}\right) = f\left(\frac{y}{n}\right)f\left(\frac{x}{n}\right),$$

which shows that $f\left(\frac{x}{n}\right)$ and $f\left(\frac{y}{n}\right)$ commute; by the definition of f_1, we have therefore $f_1(x+y) = f_1(x)f_1(y)$. If $\mathrm{J} = \mathbf{R}$, the proof is complete; if not, say $\mathrm{J} = [0, +\infty[$, and for each $x < 0$ define $f_1(x)$ to be $(f_1(-x))^{-1}$. Then the relation $f_1(x+y) = f_1(x)f_1(y)$ remains valid for all $x \in \mathbf{R}$ and all $y \in \mathbf{R}$. This is clear if $x < 0$ and $y < 0$; if $x \geqslant 0$, $y < 0$ and $x + y \geqslant 0$ it follows from $f_1(x) = f_1(x+y)f_1(-y)$; similarly if $x \geqslant 0, y < 0$ and $x + y < 0$, for then we have

$$f_1(-y) = f_1(-x-y)f_1(x);$$

analogous proofs for $x < 0$ and $y \geqslant 0$. We see therefore that f_1 is a homomorphism of $\mathbf{R}$ into G, so that $f_1(0) = e$, the identity element of G; and since f_1 is continuous with respect to J, it has a limit on the right at 0, equal to e; since $f_1(-x) = (f_1(x))^{-1}$, f_1 also has a limit at 0 on the left, equal to e; thus f_1 is continuous at 0, and the proof is complete.

COROLLARY. *Let f be a local isomorphism of* $\mathbf{R}$ *with a topological group* G. *Then there is a unique strict morphism of* $\mathbf{R}$ *onto an open subgroup of* G *which coincides with f at all points of some neighbourhood of* 0.

Let $\bar{f}$ be the continuous homomorphism of $\mathbf{R}$ into G which coincides with f at all points of an open interval I, which contains 0 and is contained in the set on which f is defined; $\bar{f}(\mathbf{R})$ by hypothesis contains a neighbourhood of the identity element of G, hence (Chapter III, § 2, no. 1, Corollary to Proposition 4) is an open subgroup of G; and $\bar{f}$ is a strict morphism of $\mathbf{R}$ onto $\bar{f}(\mathbf{R})$, by Chapter III, § 2, no. 8, Proposition 24.

PROPOSITION 7. *Every connected group* G *which is locally isomorphic to* $\mathbf{R}$ *is isomorphic to either* $\mathbf{R}$ or $\mathbf{T}$.

For a local isomorphism of $\mathbf{R}$ with G extends to a strict morphism of $\mathbf{R}$ onto an open subgroup of G (Corollary to Proposition 6), hence onto G itself since G is connected. Hence G is isomorphic to a quotient group of $\mathbf{R}$; since G is Hausdorff and does not consist of the identity element alone (because it is locally isomorphic to $\mathbf{R}$), it is isomorphic to either $\mathbf{R}$ or $\mathbf{T}$ by Proposition 3 of no. 2.

2. MEASUREMENT OF MAGNITUDES

We have seen (cf. the historical note to Chapter IV) that the problem of the *measurement of magnitudes* is at the origin of the concept of real number; more precisely, the various types of magnitudes which gradually came to be studied, for practical or theoretical purposes, were first considered separately from each other, and the possibility of measuring all types by the same number system had appeared as an experimental fact well before the Greeks had conceived the bold idea of giving a rigorous demonstration. In the axiomatic theory they established, the idea of magnitude was related to a law of composition (" addition " of magnitudes of the same type) and an order-relation (the relation " A is smaller than B ", called the relation of comparison of magnitudes). In what follows we shall examine the same problem, that is to say we shall investigate the conditions that must be imposed on an internal law of composition and an order-relation on a set E in order that it should be *isomorphic* to a subset E′ of **R**, endowed with the structure induced by addition and the relation $\leqslant$ in **R**. As we shall not assume *a priori* that the given law of composition on E is commutative, we shall use the multiplicative notation; but apart from this we shall scarcely depart from the classical arguments.

Let E be a set *linearly ordered* by an order relation written $x \leqslant y$, and suppose that E has a smallest element ω. Let I be a subset of E such that $\omega \in \mathrm{I}$, and such that the relations $x \in \mathrm{I}$ and $y \leqslant x$ imply $y \in \mathrm{I}$; suppose also that we are given a law of composition $(x, y) \to xy$ on E, the product xy being defined for all pairs of elements of I (xy belongs to E, but not necessarily to I). Furthermore, we make the following assumptions:

($\mathrm{GR_I}$) *ω is the identity element* [$\omega x = x\omega = x$ for all $x \in \mathrm{I}$] *and the law of composition is associative* [in the following sense: whenever $x \in \mathrm{I}$, $y \in \mathrm{I}$, $z \in \mathrm{I}$, $xy \in \mathrm{I}$ and $yz \in \mathrm{I}$, then $x(yz) = (xy)z$].

($\mathrm{GR_{II}}$) *The relation $x < y$ between elements of* I *implies, for all $z \in \mathrm{I}$, the relations $xz < yz$ and $zx < zy$.*

($\mathrm{GR_{III}}$) *The set of elements $> \omega$ in* I *is not empty and has no smallest element; and given any elements x, y of* I *such that $x < y$, there exists $z > \omega$ such that $xz \leqslant y$.*

The condition ($\mathrm{GR_{II}}$) implies that inequalities between elements of I may be multiplied term by term: $x < y$ and $x' < y'$ imply $xx' < yy'$ (for $xx' < yx'$ and $yx' < yy'$). In particular we have $y < yx$ for all $x > \omega$ ($x \in \mathrm{I}$, $y \in \mathrm{I}$).

Given a finite sequence $(x_i)_{1 \leqslant i \leqslant p}$ of elements of I, we may define by induction on p the product of this sequence $\prod_{i=1}^{p} x_i$ as being equal to $\left(\prod_{i=1}^{p-1} x_i\right)x_p$, provided that the product $\prod_{i=1}^{p-1} x_i$ is defined and belongs to I: thus if $\prod_{i=1}^{p} x_i$ is defined, each of the products $\prod_{i=1}^{q} x_i$ is defined and belongs to I, for $2 \leqslant q \leqslant p-1$. By taking all the x_i equal to the same element $x \in I$, we see in particular that if x^p is defined, then x^q is defined and belongs to I for $2 \leqslant q \leqslant p-1$. Conventionally we define x^0 to be equal to ω, for all $x \in I$. By (GR_{II}), if $x > \omega$, we have $\omega < x^q < x^p$ for $1 \leqslant q \leqslant p-1$ if x^p is defined; if $x < y$ and if y^p is defined, then we see (by induction on p) that x^p is defined and $x^p < y^p$. On the other hand, the associativity condition (GR_I) implies by induction on n that, if x^{m+n} is defined, then so is $x^m x^n$, and that $x^{m+n} = x^m x^n$. Conversely, by virtue of (GR_I) and (GR_{II}), if $x^m x^n$ is defined and belongs to I, then x^{m+n} is defined and we have $x^{m+n} = x^m x^n$; again this is proved by induction on n, for we have $x^{n-1} \leqslant x^n$, therefore $x^m x^{n-1}$ is defined and belongs to I; by hypothesis, $x^m x^{n-1} = x^{m+n-1} \in I$, hence $(x^{m+n-1})x = x^{m+n}$ is defined and equal to $x^m x^n$ by the previous result. One shows likewise by induction on n that, if x^{mn} is defined, then $(x^m)^n$ is defined and that $x^{mn} = (x^m)^n$; and that conversely if $(x^m)^n$ is defined and belongs to I, then x^{mn} is defined and equal to $(x^m)^n$.

Finally, the axiom (GR_{III}) implies that, for all $x \in I$ such that $x > \omega$, there exists $y > \omega$ such that $y^2 \leqslant x$. For if $x > \omega$ there exists $z > \omega$ such that $z < x$, and then $t > \omega$ such that $zt \leqslant x$; take y to be the smaller of the two elements z, t. By induction on n, we deduce that there exists $u > \omega$ such that $u^{2^n} \leqslant x$.

Let us now introduce the following assumption:

(GR_{IV}) (" Archimedes' axiom "). *For all* $x \in I$ *and* $y \in I$ *such that* $x > \omega$, *there exists an integer* $n > 0$ *such that* x^n *is defined and* $x^n > y$.

If we take E to be a set of real numbers $\geqslant 0$ which contains 0 and arbitrarily small numbers > 0, I to be the intersection of E with an interval of **R** which has 0 as its left-hand end-point and contains at least one other number, the law of composition to be addition of elements of I, and if we suppose that $x + y \in E$ whenever $x \in I$ and $y \in I$ then it is clear that the axioms (GR_I), (GR_{II}), (GR_{III}) and (GR_{IV}) are satisfied (*).

(*) In the sets of " magnitudes " which arise in the experimental sciences, the axioms (GR_I) and (GR_{II}) are in general capable of experimental verification, at any rate approximately. On the other hand, axiom (GR_{III}), which postulates the existence of magnitudes " as small as we please ", clearly cannot be established in the same way; it is a purely *a priori* assumption. As to axiom (GR_{IV}), it can be considered as an " extrapolation " of a fact which can be verified by experiment for magnitudes which are not " too small ".

Conversely:

PROPOSITION 1. *Let* E *be a linearly ordered set with a smallest element* ω; *let* I *be a subset of* E *such that* $\omega \in I$ *and such that the relations* $x \in I, y \leqslant x$ *imply* $y \in I$; *let* $(x, y) \to xy$ *be a law of composition on* E, *defined for* $x \in I$ *and* $y \in I$. *Then, if the axioms* (GR_I), (GR_{II}), (GR_{III}) *and* (GR_{IV}) *are satisfied, there exists a strictly increasing mapping* f *of* I *into the set* $\mathbf{R}_+$ *of real numbers* $\geqslant 0$, *such that*

$$f(xy) = f(x) + f(y)$$

whenever $x \in I, y \in I$ *and* $xy \in I$; *moreover, the intersection of* $f(I)$ *with every interval* $[0, f(b)]$ *of* $\mathbf{R}$ *is dense in this interval, where* b *denotes any element of* I.

Given any two elements x, y of I such that $y \neq \omega$, let us denote by $(x : y)$ the largest integer $n \geqslant 0$ such that y^n is defined and $\leqslant x$ (*); this integer exists by (GR_{IV}); if $(x : y) = p$, then y^{p+1} is defined and $> x$. If $x \in I, y \in I$ and $xy \in I$, we have

$$(x : z) + (y : z) \leqslant (xy : z) \leqslant (x : z) + (y : z) + 1. \tag{1}$$

For let $(x : z) = p$, $(y : z) = q$; then we have $z^p \leqslant x$, $z^q \leqslant y$; since $xy \in I$, $z^p z^q$ is defined and belongs to I, therefore z^{p+q} is defined and $z^{p+q} = z^p z^q \leqslant xy$; moreover, if z^{p+q+2} is defined, we have $z^{p+q+2} > xy$ because $z^{p+1} > x$ and $z^{q+1} > y$.

Next we establish the inequalities

$$\begin{cases} (x : y)\,(y : z) \leqslant (x : z), \\ ((x : y) + 1)\,((y : z) + 1) \geqslant (x : z) + 1. \end{cases} \tag{2}$$

Let $(x : y) = p$ and $(y : z) = q$; then $y^p \leqslant x$ and $z^q \leqslant y$, so that $(z^q)^p$ is defined and $\leqslant x$; it belongs therefore to I; consequently z^{pq} is defined and we have $z^{pq} = (z^q)^p \leqslant x$, from which the first inequality follows. On the other hand, if $z^{(p+1)(q+1)}$ is defined, we have $z^{(p+1)(q+1)} > x$, because $y^{p+1} > x$ and $z^{q+1} > y$; hence the second inequality.

Let $\mathfrak{F}$ denote the filter of sections of the ordered set of elements $> \omega$ in I, with respect to the relation $\geqslant$; the intervals $]\omega, z]$, where z runs through the set of all elements $> \omega$, form a base of $\mathfrak{F}$. Given two elements a and x of I such that $a > \omega$, we shall show that the ratio $\dfrac{(x : z)}{(a : z)}$, which is defined for $z \leqslant a$ and is a rational number > 0,

(*) When $E = I$ is the set of natural integers, the law of composition being addition, $(x : y)$ is the integral part of x/y.

is a function of z which has a *limit* with respect to $\mathfrak{F}$. This is obvious if $x = \omega$, for then $(x : z) = 0$ for all z. If $x > \omega$, we shall show that the image $\mathfrak{G}$ of $\mathfrak{F}$ under the mapping $z \to \frac{(x : z)}{(a : z)}$ (restricted to the set of those $z > \omega$ which are $\leqslant x$ and $\leqslant a$) is a Cauchy filter base for the uniform structure of the *multiplicative* group $\mathbf{R}_+^*$, and therefore converges to a real number > 0. Note first that, $u > \omega$ being given, $(u : z)$ has limit $+\infty$ with respect to $\mathfrak{F}$: for there exists $z > \omega$ such that $z^{2^n} \leqslant u$, so that $(u : z) \geqslant 2^n > n$. Now take a number $\varepsilon > 0$ arbitrarily; there exists $t > \omega$ such that $(x : t) \geqslant 1/\varepsilon$ and $(a : t) \geqslant 1/\varepsilon$. Consider the double inequality

$$\frac{(x : t)}{(a : t) + 1} \cdot \frac{(t : z)}{(t : z) + 1} \leqslant \frac{(x : z)}{(a : z)} \leqslant \frac{(x : t) + 1}{(a : t)} \cdot \frac{(t : z) + 1}{(t : z)},$$

which follows immediately from the inequalities (2). There exists $z_0 > \omega$ such that $z \leqslant z_0$ implies $(t : z) \geqslant 1/\varepsilon$, so that

$$\frac{1}{(1 + \varepsilon)^2} \frac{(x : t)}{(a : t)} \leqslant \frac{(x : z)}{(a : z)} \leqslant (1 + \varepsilon)^2 \frac{(x : t)}{(a : t)},$$

which shows that $\mathfrak{G}$ is a Cauchy filter base for the multiplicative uniformity.

Fix once and for all the element $a > \omega$ (the "unit of measure") and for each $x \in I$ put

$$f(x) = \lim_{\mathfrak{F}} \frac{(x : z)}{(a : z)}.$$

From what has already been proved, we have $f(\omega) = 0$, $f(x) > 0$ for $x > \omega$, and $f(a) = 1$. If we divide the inequality (1) throughout by $(a : z)$ and pass to the limit with respect to $\mathfrak{F}$, we see that

$$f(xy) = f(x) + f(y)$$

whenever $x \in I, y \in I$ and $xy \in I$. Likewise, the relation $x \leqslant y$ implies $(x : z) \leqslant (y : z)$, whence by dividing by $(a : z)$ and passing to the limit we have $f(x) \leqslant f(y)$, so that f is *increasing* on I. We deduce that f is *strictly increasing* on I; for if $x < y$, there exists $z > \omega$ such that $xz \leqslant y$, whence $f(xz) \leqslant f(y)$; and since $xz \in I$,

$$f(x) + f(z) = f(xz) \leqslant f(y);$$

but $f(z) > 0$, so that indeed $f(x) < f(y)$.

Finally, if $b \in \mathrm{I}$, the intersection of $f(\mathrm{I})$ and the interval $[0, f(b)]$ of $\mathbf{R}$ is dense in this interval. For if n is any integer > 0, there exists $x > \omega$ such that $f(x) \leqslant 2^{-n}$ (take x such that $x^{2^n} \leqslant a$); if p is the smallest integer such that $x^{p+1} > b$, we have $(p+1) f(x) > f(b)$ and $qf(x) \leqslant f(b)$ for $1 \leqslant q \leqslant p$; therefore every interval contained in $[0, f(b)]$ and of length $> 2^{-n}$ contains at least one point of the form $qf(x) = f(x^q) \in f(\mathrm{I})$. The proof of Proposition 1 is therefore complete.

Remarks. 1) The relations $x \in \mathrm{I}, y \in \mathrm{I}, xy \in \mathrm{I}, yx \in \mathrm{I}$ imply

$$f(xy) = f(x) + f(y) = f(yx),$$

and hence $yx = xy$ since f is strictly increasing; in other words, the law induced by the law of composition of E on an interval $[0, b]$ suitably chosen (e.g., such that $b^2 \leqslant a$) is *commutative*.

2) Every mapping g of I into $\mathbf{R}_+$ which satisfies the same conditions as f is of the form $x \to \lambda f(x)$ where $\lambda > 0$. For if $\lambda = g(a) > 0$, the relations $z^p \leqslant x \leqslant z^{p+1}$, $z^q \leqslant a \leqslant z^{q+1}$ imply, by hypothesis,

$$pg(z) \leqslant g(x) \leqslant (p+1)g(z), \qquad qg(z) \leqslant g(a) \leqslant (q+1)g(z),$$

whence

$$\lambda \frac{(x:z)}{(q:z)+1} \leqslant g(x) \leqslant \lambda \frac{(x:z)+1}{(q:z)},$$

and therefore, passing to the limit with respect to $\mathfrak{F}$, we have $g(x) = \lambda f(x)$.

Let us seek conditions under which $f(\mathrm{I})$ is an *interval* of $\mathbf{R}_+$. Clearly the following two conditions are necessary:

($\mathrm{GR}_{\mathrm{III}a}$) *The set of elements* $> \omega$ *in* I *is not empty and has no smallest element, and given any two elements* x, y, *of* I *such that* $x < y$, *there exists* $z \in \mathrm{I}$ *such that* $xz = y$ ("subtraction" of magnitudes).

($\mathrm{GR}_{\mathrm{IV}a}$) *Every increasing sequence of elements of* I, *which is bounded above by an element of* I *has a least upper bound in* I.

We shall show that these conditions are also sufficient, and moreover that they allow us to dispense with axiom ($\mathrm{GR}_{\mathrm{IV}}$) (Archimedes' axiom). To be precise, we shall prove the following proposition:

PROPOSITION 2. *If a linearly ordered set* E *and a subset* I *of* E *satisfy the axioms* (GR_{I}), ($\mathrm{GR}_{\mathrm{II}}$), ($\mathrm{GR}_{\mathrm{III}\,a}$) *and* ($\mathrm{GR}_{\mathrm{IV}\,a}$), *there exists a strictly increasing mapping* f *of* I *onto an interval of* $\mathbf{R}$, *with* 0 *as its left-hand end-point, such that* $f(\omega) = 0$ *and* $f(xy) = f(x) + f(y)$ *whenever* x, y *and* xy *belong to* I.

Let us first show that axiom ($\mathrm{GR}_{\mathrm{IV}}$) is satisfied. We argue by contradiction: suppose that there exist $x, y \in \mathrm{I}$ such that $x > \omega$, x^n is defined and $x^n \leqslant y$ for all integers $n > 0$. The increasing sequence (x^n) has a least upper bound $b \in \mathrm{I}$ by ($\mathrm{GR}_{\mathrm{IV}\,a}$). Since $x < b$, there exists $c \in \mathrm{I}$ such that $xc = b$ by ($\mathrm{GR}_{\mathrm{III}\,a}$), and we have $c < b$ since $x > \omega$. Now, for every n, we have $x^{n+1} \leqslant b = xc$, whence $x^n \leqslant c$ by ($\mathrm{GR}_{\mathrm{II}}$): the upper bound b of the x^n is therefore $\leqslant c$, which is a contradiction.

We are therefore in a position to apply Proposition 1. It remains to show that, if $\gamma = f(c)$ $(c > \omega)$ is any element of $f(\mathrm{I})$, and if β is any real number such that $0 < \beta < \gamma$, there exists $b \in \mathrm{I}$ such that $f(b) = \beta$ (Chapter IV, § 2, no. 4, Proposition 1). Since the intersection of $f(\mathrm{I})$ and $[0, \gamma]$ is dense in $[0, \gamma]$, there exists an increasing sequence (x_n) of elements of I such that $f(x_n)$ has β as limit. Let b be the least upper bound of the sequence (x_n) in I; we have $f(b) \geqslant f(x_n)$ for all n, hence $f(b) \geqslant \beta$; but $f(b) > \beta$ is impossible, otherwise there would exist $y \in \mathrm{I}$ such that $\beta < f(y) < f(b)$, and since β is the least upper bound of the sequence $(f(x_n))$, we should have $f(x_n) < f(y) < f(b)$ for all n, whence $x_n < y < b$ for all n, which is absurd. Hence $f(b) = \beta$, and Proposition 2 is therefore proved.

Remark. When $\mathrm{I} = \mathrm{E}$, the image $f(\mathrm{I}) = f(\mathrm{E})$ is the whole of $\mathbf{R}_+$, for if $b > \omega$, then b^n is defined for all n, and therefore $n.f(b)$ belongs to $f(\mathrm{E})$ for all n; this implies that $f(\mathrm{E})$ is not bounded above, because $f(b) > 0$.

3. TOPOLOGICAL CHARACTERIZATION OF THE GROUPS **R** AND **T**

THEOREM 1. *A topological group* G *in which there exists a neighbourhood of the identity element homeomorphic to an open interval of* **R** *is locally isomorphic to* **R**.

The significance of this theorem is that it allows us to deduce, from a purely topological property of a group G, a property of the *group structure* of G.

We are concerned here with a phenomenon which is peculiar to the group **R** and has no analogue for the groups $\mathbf{R}^n$ when $n > 1$ (cf. Chapter VIII, § 1, no. 4). Groups locally isomorphic to **R** are sometimes called *one-parameter groups*.

To prove Theorem 1 we shall reduce it to Proposition 2 of § 2. By hypothesis, there is a homeomorphism φ of an open neighbourhood U of the identity element e of G onto an open interval in **R**. By means of the inverse of the mapping φ we can transport to U the linear order structure of the interval $\varphi(\mathrm{U})$; the topology of U (induced by that of G) then has a base consisting of all the open intervals of U (Chapter

IV, § 1, no. 4, Proposition 5). We can find a *symmetric* neighbourhood V of e such that $V.V \subset U$ and such that V is an open interval; for there exists an open interval V' containing e such that $V'.V' \subset U \cap U^{-1}$, $V'.V'^{-1} \subset U$ and $V'^{-1}.V' \subset U$; taking $V = V' \cup V'^{-1}$, V is open and symmetric, satisfies $V.V \subset U$ and is connected, hence is an interval (Chapter IV, § 2, no. 5, Theorem 4).

We show that, if x, y, z belong to V, the relation $x < y$ implies $xz < yz$ and $zx < zy$. Indeed, the functions $f_1(z) = \varphi(yz) - \varphi(xz)$ and $f_2(z) = \varphi(zy) - \varphi(zx)$ are continuous on V; they are > 0 for $z = e$ and do not vanish in V [e.g., if we had $\varphi(yz) = \varphi(xz)$, we should have $yz = xz$ and therefore $y = x$]. Since $f_1(V)$ and $f_2(V)$ are connected (Chapter I, § 11, no. 2, Proposition 4) and are therefore intervals in **R** (Chapter IV, § 2, no. 5, Theorem 4), and since these intervals each contain a number > 0 and do not contain 0, they are contained in $\mathbf{R}_+^*$: that is, we have $f_1(z) > 0$ and $f_2(z) > 0$ for all $z \in V$.

If x and y are two elements of V such that $x \geqslant e$ and $y \geqslant e$, then in particular we have $xy \geqslant e$. Let E denote the (linearly ordered) set of elements of U which are $\geqslant e$, and let I denote the set of elements of V which are $\geqslant e$; then the axioms (GR_I), (GR_{II}), (GR_{IIIa}) and (GR_{IVa}) of § 2 are satisfied (taking ω to be the element e, and the law of composition to be that of the group G). This is clear for (GR_I), (GR_{II}) and (GR_{IVa}), from what precedes. As to (GR_{IIIa}), it is enough to remark that, if $e < x < y$ $(x \in V, y \in V)$, we have $x^{-1} \in V$, hence $x^{-1} < e < x^{-1}y$ and $x^{-1}y < y$; consequently $z = x^{-1}y$ belongs to I and we have $xz = y$. By Proposition 2 of § 2 there exists therefore a strictly increasing mapping f of I onto an interval of **R**, with left-hand end-point 0, such that $f(e) = 0$ and $f(xy) = f(x) + f(y)$ whenever x, y and xy belong to I (which will be the case whenever x and y belong to $W \cap I$, W being a neighbourhood of e such that $W.W \subset V$).

For every element $x \in V$ which does not belong to I we have $x < e$, hence $x^{-1} > e$; consequently we can extend f to a *strictly increasing* mapping $\bar{f}$ of V *onto* an interval of **R** by putting $\bar{f}(x) = -f(x^{-1})$ for all $x < e$ in V. The inverse image under $\bar{f}$ of an open interval contained in $\bar{f}(V)$ is an open interval of V, so that $\bar{f}$ is continuous on V; conversely, the image under $\bar{f}$ of an open interval of V is an open interval of $\bar{f}(V)$, and therefore $\bar{f}$ is a *homeomorphism* of V onto a neighbourhood of 0 in the group **R**. On the other hand, it is easily checked (as in Proposition 6 of § 1, no. 4, by considering the various possible cases) that we have $\bar{f}(xy) = \bar{f}(x) + \bar{f}(y)$ whenever x, y and xy all belong to V; and we therefore concude that $\bar{f}$, restricted to a suitable neighbourhood of e in G, is a local isomorphism of G with **R** (Chapter III, § 1, no. 3, Proposition 3).

Q.E.D.

THEOREM 2. *A connected group* G *in which there exists a neighbourhood of the identity element homeomorphic to an open interval of* **R** *is isomorphic to either* **R** *or* **T**.

This is an immediate consequence of the preceding theorem, together with Proposition 7 of § 1, no. 4.

Remarks. 1) To decide whether a group G which satisfies the conditions of Theorem 2 is isomorphic to **T** or to **R**, it is enough to see whether G is compact or not.

2) Theorem 2 shows in particular that every topological group which is *homeomorphic* to the group **R** is necessarily *isomorphic* to **R**.

3) The preceding topological characterization of the groups **R** and **T** involves the topological space **R** as an auxiliary set. It is possible to characterize the topological group structures of **R** and **T** by means of axioms which do not involve any auxiliary set (see Exercises 4 and 5).

4. EXPONENTIALS AND LOGARITHMS

1. DEFINITION OF a^x AND $\log_a x$

THEOREM 1. *The multiplicative group* $\mathbf{R}_+^*$ *of real numbers* > 0 *is a topological group isomorphic to the additive group* **R** *of real numbers.*

For $\mathbf{R}_+^* =]0, +\infty[$ is an open interval of **R** and is therefore *homeomorphic* to **R** (Chapter IV, § 4, no. 1, Proposition 1). By Theorem 2 of § 3, it is therefore a topological group *isomorphic* to **R**.

From the Corollary to Proposition 5 of § 1, no. 3, for every number $a > 0$ there is a unique continuous homomorphism f_a of **R** into $\mathbf{R}_+^*$ such that $f_a(1) = a$. Hence, for all $x \in \mathbf{R}$ and all $y \in \mathbf{R}$ we have

$$f_a(x+y) = f_a(x)f_a(y), \qquad f_a(-x) = \frac{1}{f_a(x)},$$

and hence, in particular, for all $n \in \mathbf{Z}$,

$$f_a(n) = a^n.$$

For this reason we denote $f_a(x)$ by a^x for all $x \in \mathbf{R}$; the functions a^x (for all values of $a > 0$) are called *exponential functions*. We have $1^x = 1$ for all $x \in \mathbf{R}$; if $a \neq 1$, $x \to a^x$ is an *isomorphism* of the group **R** onto the group $\mathbf{R}_+^*$.

If $a \neq 1$, the isomorphism of $\mathbf{R}_+^*$ onto **R** which is the inverse of $x \to a^x$ is called the *logarithm to base* a, and its value at $x \in \mathbf{R}_+^*$ is denoted by $\log_a x$. Thus, with this notation, we have

(1) $$a^{x+y} = a^x a^y \qquad (x \in \mathbf{R}, y \in \mathbf{R}, a > 0);$$

(2) $$a^{-x} = \frac{1}{a^x} \qquad (x \in \mathbf{R}, a > 0);$$

(3) $$\log_a 1 = 0, \quad \log_a a = 1 \qquad (a > 0, a \neq 1);$$

(4) $$\log_a (xy) = \log_a x + \log_a y \qquad (x > 0, y > 0);$$

(5) $$\log_a\left(\frac{1}{x}\right) = -\log_a x \qquad (x > 0);$$

(6) $$a^{\log_a x} = x \qquad (x > 0);$$

(7) $$\log_a a^x = x \qquad (x \in \mathbf{R}).$$

By Proposition 5 of § 1, no. 3, every continuous homomorphism of $\mathbf{R}$ into $\mathbf{R}_+^*$ is of the form $y \to a^{xy}$, where $x \in \mathbf{R}$; since its value when $y = 1$ is a^x, we have identically

(8) $$(a^x)^y = a^{xy} \qquad (x \in \mathbf{R}, y \in \mathbf{R}, a > 0),$$

or, changing the notation,

(9) $$x^y = a^{y.\log_a x} \qquad (x > 0, y \in \mathbf{R}, a > 0, a \neq 1).$$

The formula (8) shows that, for every integer $n > 0$, we have $(a^{1/n})^n = a$, which justifies the notation $a^{1/n}$ introduced for the *nth root* $\sqrt[n]{a}$, defined in Chapter IV, § 3, no. 3.

Formulas (7) and (9) show that

(10) $$\log_a (x^y) = y.\log_a x \qquad (x > 0, y \in \mathbf{R}),$$

or, changing the notation,

(11) $$\log_a x = \log_a b.\log_b x \qquad (x > 0, a > 0, b > 0, a \neq 1, b \neq 1),$$

which is the formula for " change of base ".

Finally, let us obtain all the *continuous homomorphisms* of the topological group $\mathbf{R}^*$ into itself; if g is such a continuous homomorphism,

$$\log_a (g(a^x))$$

is a continuous homomorphism of $\mathbf{R}$ into $\mathbf{R}$, therefore (§ 1, no. 3, Proposition 5) there exists $\alpha \in \mathbf{R}$ such that $\log_a (g(a^x)) = \alpha x$ for all $x \in \mathbf{R}$; hence, by (8), $g(x) = x^\alpha$ for all $x > 0$. Hence we have identically

(12) $$(xy)^\alpha = x^\alpha y^\alpha \text{ for all } x > 0, y > 0 \text{ and } \alpha \in \mathbf{R}.$$

By reason of formula (4), which reduces every multiplication to an addition (the only operation to which the customary system of numeration is well adapted), logarithms have long been an indispensable instrument in numerical calculations (see the Historical Note to this Chapter).

When used for this purpose, the base chosen is $a = 10$; and there are tables giving the values of the function $\log_{10} x$ (to a certain approximation). In analysis, one is led to choose a different base (denoted by e) which is such that $\lim_{x \to 1, x \neq 1} \frac{\log_e x}{(x - 1)} = 1$ (cf. Exercise 1).

2. BEHAVIOUR OF THE FUNCTIONS a^x AND $\log_a x$

By Theorem 5 of Chapter IV, § 2, no. 6, if $a \neq 1$, $x \to a^x$ is a *strictly monotone* mapping of $\mathbf{R}$ onto the interval $\mathbf{R}_+^* =]0, +\infty[$. If $a > 1$, we have $a^1 = a \geqslant 1 = a^0$, hence a^x is *strictly increasing*; moreover, since $\mathbf{R}_+^*$ is not bounded above, a^x is not bounded above in $\mathbf{R}$, so that

(13) $$\lim_{x \to +\infty} a^x = +\infty \qquad (a > 1)$$

and, by (2),

(14) $$\lim_{x \to -\infty} a^x = 0 \qquad (a > 1).$$

On the other hand, if $a < 1$, the function a^x is strictly decreasing, tends to 0 as x tends to $+\infty$, and tends to $+\infty$ as x tends to $-\infty$ (Fig. 1).

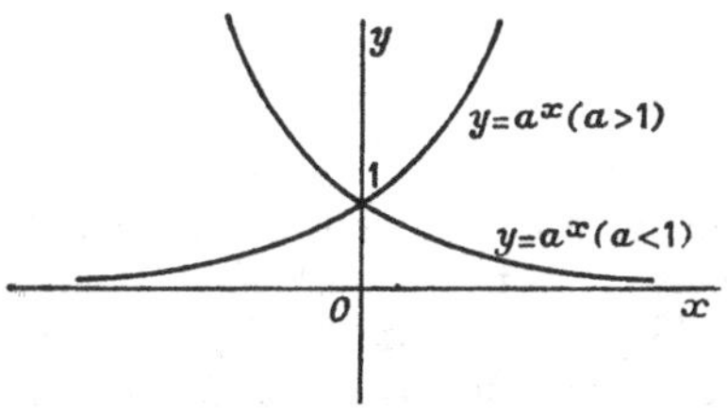

Figure 1.

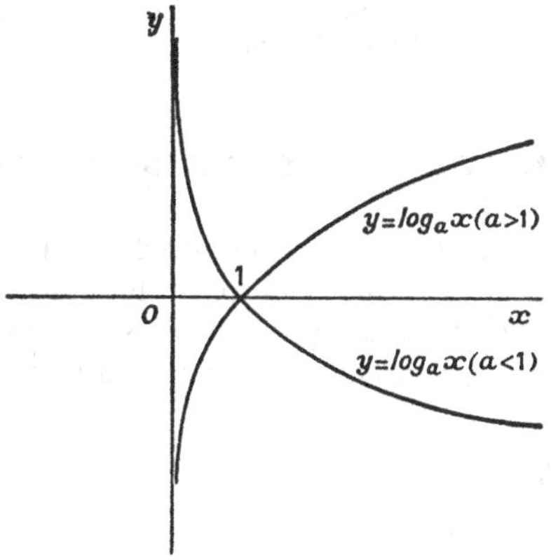

Figure 2.

From these properties and from (12), we deduce that if $0 < a < b$, we have $a^x < b^x$ for $x > 0$, and $a^x > b^x$ for $x < 0$; for $\left(\frac{b}{a}\right)^x > 1$ if $x > 0$, and $\left(\frac{b}{a}\right)^x < 1$ if $x < 0$.

The behaviour of $\log_a x$ in $\mathbf{R}_+^*$ is deduced from that of a^x in $\mathbf{R}$; if $a > 1$, the function $\log_a x$ is strictly increasing, tends to $-\infty$ as x tends to 0, and tends to $+\infty$ as x tends to $+\infty$; if $a < 1$, the function $\log_a x$ is strictly decreasing, tends to $+\infty$ as x tends to 0, and to $-\infty$ as x tends to $+\infty$ (Fig. 2).

The function a^x (resp. $\log_a x$), considered as defined on a subset of the extended line $\overline{\mathbf{R}}$ and taking its values in $\overline{\mathbf{R}}$, can be *extended by continuity* to $\overline{\mathbf{R}}$ (resp. to the interval $[0, +\infty]$ of $\overline{\mathbf{R}}$) by assigning to it its limiting values at the points $+\infty$ and $-\infty$ (resp. 0 and $+\infty$).

More generally, formula (9) shows that the function x^y is continuous on the subspace $\mathbf{R}_+^* \times \mathbf{R}$ of $\overline{\mathbf{R}}^2$ and tends to a limit when (x, y) tends to any point (a, b) of $\overline{\mathbf{R}}^2$ which lies in the closure of $\mathbf{R}_+^* \times \mathbf{R}$, with the exception of the points $(0, 0)$, $(+\infty, 0)$, $(1, +\infty)$, $(1, -\infty)$. We can therefore extend x^y by continuity to those points of $\overline{\mathbf{R}}^2$ at which the limit exists; by the principle of extension of identities (Chapter I, § 8, no. 1, Proposition 2, Corollary 1), formulas (1), (4) and (8) remain valid whenever both sides have a meaning.

Note that the extension by continuity of x^y does not allow us to obtain the formula $0^0 = 1$.

Note also that the definition of the exponential allows us to extend to $\mathbf{R}$ the function $n \to a^n$ defined on $\mathbf{Z}$, for all $a > 0$; but we do not obtain in this way any extension of this function when $a < 0$; a "natural" extension of this function can be defined only in terms of the theory of analytic functions.

3. MULTIPLIABLE FAMILIES OF NUMBERS > 0

The isomorphism of the topological groups $\mathbf{R}$ and $\mathbf{R}_+^*$ shows immediately that for a family (x_ι) of finite real numbers > 0 to be *multipliable* (Chapter IV, § 7, no. 4) it is necessary and sufficient that the family $(\log_a x_\iota)$ should be *summable* (a being any number > 0 and $\neq 1$); and we have

$$\prod_\iota x_\iota = a^{\sum_\iota \log_a x_\iota}. \tag{15}$$

Likewise, an infinite product defined by a sequence $(1 + u_n)$ of finite numbers > 0 is *convergent* (Chapter IV, § 7, no. 6) if and only if the series whose general term is $\log_a (1 + u_n)$ is *convergent*, and then we have

$$(16) \qquad \prod_{n=0}^{\infty} (1 + u_n) = a^{\sum_{n=0}^{\infty} \log_a (1+u_n)}.$$

The study of infinite products of real numbers > 0 is thus reduced to that of infinite series of real numbers whose terms appear in the form of logarithms; we shall see later how sums of this nature can be easily studied by means of the differential properties of the logarithm.

EXERCISES

§ 1

1) * *a*) Let f be a homomorphism of the additive group $\mathbf{R}$ into itself. Show that, if the graph of f is not dense in $\mathbf{R}^2$, f is of the form $x \to ax$ [consider the closure in $\mathbf{R}^2$ of the graph of f, and use the structure theorem for closed subgroups of $\mathbf{R}^2$ (Chapter VII, § 1, no. 2, Theorem 2)]. * [Compare with Chapter VI, § 1, Exercise 12 *b*) and Chapter IV, § 6, Exercise 2.]

b) If the graph of f is dense in $\mathbf{R}^2$, the inverse image of the topology of $\mathbf{R}^2$ under the mapping $x \to (x, f(x))$ is compatible with the group structure of $\mathbf{R}$ and is strictly finer than the usual topology of $\mathbf{R}$. If moreover f is injective, the inverse image under f of the usual topology of $\mathbf{R}$ is compatible with the group structure of $\mathbf{R}$ and is not comparable with the usual topology of $\mathbf{R}$.

¶ 2) Let $\mathfrak{T}$ be a Hausdorff topology on $\mathbf{R}$, compatible with the group structure of $\mathbf{R}$ and *strictly coarser* than the usual topology $\mathfrak{T}_0$.

a) Show that every open neighbourhood of 0 with respect to $\mathfrak{T}$ is unbounded in $\mathbf{R}$ (note that a Hausdorff topology which is coarser than the topology of a compact space must coincide with the latter).

b) Let $V \neq \mathbf{R}$ be a symmetric open neighbourhood of 0 with respect to $\mathfrak{T}$, and let W be a symmetric open neighbourhood of 0 with respect to $\mathfrak{T}$ such that $W + W \subset V$. Show that if a is the length of the component of V (with respect to $\mathfrak{T}_0$) which contains 0, then every component (with respect to $\mathfrak{T}_0$) of W has length $\leqslant a$. Moreover, the set of lengths of components of the interior of $\mathbf{R} - W$ (with respect to $\mathfrak{T}_0$) is bounded [use *a*) and the fact that there exists a symmetric open neighbourhood W_1 of 0 with respect to $\mathfrak{T}$ such that $W_1 + W_1 \subset W$].

c) Deduce from *b*) that $\mathbf{R}$ is *precompact* and not locally compact in the topology $\mathfrak{T}$.

d) Show likewise that $\mathbf{Z}$ is precompact and not locally compact in any topology $\mathfrak{T}$ compatible with its group structure and distinct from the discrete topology.

e) Let f be a continuous homomorphism of a subgroup Γ of $\mathbf{R}$, not consisting of 0 alone and endowed with the topology induced by $\mathfrak{T}_0$, into a complete Hausdorff group G. Show that if f is not an isomorphism of Γ onto the subgroup $f(\Gamma)$ of G, then $f(\Gamma)$ is relatively compact in G [use *c*) and *d*)].

$_*$ *f*) For every integer $n \geqslant 2$, give examples of continuous injective homomorphisms f of **R** into $\mathbf{T}^n$ such that $f(\mathbf{R})$ is dense in $\mathbf{T}^n$ (cf. Chapter VII, § 1, no. 4, Corollary 1 of Proposition 7). $_*$

3) Show that the group **T** is algebraically isomorphic to the product $\mathbf{R} \times (\mathbf{Q}/\mathbf{Z})$ (take a suitable Hamel base in **R**). Deduce that there exists a Hausdorff topology on **R**, compatible with its group structure, not comparable with the usual topology, and with respect to which **R** is precompact.

§ 2

1) Let E be a linearly ordered set with a smallest element ω, and let I be an interval of E, containing ω, with ω as its left-hand end-point. Suppose that E and I satisfy axioms (GR_{I}), (GR_{II}), (GR_{IV}) and the following axiom:

(GR_{IIIb}) *The set of elements* $x > \omega$ *in* I *is not empty, and if* x *is any element* $> \omega$ *in* I, *there exists* $y > \omega$ *in* I *such that* $y^2 \leqslant x$.

Show that there exists an increasing mapping f of I into **R**, such that $f(\omega) = 0$ and $f(xy) = f(x) + f(y)$ whenever x, y and xy are in I; also that $f(\mathrm{I}) \cap [0, f(b)]$ is dense in $[0, f(b)]$ for all $b \in \mathrm{I}$.

2) Let G be a non-commutative linearly ordered group (for example the multiplicative group of a non-commutative ordered division ring). In the set $\mathbf{R}_+ \times \mathrm{G}$ consider the set E consisting of $(0, e)$ (e being the identity element of G) and all pairs (x, y) where x is any real number > 0 and y is any element of G. Define a law of composition on E by the rule $(x, y)(x', y') = (x + x', yy')$, and order E lexicographically [i.e. $(x, y) < (x', y')$ if $x < x'$ or if $x = x'$ and $y < y'$]. If we take I = E, show that axioms (GR_{I}), (GR_{II}), (GR_{IIIb}) (Exercise 1) and (GR_{IV}) are satisfied; but that if f is an increasing mapping of E into $\mathbf{R}_+$ such that $f(zz') = f(z) + f(z')$ for all a and z' in E, then f is not strictly increasing.

§ 3

1) A linearly ordered group G (not necessarily commutative) is said to be *Archimedean* if the set I of elements $\geqslant e$ (the identity element of G) satisfies axiom (GR_{IV}) of § 2. Show that a linearly ordered group G is isomorphic to a subgroup of the additive group **R** if and only G is Archimedean (distinguish two cases according as the set of elements $> e$ in G has or has not a smallest element, and use Proposition 1 of § 2).

2) Let G be a linearly ordered group (not necessarily commutative), with more than one element. Then the topology $\mathcal{T}_0(G)$ (Chapter I, § 2, Exercise 5) is compatible with the group structure of G. If G is *connected* in this topology, show that G is isomorphic to the additive group **R** (use Exercise 7 of Chapter IV, § 2, and Proposition 2 of § 2).

3) Let G be a (not necessarily commutative) linearly ordered group. If G, endowed with the topology $\mathcal{T}_0(G)$, is *locally compact* and not discrete, then G is locally isomorphic to **R**, and the identity component of G is an open subgroup isomorphic to **R** (use Exercise 6 of Chapter IV, § 2, and Proposition 2 of § 2).

¶ 4) Let G be a topological group which satisfies the following conditions:

(R_I) G is connected.

(R_{II}) The complement G^* of the identity element e of G is not connected.

Then there exists a *continuous bijective homomorphism* of G onto **R** (in other words, G is algebraically isomorphic to **R** and its topology is finer than that of **R**). The proof is in several steps:

a) Let $(U_i)_{1 \leqslant i \leqslant n}$ be a partition of G^* into open sets *in* G^* ($n \geqslant 2$). Show that each U_i is open in G, that e lies in each $\overline{U}_i$, and that G is Hausdorff. Deduce that the closures $\overline{U}_i = U_i \cup \{e\}$ in G are connected (Chapter I, § 11, Exercise 4).

b) Let A be a connected component of U_i. Show that for each index $j \neq i$, we have $A^{-1}\overline{U}_j = A^{-1}$ (observe that $A^{-1}\overline{U}_j$ is connected and contains A^{-1}, and that $A^{-1}\overline{U}_j \subset G^*$ if $j \neq i$).

c) Show that n must be equal to 2 [for each index $i = 1, 2, \ldots, n$ take a component A_i of U_i; if $j \neq i$ we have, by *b*), $A_i^{-1}A_j \subset A_i^{-1}$, $A_j^{-1}A_i \subset A_j^{-1} \subset U_j^{-1}$, whence $A_i^{-1} \subset U_j$]. Deduce that G^* has exactly *two* components A and B, that $B = A^{-1}$ and $\overline{A} = \complement(A^{-1})$.

d) The relation $yx^{-1} \in \overline{A}$ is an order relation, which makes G into a linearly ordered group (show that $\overline{A}^2 = \overline{A}$ and that $x\overline{A}x^{-1} = \overline{A}$ for all $x \in G$).

e) The topology $\mathcal{T}_0(G)$ is coarser than the given topology $\mathcal{T}$ on G (observe that A is open with respect to $\mathcal{T}$).

Complete the proof by using Exercise 2 above and Chapter I, § 11, no. 2, Proposition 4. [Cf. Chapter VI, § 1, Exercise 12 *b*).]

f) Show that if in addition we suppose G to be either *locally compact* or *locally connected*, then G is isomorphic to **R**.

* 5) Give an example of a topology on **R**, compatible with the group structure, for which **R** is connected, locally compact and locally connected, and for which the complement of $\{0\}$ in **R** is connected (use the fact that **R** and $\mathbf{R}^n$ are algebraically isomorphic). *

¶ 6) Let G be a topological group which satisfies the following conditions:

(LR_I) G is Hausdorff and locally connected.

(LR_{II}) There exists a connected neighbourhood U of the identity element e of G such that the complement of e in U is not connected.
Show that, under these conditions, G is *locally isomorphic to* **R**.

[Prove first that if V is any connected neighbourhood of e in G contained in U, then $V \cap \complement\{e\}$ is not connected, that e lies in the closure of every component of $V \cap \complement\{e\}$, and that $V \cap \complement\{e\}$ has exactly two components, by reasoning as in Exercise 4. Then take the connected neighbourhood V sufficiently small and define a linear ordering on V such that $\mathfrak{T}_0(V)$ is coarser than the topology induced on V by the topology of G, and such that Proposition 2 of § 2 applies to the set E of elements $\geqslant e$ of V.]

7) Let G be a non-discrete locally compact group with a countable base of open sets, and let V_0 be a compact neighbourhood of the identity element e of G which contains *no subgroup* of G other than $\{e\}$.

a) Let (x_n) be a sequence of points of V_0 which converges to e. For each n, there exists an integer $p(n) > 0$ such that $x_n^k \in V_0$ whenever $k \leqslant p(n)$ and $x_n^{p(n)+1} \notin V_0$, and we have $\lim_{n\to\infty} p(n) = +\infty$. Show that there exists a subsequence (x_{k_n}) of (x_n) such that, for every rational number r with $|r| \leqslant 1$, the sequence $(x_{k_n}^{[rp(k_n)]})$ has a limit $f(r)$ in V_0 (use the diagonal technique); and that if r, r' are rational numbers such that $|r| \leqslant 1$, $|r'| \leqslant 1$ and $|r + r'| \leqslant 1$ we have

$$f(r + r') = f(r)f(r').$$

b) Show that f is continuous in a neighbourhood of 0 in **Q**. [Argue by contradiction: if there were a sequence (r_n) of rational numbers tending to 0 and such that $f(r_n)$ tended to an element $y \neq e$ in V_0, show that we should have $y^m \in V_0$ for all integers m.]

c) Deduce from *b*) that there exists a non-trivial continuous homomorphism of **R** into G (use Proposition 6 of § 1).

d) Let H be a closed normal subgroup of G other than G itself, and suppose that there is a neighbourhood of the identity element in G/H which contains no non-trivial subgroup. Show that there exists a continuous homomorphism $f : \mathbf{R} \to G$ such that the composition

$$\mathbf{R} \xrightarrow{f} G \to G/H$$

is non-trivial.

8) Let G be a topological group and let H be a closed subgroup of G, contained in the centre of G, and such that there is a continuous

surjective homomorphism $\varphi : \mathbf{R} \to G/H$. Show that G is abelian [if a_0 is an element of G which is not in H, observe that there exists a sequence (a_n) of elements of G and a sequence (c_n) of elements of H such that $a_n = c_n a_{n+1}$ and that the subgroup of G generated by H and the a_n is dense in G].

§ 4

1) *a*) Let $(x_n, y_n)_{n \in \mathbf{Z}}$ be a family of points of $\mathbf{R}^2$ such that for all $n \in \mathbf{Z}$ we have $x_n < x_{n+1}$ and

$$\frac{y_{n+1} - y_n}{x_{n+1} - x_n} < \frac{y_{n+2} - y_{n+1}}{x_{n+2} - x_{n+1}}.$$

Show that, for all integers $n \in \mathbf{Z}$, $m > 0$, $p > 0$, we have

$$\frac{y_{n+m} - y_n}{x_{n+m} - x_n} < \frac{y_{n+m+p} - y_n}{x_{n+m+p} - x_n}.$$

b) If $a > 0$ and $x \neq 0$, put

$$f_a(x) = \frac{a^x - 1}{x}.$$

Show that, if $x < y$, $x \neq 0$, $y \neq 0$ and $a \neq 1$, we have $f_a(x) < f_a(y)$ [prove the relation first for x, y integral, by means of *a*), then for x, y rational, and finally in general].

c) Deduce that, for all $a > 0$, the function $f_a(x)$, defined for $x \neq 0$, has a limit on the right and a limit on the left as $x \to 0$; show that these two limits have a common value $\varphi(a)$, and that $\varphi(a) \neq 0$ if $a \neq 1$. Show that if $a \neq 1$, the function $\log_a x/x - 1$, defined for $x \neq 1$, tends to the limit $1/\varphi(a)$ as $x \to 1$.

d) Show that, for all $a > 0$ and $b > 0$, we have $\varphi(b) = \varphi(a) \log_a b$ (if $a \neq 1$). Deduce that there exists a number e between 2 and 4 such that $\varphi(e) = 1$, and that $\varphi(a) = \log_e a$ for all $a > 0$.

2) Show, by induction on n, that for all integers $n > 0$ we have

$$2^n > \frac{1}{2} n(n+1).$$

Deduce that, if $a > 1$ and $\alpha > 0$,

$$\lim_{x \to +\infty} \frac{a^x}{x^\alpha} = +\infty, \qquad \lim_{x \to +\infty} \frac{\log_a x}{x^\alpha} = 0$$

(begin by proving the first of these relations for $a = 2$ and $\alpha = 1$).

HISTORICAL NOTE

(Numbers in brackets refer to the bibliography at the end of this note.)

The history of the theory of the multiplicative group $\mathbf{R}_+^*$ of real numbers > 0 is closely related to that of the development of the notion of *powers* of a number > 0 and the notations employed to denote them. The idea of the " geometric progression " formed by successive powers of the same number goes back to the Egyptians and the Babylonians; the Greek mathematicians were familiar with it, and as early as Euclid [1] we find a general statement equivalent to the rule $a^m a^n = a^{m+n}$ for integral exponents $m, n > 0$. In the Middle Ages, the French mathematician N. Oresme (14 th century) rediscovered this rule. He was the first to have the idea of a fractional exponent > 0, with a notation similar to our own and to our relevant rules of calculation, stated in general terms; for example, he used the two rules which we should write as

$$(ab)^{1/n} = a^{1/n}b^{1/n}, \qquad (a^m)^{p/q} = (a^{mp})^{1/q} \ [2].$$

But the ideas of Oresme were too far ahead of the mathematics of his time to exercise an influence on his contemporaries, and his treatise soon sank into oblivion. A century later, N. Chuquet stated Euclid's rule anew; furthermore, he introduced an exponential notation for the powers of unknowns in his equations, and did not hesitate to use the exponent 0 and integral exponents < 0 (*). This time, even though Chuquet's work remained in manuscript and seems not to have been widely circulated, the idea of the isomorphism between the " arithmetic progression " of the exponents and the " geometric progression " of the powers was not lost sight of again; it was extended to negative and fractional exponents by Stifel [4], and led finally to the definition of logarithms and the construction of the first tables, independently undertaken by the Scotsman J. Napier in 1614-1620 [5] and the Swiss J. Bürgi (whose work did not appear until 1620, although his ideas went back to the first years of the 17th

(*) Chuquet writes for example 12^1, 12^2, 12^3, etc., for $12\,x$, $12\,x^2$, $12\,x^3$, etc., 12^0 for the number 12, and $12^{2\tilde{m}}$ for $12\,x^{-2}$ [3].

century). Bürgi implicitly assumes the continuity of the isomorphism established between $\mathbf{R}$ and $\mathbf{R}_+^*$ by means of interpolation in the use of his tables; Napier on the other hand formulates it in his definition explicitly (or at any rate as explicitly as the hazy notion of continuity current at that time allowed) (*).

It is not our purpose here to dwell on the services rendered by logarithms to numerical calculation; from the theoretical point of view, their importance dates from the beginnings of the infinitesimal calculus, with the discovery of the expansions in series of $\log(1+x)$ and e^x, and the differential properties of these functions. As far as the definition of logarithms and exponentials was concerned, it was assumed intuitively, up to the middle of the 19th century, that the function a^x defined for all rational x could be extended by continuity to the set of all real numbers; and it was not until the notion of real number had been definitively clarified and deduced from that of rational number that a rigorous justification was sought for this extension by continuity. An analogous principle of extension, suitably applied, is again at the base of the proofs of Propositions 1 and 2 of § 2; from it there follows not only the definition of exponentials and logarithms, but also, as we shall see in Chapter VIII, angular measure.

BIBLIOGRAPHY

[1] *Euclidis Elementa*, 5 vols., ed. J. L. Heiberg, Leipzig (Teubner), 1883-88. IX, II.

[2] M. Curtze, *Zeitschr. Math. Phys.*, **13**, supplement (1868), p. 65.

[3] N. Chuquet, *Bull. bibl. storia math.*, **13** (1880), pp. 737-738.

[4] M. Stifel, *Ari.hme.ica in.egra*, Nuremberg (1544), fol. 35 and 249-250.

[5] J. Napier, *Mirifici logarithmorum canonis constructio*, Lyon, 1620.

(*) Napier considers two points M, N moving simultaneously on two lines, the movement of M being uniform, and that of N being such that its velocity is proportional to its abscissa; the abscissa of M is then by definition the logarithm of the abscissa of N ([2], p. 20-21).

CHAPTER VI

Real number spaces and projective spaces

1. REAL NUMBER SPACE $\mathbf{R}^n$

1. THE TOPOLOGY OF $\mathbf{R}^n$

DEFINITION 1. *The topological product of* n *spaces identical with the real line is called real number space of* n *dimensions, and is denoted by* $\mathbf{R}^n$.

Remark. The space $\mathbf{R}^0$ consists of a single point.

From *Set Theory*, Chapter III, § 6, no. 3, Theorem 2, Corollary 1, we know that, if E is an infinite set, E^n is equipotent with E for all integers $n > 0$; hence, if $n > 0$, $\mathbf{R}^n$ is equipotent with $\mathbf{R}$, that is, $\mathbf{R}^n$ *has the power of the continuum* (cf. Exercises 1 and 2).

DEFINITION 2. *Any subset of* $\mathbf{R}^n$ *which is the product of* n *open* (resp. *closed*) *intervals* of $\mathbf{R}$ *is called an open* (resp. *closed*) *box in* $\mathbf{R}^n$. [For $n = 2$ it is called an *open* (resp. *closed*) *rectangle*.]

The open boxes in $\mathbf{R}^n$ form a *base* of the topology of $\mathbf{R}^n$ (Chapter I, § 4, no. 1); the open boxes which contain a point $x = (x_i)_{1 \leqslant i \leqslant n}$ of $\mathbf{R}^n$ form a fundamental system of neighbourhoods of x, and so do the closed boxes of $\mathbf{R}^n$ for which x is an interior point.

Every non-empty open box in $\mathbf{R}^n$ is *homeomorphic* to $\mathbf{R}^n$ (Chapter IV, § 4, no. 1, Proposition 1).

It follows that, when $n \geqslant 1$, every non-empty open set in $\mathbf{R}^n$ has the power of the continuum.

An *open* (resp. *closed*) *cube* of $\mathbf{R}^n$ is an open (resp. closed) box which is the product of n *bounded* intervals *of equal length* [for $n = 2$, it is called an *open* (resp. *closed*) *square*]; the common length of these intervals is called the *side* (or *side-length*) of the cube. The open cubes

$$K_m = \prod_{1 \leqslant i \leqslant n} \left]x_i - \frac{1}{m}, x_i + \frac{1}{m}\right[$$

form a countable fundamental system of neighbourhoods of the point $x = (x_i)$, as m runs through the set of all integers > 0 or through any sequence of integers increasing to infinity.

Every open (or closed) box in $\mathbf{R}^n$ is *connected* (Chapter I, § 11, no. 4, Proposition 8); in particular, $\mathbf{R}^n$ is *connected* and *locally connected*.

If A is a non-empty open set in $\mathbf{R}^n$, its components are therefore *open* sets (Chapter I, § 11, no. 6, Proposition 11); and the set of these components is *countable*, for $\mathbf{R}^n$ has a countable dense subset (for example $\mathbf{Q}^n$).

Consider under what conditions a subset A of $\mathbf{R}^n$ will be *relatively compact*. By Tychonoff's theorem (Chapter I, § 9, no. 5, Theorem 3) it is necessary and sufficient that the projections of A on the factors of $\mathbf{R}^n$ should be relatively compact; by the Borel-Lebesgue theorem (Chapter IV, § 2, no. 2, Theorem 2) this is equivalent to saying that these projections are *bounded* subsets of $\mathbf{R}$. We say that a subset A of $\mathbf{R}^n$ is *bounded* if all its projections are bounded subsets of $\mathbf{R}$; thus we have proved:

PROPOSITION 1. *A subset* A *of* $\mathbf{R}^n$ *is relatively compact if and only if it is bounded.*

COROLLARY. *The space* $\mathbf{R}^n$ *is locally compact, but is not compact if* $n \geqslant 1$.

2. THE ADDITIVE GROUP $\mathbf{R}^n$

The set $\mathbf{R}^n$, endowed with the group structure which is the *product* of the additive group structures of the n factors of $\mathbf{R}^n$, is an abelian group; we use the additive notation, the sum of $x = (x_i)$ and $y = (y_i)$ being therefore $x + y = (x_i + y_i)$. The topology of the number space is compatible with this group structure; endowed with these two structures, $\mathbf{R}^n$ is a topological group called the *additive group of n-dimensional real number space.* If $n = 0$, we make the convention that $\mathbf{R}^0$ denotes a group consisting only of the identity element.

The uniform structure of this group, called the *additive uniformity* of $\mathbf{R}^n$, is the product of the uniformities of the factors of $\mathbf{R}^n$ (Chapter III, § 3, no. 2). If, for each integer $p > 0$, V_p denotes the set of pairs (x, y) of $\mathbf{R}^n$ such that $\max_{1 \leqslant i \leqslant n} |x_i - y_i| \leqslant 1/p$, the sets V_p form a *fundamental system of entourages* of this uniformity. Whenever we consider $\mathbf{R}^n$ as a uniform space we shall always have in mind the additive uniformity just defined, unless the contrary is expressly stated. Endowed with this uniform structure, $\mathbf{R}^n$ is a *complete* uniform space (Chapter II, § 3, no. 5, Proposition 10).

3. THE VECTOR SPACE $\mathbf{R}^n$

Since $\mathbf{R}$ is a *field*, we can define on $\mathbf{R}^n$ a *vector space structure* over the field $\mathbf{R}$, the product tx of a scalar $t \in \mathbf{R}$ and a point (or vector) $x = (x_i)$ of $\mathbf{R}^n$ being the point (tx_i). Note that the homothety $(t, x) \to tx$ is *continuous* on $\mathbf{R} \times \mathbf{R}^n$. If e_i denotes the vector of $\mathbf{R}^n$ all of whose coordinates are zero, except for that of index i, which is equal to 1, then the e_i form a *basis* of the vector space $\mathbf{R}^n$, called the *canonical basis* of this space. Every vector $x = (x_i) \in \mathbf{R}^n$ can be written as $x = \sum_{i=1}^{n} x_i e_i$, and the relation $\sum_{i=1}^{n} t_i e_i = 0$ implies that $t_i = 0$ for $1 \leqslant i \leqslant n$.

The vector space $\mathbf{R}^n$ is therefore of *dimension* n over the field $\mathbf{R}$, in the sense of algebra (*Algebra*, Chapter II, § 7, no. 2); hence its name of *n-dimensional* real number space.

Let f be an *affine* mapping of the vector space $\mathbf{R}^n$ into the vector space $\mathbf{R}^m$ (m and n being integers > 0). If we put $g(x) = f(x) - f(0)$, g is a *linear* mapping of $\mathbf{R}^n$ into $\mathbf{R}^m$. Let a_{ij} $(1 \leqslant j \leqslant m)$ be the coordinates of $g(e_i)$ in $\mathbf{R}^m$ and let b_j $(1 \leqslant j \leqslant m)$ be those of $f(0)$; if x_i $(1 \leqslant i \leqslant n)$ is the ith coordinate of $x \in \mathbf{R}^n$ and if y_j is the jth coordinate of $y = f(x)$, we have

$$y_j = \sum_{i=1}^{n} a_{ij} x_i + b_j \qquad (1 \leqslant j \leqslant m).$$

Since every linear polynomial in $x_1, x_2, \ldots, x_n$ is uniformly continuous on $\mathbf{R}^n$, it follows that every affine mapping of $\mathbf{R}^n$ into $\mathbf{R}^m$ is *uniformly continuous* on $\mathbf{R}^n$ (Chapter II, § 2, no. 6, Proposition 7).

In particular, we know that every affine mapping of $\mathbf{R}^n$ *onto* itself is *bijective* and that its inverse is again an affine mapping; hence every affine mapping of $\mathbf{R}^n$ onto itself is a *homeomorphism* (and an automorphism of the uniform structure of $\mathbf{R}^n$).

Let $(a_i)_{1 \leqslant i \leqslant n}$ be a *free system* of n vectors of $\mathbf{R}^n$ [in other words a *basis* of the vector space $\mathbf{R}^n$]; if b is any point of $\mathbf{R}^n$, the set P of points $x = b + \sum_{i=1}^{n} u_i a_i$ such that $-1 \leqslant u_i \leqslant 1$ for $1 \leqslant i \leqslant n$ is a *compact neighbourhood* of b; for there exists a bijective affine mapping f of $\mathbf{R}^n$ onto itself such that $f(b) = 0$, $f(b + a_i) = e_i$ for $1 \leqslant i \leqslant n$; and $f(\mathrm{P})$ is the cube which is the product of the n intervals $[-1, +1]$ in the n factor spaces. P is called the *closed parallelotope* with centre b and basis vectors a_i. The interior of P consists of the points $b + \sum_{i=1}^{n} u_i a_i$ such that $-1 < u_i < 1$ for $1 \leqslant i \leqslant n$; it is called the *open parallelotope* with centre b and basis vectors a_i.

4. AFFINE LINEAR VARIETIES IN $\mathbf{R}^n$

Given a p-dimensional *affine linear variety* V in $\mathbf{R}^n$, there exists an affine mapping f of $\mathbf{R}^n$ onto itself which transforms V into a p-dimensional *coordinate* variety, that is to say a vector subspace V′ generated by p of the vectors of the canonical basis (e_i) of $\mathbf{R}^n$. There exists a mapping of V′ onto $\mathbf{R}^p$ which is an isomorphism of the vector space structure and the topology of V′ onto the corresponding structures of $\mathbf{R}^p$ (it is moreover often convenient to *identify* $\mathbf{R}^p$ with such a coordinate variety V′, e.g., with the vector subspace generated by $e_1, e_2, \ldots, e_p$). In addition, V′ is a *closed* subset of $\mathbf{R}^n$ (Chapter I, § 4, no. 3, Corollary to Proposition 7); hence:

PROPOSITION 2. *Every p-dimensional linear affine variety in* $\mathbf{R}^n$ *is a closed subset of* $\mathbf{R}^n$, *homeomorphic to* $\mathbf{R}^p$.

It is this result which is the origin of the names *line* and *plane* given to affine linear varieties of *one* and *two* dimensions in a vector space over an arbitrary division ring. We recall also that, for $n \geqslant 1$, the affine linear varieties of $n - 1$ dimensions of $\mathbf{R}^n$ are called *hyperplanes* (*loc. cit.*).

The n one-dimensional coordinate varieties, that is to say the n lines passing through 0 and the n points e_i respectively, are called the *coordinate axes*. For $n = 2$ the axis through e_1 is sometimes called the *axis of abscissas* and the axis through e_2 is called the *axis of ordinates*; the first coordinate of a point $x \in \mathbf{R}^2$ is called its *abscissa*, the second coordinate its *ordinate*.

Every line D passing through a point a has a parametric representation $t \to a + tb$, where t runs through $\mathbf{R}$ and $b \neq 0$; this mapping is a homeomorphism of $\mathbf{R}$ onto D. The vector b is called a *direction vector* of D, and its components b_i ($1 \leqslant i \leqslant n$) are called *direction ratios* of D. If $\boldsymbol{b}'$ is another direction vector of D, there exists $h \neq 0$ in $\mathbf{R}$ such that $\boldsymbol{b}' = h\boldsymbol{b}$.

The set of points $\boldsymbol{a} + t\boldsymbol{b}$, where t runs through the set of real numbers $\geqslant 0$, is called the *closed ray* (or simply *ray*, or *half-line*) with *origin* $\boldsymbol{a}$ and *direction vector* $\boldsymbol{b}$ (or with *direction ratios* b_i). It is a *closed* subset of $\mathbf{R}^n$, homeomorphic to the interval $[0, +\infty[$ of $\mathbf{R}$, and therefore *connected.* The line D is the union of the two rays with origin $\boldsymbol{a}$ and direction vectors $\boldsymbol{b}$ and $-\boldsymbol{b}$ respectively; these rays are said to be *opposite.*

By abuse of language, the set of points $\boldsymbol{a} + t\boldsymbol{b}$, where t runs through the set of real numbers > 0, is called the *open ray* with origin $\boldsymbol{a}$ and direction vector $\boldsymbol{b}$; it is homeomorphic to the interval $]0, +\infty[$ (and therefore homeomorphic to $\mathbf{R}$), but is not open in $\mathbf{R}^n$ if $n > 1$, although it is open in the line which contains it.

A line passing through two distinct points $\boldsymbol{x}$ and $\boldsymbol{y}$ also has a parametric representation $(u, v) \rightarrow u\boldsymbol{x} + v\boldsymbol{y}$, where (u, v) runs through the set of pairs of real numbers such that $u + v = 1$. Given any two points $\boldsymbol{x}, \boldsymbol{y}$ (distinct or not) the set of points $u\boldsymbol{x} + u\boldsymbol{y}$, where (u, v) runs through the set of pairs of real numbers such that $u \geqslant 0, v \geqslant 0$ and $u + v = 1$, is called the *closed segment* (or simply *segment*) with end-points $\boldsymbol{x}, \boldsymbol{y}$. A closed segment is *compact* and *connected,* for if its end-points are distinct it is homeomorphic to the interval $[0, 1]$ of $\mathbf{R}$.

If $\boldsymbol{x} \neq \boldsymbol{y}$, the set of points $u\boldsymbol{x} + v\boldsymbol{y}$ such that $u > 0, v > 0$ and $u + v = 1$ is called (by abuse of language) the *open segment* with end-points $\boldsymbol{x}, \boldsymbol{y}$; it is homeomorphic to the open interval $]0, 1[$ (and hence also homeomorphic to $\mathbf{R}$). Finally the union of $\{\boldsymbol{y}\}$ and the open segment with end-points $\boldsymbol{x}, \boldsymbol{y}$ is sometimes called the *segment open at* $\boldsymbol{x}$ *and closed at* $\boldsymbol{y}$; it is homeomorphic to the interval $[0, 1[$. All the segments with $\boldsymbol{x}$ and $\boldsymbol{y}$ as end-points are connected, and the closure of each of them is the closed segment with the same end-points.

PROPOSITION 3. *Let $f(\boldsymbol{x}) = f(x_1, x_2, \ldots, x_n)$ be a polynomial with real coefficients, not identically zero, defined on $\mathbf{R}^n$. Then the complement of the set $\overset{-1}{f}(0)$ is dense in $\mathbf{R}^n$.*

Let $\boldsymbol{x}$ be any point of $\mathbf{R}^n$ and let $\boldsymbol{y} \in \mathbf{R}^n$ be such that $f(\boldsymbol{y}) \neq 0$; $\varphi(t) = f(\boldsymbol{x} + t(\boldsymbol{y} - \boldsymbol{x}))$ is a polynomial in the real variable t, not identically zero; hence there exist arbitrarily small values of t such that $\varphi(t) \neq 0$. This shows that $\boldsymbol{x}$ lies in the closure of the complement of $\overset{-1}{f}(0)$.

COROLLARY. *The complement of an affine linear variety of dimension $p < n$ is dense in $\mathbf{R}^n$.*

Since every affine linear variety of dimension $p < n$ is contained in a hyperplane, it is enough to prove the corollary for a hyperplane; but a hyperplane is defined by an equation $g(\boldsymbol{x}) = 0$, where g is a linear polynomial not identically zero.

PROPOSITION 4. *In* $\mathbf{R}^n$ $(n \geqslant 1)$ *the complement of every hyperplane has two connected components.*

Let $g(\boldsymbol{x}) = 0$ be an equation of a hyperplane H in $\mathbf{R}^n$, g being a linear polynomial. The set $\complement H$ is the union of the set E_1 of all points $\boldsymbol{x}$ such that $g(\boldsymbol{x}) > 0$ and the set E_2 of all points x such that $g(\boldsymbol{x}) < 0$. E_1 and E_2 are connected, for if $g(\boldsymbol{x}) > 0$ and $g(\boldsymbol{y}) > 0$ we have $g(u\boldsymbol{x} + v\boldsymbol{y}) = ug(\boldsymbol{x}) + vg(\boldsymbol{y}) > 0$ whenever $u \geqslant 0$, $v \geqslant 0$ and $u + v = 1$; in other words, the segment with end-points $\boldsymbol{x}$ and $\boldsymbol{y}$ is contained in E_1. Similarly for E_2. On the other hand, $\complement H$ is not connected, because its image in $\mathbf{R}$ under g is the union of the intervals $]0, +\infty[$ and $]-\infty, 0[$.

The components E_1, E_2 of the complement of a hyperplane H are called the *open half-spaces* determined by H.

The closures of E_1 and E_2, which are respectively $E_1 \cup H$ and $E_2 \cup H$, are called the *closed half-spaces* determined by H.

Observe that an affine mapping of $\mathbf{R}^n$ onto itself which transforms H into a " coordinate " hyperplane, e.g., the hyperplane whose equation is $x_n = 0$, also transforms the open half-spaces determined by H into the open half-spaces defined respectively by the relations $x_n > 0$ and $x_n < 0$; the latter are open boxes and therefore *homeomorphic to* $\mathbf{R}^n$.

5. TOPOLOGY OF VECTOR SPACES AND ALGEBRAS OVER THE FIELD R

Let E be a vector space of dimension n over the field $\mathbf{R}$; if $(\boldsymbol{a}_i)_{1 \leqslant i \leqslant n}$ is a *basis* of E, then every point $\boldsymbol{x} \in E$ can be written uniquely in the form $\boldsymbol{x} = \sum_{i=1}^{n} x_i \boldsymbol{a}_i$, where the x_i are real numbers; the mapping $(x_i) \rightarrow \sum_{i=1}^{n} x_i \boldsymbol{a}_i$ is therefore a bijective linear mapping of $\mathbf{R}^n$ onto E. If we *transport* to E the topology of $\mathbf{R}^n$ by this mapping, E is endowed with a topology compatible with its additive group structure, and the mapping $(t, \boldsymbol{x}) \rightarrow t\boldsymbol{x}$ of $\mathbf{R} \times E$ into E is continuous with respect to this topology. This topology is *independent of the basis* chosen in E; for if $(\boldsymbol{a}')$ is another basis of E, and if $\boldsymbol{x} = \sum_{i=1}^{n} x'_i \boldsymbol{a}'_i = \sum_{i=1}^{n} x_i \boldsymbol{a}_i$, the mapping $(x_i) \rightarrow (x'_i)$ of $\mathbf{R}^n$ onto itself is linear and therefore a homeomorphism.

This fact leads one to suspect that the topology so defined on E should be capable of characterization without the help of a basis of E. In fact, we shall see later that this is the *only* Hausdorff topology on E for which the functions $x - y$ (on $E \times E$) and tx (on $\mathbf{R} \times E$) are continuous.

If now A is an *algebra* of finite rank n over the field $\mathbf{R}$, the above topology on A (considered as an n-dimensional vector space over $\mathbf{R}$) is compatible not only with the additive group structure of A, but also with its *ring* structure. This is a consequence of the following more general result:

PROPOSITION 5. *Let* E, F, G *be three finite-dimensional vector spaces over the field* $\mathbf{R}$. *Then every bilinear mapping* (*) f *of* $E \times F$ *into* G *is continuous.*

We may suppose that $E = \mathbf{R}^m$, $F = \mathbf{R}^n$, $G = \mathbf{R}^p$; it is enough to show that the coordinates in $\mathbf{R}^p$ of $f(x, y)$ are continuous functions of $(x, y) \in E \times F$ (Chapter I, § 4, no. 1, Proposition 1). In other words, it is enough to show that every *bilinear form* g is continuous on $E \times F$; and this is immediate, since $g(x, y)$ is a polynomial in the coordinates of x and y.

6. TOPOLOGY OF MATRIX SPACES OVER R

An important example of a vector space over $\mathbf{R}$ is the space $\mathbf{M}_{m,n}(\mathbf{R})$ of *matrices with* m *rows and* n *columns* whose elements belong to $\mathbf{R}$; this is a space of dimension mn over $\mathbf{R}$, hence is homeomorphic to $\mathbf{R}^{mn}$. By Proposition 5 of § 5, the product $X.Y$ of two matrices $X \in \mathbf{M}_{m,n}(\mathbf{R})$, $Y \in \mathbf{M}_{n,p}(\mathbf{R})$ is a continuous function of (X, Y). In particular, the topology of the space $\mathbf{M}_n(\mathbf{R})$ of *square* matrices of order n is compatible with the ring structure on $\mathbf{M}_n(\mathbf{R})$. Furthermore:

PROPOSITION 6. *In the ring* $\mathbf{M}_n(\mathbf{R})$, *the group* $\mathbf{GL}_n(\mathbf{R})$ *of non-singular matrices is a dense open subset, and the topology induced on this set is compatible with its group structure.*

(*) If E, F, G are three vector spaces over a field K, a mapping f of $E \times F$ into G is said to be *bilinear* if we have identically

$$f(x + x', y) = f(x, y) + f(x', y), \quad f(x, y + y') = f(x, y) + f(x, y'),$$
$$f(\lambda x, y) = f(x, \lambda y) = \lambda f(x, y)$$

for all $x, x' \in E$, all $y, y' \in F$ and all $\lambda \in K$.

If X is a nonsingular square matrix, the elements of X^{-1} are rational functions of the elements of X; these functions are therefore defined and continuous in a neighbourhood of X, so that every matrix Y in this neighbourhood is non-singular, and the mapping $Y \to Y^{-1}$ is continuous at the point X; hence $\mathbf{GL}_n(\mathbf{R})$ is open in $\mathbf{M}_n(\mathbf{R})$ and the topology of $\mathbf{GL}_n(\mathbf{R})$ is compatible with its group structure.

Finally, $\mathbf{GL}_n(\mathbf{R})$ is the complement of the set of square matrices X whose determinant is zero; since the determinant of X is a polynomial in the elements of X, Proposition 3 of no. 4 shows that $\mathbf{GL}_n(\mathbf{R})$ is *dense* in $\mathbf{M}_n(\mathbf{R})$.

2. EUCLIDEAN DISTANCE; BALLS AND SPHERES

1. EUCLIDEAN DISTANCE IN $\mathbf{R}^n$

In conformity with the general definitions the *Euclidean distance* between two points $x = (x_i)$ and $y = (y_i)$ is the number

$$d(x, y) = \sqrt{\sum_{i=1}^{n} (x_i - y_i)^2} \geqslant 0.$$

We recall its principal properties. The relation $d(x, y) = 0$ is equivalent to $x = y$. We have $d(x, y) = d(y, x)$; for all scalars $t \in \mathbf{R}$, $d(tx, ty) = |t| d(x, y)$; for all $z \in \mathbf{R}^n$, $d(x + z, y + z) = d(x, y)$; in other words, the distance between two points is *invariant under translation*. The distance $d(0, x)$ from the origin 0 to a point x is denoted also by $\|x\|$ and is called the *Euclidean norm* of x (or simply the *norm* of x, when there is no likelihood of confusion; cf. Chapter IX, § 3, no. 3). Then $d(x, y) = \|y - x\|$.

For $n = 1$, the Euclidean distance between the points x, y of $\mathbf{R}$ reduces to the length $|y - x|$ of the intervals with x and y as end-points. For any n, we say that $d(x, y) = \|y - x\|$ is the *length* of the segments with x and y as end-points.

The Euclidean distance satisfies the *triangle inequality*

$$d(x, y) \leqslant d(x, z) + d(z, y) \tag{1}$$

for all x, y, z in $\mathbf{R}^n$.

We recall that the proof of (1) reduces to that of the inequality

$$\left(\sum_{i=1}^{n} (x_i + y_i)^2\right)^{\frac{1}{2}} \leqslant \left(\sum_{i=1}^{n} x_i^2\right)^{\frac{1}{2}} + \left(\sum_{i=1}^{n} y_i^2\right)^{\frac{1}{2}};$$

this in turn is equivalent to the Cauchy-Schwarz inequality

$$\left(\sum_{i=1}^{n} x_i y_i\right)^2 \leqslant \left(\sum_{i=1}^{n} x_i^2\right)\left(\sum_{i=1}^{n} y_i^2\right),$$

which is an immediate consequence of Lagrange's identity

$$\left(\sum_{i=1}^{n} x_i^2\right)\left(\sum_{i=1}^{n} y_i^2\right) - \left(\sum_{i=1}^{n} x_i y_i\right)^2 = \frac{1}{2} \sum_{i,j} (x_i y_j - x_j y_i)^2.$$

This proof shows at the same time that the two sides of (1) can be equal only if z is a point of the segment with x and y as end-points.

From (1) we deduce the inequality

$$d(x, y) \geqslant |d(x, z) - d(y, z)|. \tag{2}$$

Finally, if $x = (x_i)$, $y = (y_i)$, we have

$$\sup_{1 \leqslant i \leqslant n} |x_i - y_i| \leqslant d(x, y) \leqslant \sqrt{n} . \sup_{1 \leqslant i \leqslant n} |x_i - y_i|. \tag{3}$$

Hence a subset A of $\mathbf{R}^n$ is *bounded* (§ 1, no. 1) if and only if

$$\sup_{x \in A} \|x\| < +\infty.$$

2. DISPLACEMENTS

We recall again that the affine transformations f of $\mathbf{R}^n$ onto itself which leave *invariant* the distance between any two points [that is to say, such that $d(f(x), f(y)) = d(x, y)$ for all x, y] are called *Euclidean displacements* (or simply *displacements*) (*); they form a group, called the *group of displacements* of $\mathbf{R}^n$. This group operates transitively on $\mathbf{R}^n$; more generally, if V and V′ are any two affine linear varieties of the same dimension in $\mathbf{R}^n$, there exists a displacement which transforms V into V′. The displacements which leave the origin fixed, called *orthogonal transformations*, form a subgroup of the group of all displacements. This subgroup is called the *orthogonal group* on n real variables; the linear mappings which belong to this group are characterized by the fact that they leave invariant the *norm* $\|x\|$ of every point $x \in \mathbf{R}^n$, or, equivalently, the *quadratic*

(*) If f is subjected only to the condition $d(f(x), f(y)) = d(x, y)$ for all x, y, then in fact f must be affine and linear, and therefore a displacement.

form $||x||^2 = \sum_{i=1}^{n} x_i^2$. The *scalar product* of two vectors $x = (x_i)$ and $y = (y_i)$ of $\mathbf{R}^n$ is the value $\sum_{i=1}^{n} x_i y_i$ of the bilinear form associated with the quadratic form $\frac{1}{2} \sum_{n=1}^{n} x_i^2$; it is denoted by $(x|y)$, or simply by xy if there is no likelihood of confusion. Every orthogonal transformation leaves invariant the scalar product of any two vectors. Two vectors x, y are said to be *orthogonal* if $(x|y) = 0$; two vector subspaces V, V′ of $\mathbf{R}^n$ are said to be *orthogonal* if each $x \in V$ is orthogonal to each $y \in V'$; and two affine linear varieties P, P′ are said to be *orthogonal* if the vector subspaces parallel respectively to P and P′ are orthogonal.

3. EUCLIDEAN BALLS AND SPHERES

For each integer $p > 0$, let U_p denote the set of all pairs (x, y) of points of $\mathbf{R}^n$ such that $d(x, y) < 1/p$; the inequalities (3) show that the sets U_p form a *fundamental system of entourages* of the uniformity of $\mathbf{R}^n$ (cf. Chapter IX, § 2).

From this fact and from the inequality

$$|d(x, y) - d(x', y')| \leqslant d(x, x') + d(y, y'),$$

which is a consequence of (1), we infer that $d(x, y)$ is *uniformly continuous* on $\mathbf{R}^n \times \mathbf{R}^n$; consequently the norm $||x|| = d(0, x)$ is *uniformly continuous* on $\mathbf{R}^n$.

DEFINITION 1. *Given a point* $x_0 \in \mathbf{R}^n$ *and a real number* $r > 0$, *the open* (resp. *closed*) *Euclidean ball of* n *dimensions with centre* x_0 *and radius* r *is the set of all points* $x \in \mathbf{R}^n$ *such that* $d(x_0, x) < r$ [resp. $d(x_0, x) \leqslant r$]; *the Euclidean sphere of* $n - 1$ *dimensions with centre* x_0 *and radius* r *is the set of all* $x \in \mathbf{R}^n$ *such that* $d(x_0, x) = r$.

When there is no risk of confusion we say simply " ball " (resp. " sphere ") for " Euclidean ball " (resp. " Euclidean sphere "). When $n = 2$, a ball of two dimensions is called a *disc*, and a sphere of one dimension is called a *circle*. When $n = 1$, the open (resp. closed) ball with centre x_0 and radius r is the interval $]x_0 - r, x_0 + r[$ (resp. $[x_0 - r, x_0 + r]$); the sphere with centre x_0 and radius r is the set consisting of the two end-points $x_0 - r$, $x_0 + r$ of these intervals.

From what has been said, the balls (open or closed) with centre x_0 (or just those with radii $1/p$, where p runs through the set of integers > 0) form a *fundamental system of neighbourhoods* of the point x_0.

PROPOSITION 1. *Every open* (resp. *closed*) *ball of* $\mathbf{R}^n$ *is an open* (resp. *compact*) *set. The closure of an open ball is the closed ball with the same centre and the same radius; the interior of a closed ball is the open ball with the same centre and the same radius.*

The open (resp. closed) ball with centre $\boldsymbol{x}_0$ and radius r is the inverse image of the interval $]-\infty, r[$ (resp. $]-\infty, r]$) under the continuous function $d(\boldsymbol{x}_0, \boldsymbol{x})$; it is therefore open (resp. closed and bounded, hence compact). If $d(\boldsymbol{x}_0, \boldsymbol{x}) = r$, and if $\boldsymbol{y} = \boldsymbol{x}_0 + t(\boldsymbol{x} - \boldsymbol{x}_0)$ $(0 < t < 1)$ is a point of the open segment with end-points $\boldsymbol{x}_0$ and $\boldsymbol{x}$, we have $d(\boldsymbol{x}_0, \boldsymbol{y}) = tr < r$, and $d(\boldsymbol{x}, \boldsymbol{y}) = (1 - t)r$ is as small as we please; hence $\boldsymbol{x}$ lies in the closure of the open ball with centre $\boldsymbol{x}_0$ and radius r. Again, if $\boldsymbol{z} = \boldsymbol{x} + t(\boldsymbol{x} - \boldsymbol{x}_0)$ $(t > 0)$ is a point of the open ray with origin $\boldsymbol{x}$ and direction vector $\boldsymbol{x} - \boldsymbol{x}_0$, we have

$$d(\boldsymbol{x}_0, \boldsymbol{z}) = (1 + t)r > r,$$

and $d(\boldsymbol{x}, \boldsymbol{z}) = tr$ is as small as we please; hence $\boldsymbol{x}$ is not an interior point of the closed ball with centre $\boldsymbol{x}_0$ and radius r.

COROLLARY. *Every Euclidean sphere is a compact set and is the frontier of the open and closed balls with the same centre and the same radius.*

The mapping $\boldsymbol{x} \to \frac{1}{r}(\boldsymbol{x} - \boldsymbol{x}_0)$ transforms the sphere (resp. open ball, closed ball) with centre $\boldsymbol{x}_0$ and radius r into the sphere (resp. open ball, closed ball) with centre 0 and radius 1; this sphere is denoted by $\mathbf{S}_{n-1}$ and is called the *unit sphere* in $\mathbf{R}^n$. Likewise, the *closed* ball with centre 0 and radius 1 is denoted by $\mathbf{B}_n$ and is called the *unit ball* in $\mathbf{R}^n$. The topological study of a sphere of $(n - 1)$ dimensions (resp. a closed ball of n dimensions) is thus reduced to that of $\mathbf{S}_{n-1}$ (resp. $\mathbf{B}_n$). For the open balls, we have the following proposition:

PROPOSITION 2. *Every n-dimensional open ball is homeomorphic to* $\mathbf{R}^n$.

For the mapping $\boldsymbol{x} \to \dfrac{\boldsymbol{x}}{1 + \|\boldsymbol{x}\|}$ is continuous on $\mathbf{R}^n$ and maps $\mathbf{R}^n$ onto the open ball with centre 0 and radius 1; moreover, from $\boldsymbol{y} = \dfrac{\boldsymbol{x}}{1 + \|\boldsymbol{x}\|}$ we deduce $\boldsymbol{x} = \dfrac{\boldsymbol{y}}{1 - \|\boldsymbol{y}\|}$, so that the mapping is bijective and bicontinuous.

Let $\mathbf{R}_n^*$ denote the complement of 0 in $\mathbf{R}^n$.

PROPOSITION 3. *The space* $\mathbf{R}_n^*$ *is homeomorphic to the product of* $\mathbf{S}_{n-1}$ *and the space* $\mathbf{R}_+^*$ *of real numbers* > 0.

For every point $x \neq 0$ can be written uniquely in the form tz, where $t > 0$ and $\|z\| = 1$, since $x = tz$ implies $t = \|x\|$ and $z = x/\|x\|$. Since tz is continuous on the product $\mathbf{R} \times \mathbf{R}^n$ and hence *a fortiori* on $\mathbf{R}_+^* \times \mathbf{S}_{n-1}$, and since $\|x\|$ and $\frac{1}{\|x\|}$ are continuous on $\mathbf{R}_n^*$, the proposition is proved.

The mapping $x \rightarrow x/\|x\|$ is called the *central projection* of $\mathbf{R}_n^*$ onto $\mathbf{S}_{n-1}$. One defines in the same way the *central projection* of the complement of a point a onto a sphere with centre a.

COROLLARY 1. *The sphere* $\mathbf{S}_{n-1}$ *is homeomorphic to the quotient of* $\mathbf{R}_n^*$ *by the equivalence relation whose classes are the open rays with origin* 0.

> These classes can also be defined as the *classes of intransitivity*, other than $\{0\}$, of the group of homotheties of ratio > 0.

COROLLARY 2. *The space* $\mathbf{R}_n^*$ *is homeomorphic to* $\mathbf{R} \times \mathbf{S}_{n-1}$.

For $\mathbf{R}_+^* =]0, +\infty[$ is homeomorphic to $\mathbf{R}$ (Chapter IV, § 4, no. 1, Proposition 1).

> *Remark.* These propositions are not peculiar to Euclidean balls, but can be extended to a whole category of compact neighbourhoods of 0 in $\mathbf{R}^n$ (see Exercise 12).

The sets $\mathbf{S}_{n-1}$ and $\mathbf{B}_n$ are evidently invariant under all orthogonal transformations. If V is a p-dimensional vector subspace in $\mathbf{R}^n$, there exists an orthogonal transformation which transforms V into a p-dimensional coordinate variety; hence $V \cap \mathbf{S}_{n-1}$ (resp. $V \cap \mathbf{B}_n$) is homeomorphic to $\mathbf{S}_{p-1}$ (resp. $\mathbf{B}_p$).

4. STEREOGRAPHIC PROJECTION

Consider the point $e_n = (0, \ldots, 0, 1)$ of $\mathbf{S}_{n-1}$, and the hyperplane H with equation $x_n = 0$, orthogonal to the vector e_n. To every point $x = (x_i)$ of $\mathbf{S}_{n-1}$, other than e_n, let us make correspond the point y where the line through e_n and x meets the hyperplane H (Fig. 3). It is easily verified that we have

$$y = \frac{1}{1 - x_n}(x - x_n e_n)$$

and

$$x = \frac{\|y\|^2 - 1}{\|y\|^2 + 1} e_n + \frac{2}{\|y\|^2 + 1} y.$$

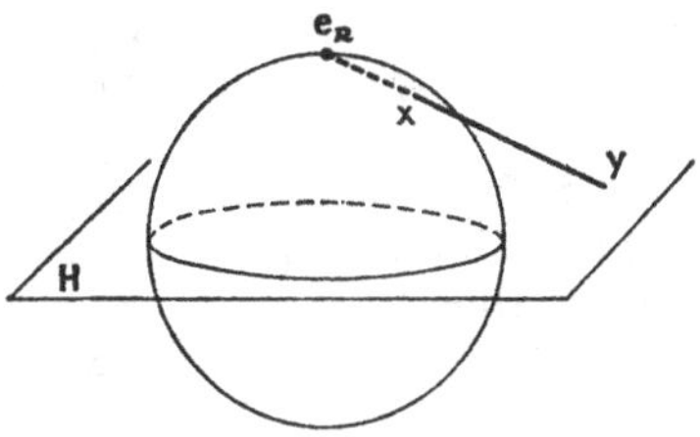

Figure 3.

If we denote by A the complement of $\{e_n\}$ in S_{n-1}, these formulas show that we have thus defined a *homeomorphism* of A *onto* the hyperplane H. This homeomorphism is called the *stereographic projection* of A onto H, or (by abuse of language) the stereographic projection of S_{n-1} onto H; e_n is the *vertex* of the projection, H the *hyperplane of projection*. More generally, if H′ is any hyperplane passing through 0 (a *diametral hyperplane* of B_n) and if a is one of the points of intersection of S_{n-1} and the line orthogonal to H′ passing through 0, we can define in the same way the stereographic projection with vertex a onto the hyperplane of projection H′; in any case this projection can be brought back to the preceding one by an orthogonal transformation which transforms H′ into H and a into e_n.

PROPOSITION 4. *If* $n > 1$, *the Euclidean sphere* S_{n-1} *is homeomorphic to the space* $\mathbf{R}^{n-1}$ *made compact by adjoining a "point at infinity"* (Chapter I, § 9, no. 8, Theorem 4).

For the stereographic projection defines a homeomorphism of the complement of a point in S_{n-1} onto a hyperplane of $\mathbf{R}^n$, which is homeomorphic to $\mathbf{R}^{n-1}$.

COROLLARY 1. *The sphere* S_n *is homeomorphic to the quotient space of the ball* B_n *obtained by identifying all the points of the sphere* S_{n-1}.

The ball B_n is a *regular* space (Chapter I, § 8, no. 4); hence the quotient space F of B_n obtained by identifying all the points of S_{n-1} is *Hausdorff* (Chapter I, § 8, no. 6, Proposition 15). Since B_n is compact, so is F, and F is therefore homeomorphic to an open ball of n dimensions made compact by adjoining a point at infinity, by Alexandroff's theorem (Chapter I, § 9, no. 8, Theorem 4). The result therefore follows from Propositions 2 and 4.

COROLLARY 2. *The circle* S_1 *is homeomorphic to the torus* $\mathbf{T}$.

In Chapter VIII, § 2, no. 1, we shall obtain this result again as a consequence of a more precise theorem.

PROPOSITION 5. *If* $n > 1$, *the Euclidean sphere* $\mathbf{S}_{n-1}$ *is connected and locally connected, and every point of it has an open neighbourhood homeomorphic to* $\mathbf{R}^{n-1}$.

The complement of a point in $\mathbf{S}_{n-1}$ is a connected dense set, and therefore (Chapter I, § 11, no. 1, Proposition 1) $\mathbf{S}_{n-1}$ is connected. To see that every point has a neighbourhood homeomorphic to $\mathbf{R}^{n-1}$, we have only to project stereographically from a vertex other than the given point.

From this proposition and from Proposition 3 of no. 3 it follows that $\mathbf{R}_n^*$, being the product of two connected spaces, is connected (Chapter I, § 11, no. 4, Proposition 8; cf. § 1, Exercise 10).

The intersection of $\mathbf{S}_{n-1}$ and a closed (resp. open) half-space determined by a diametral hyperplane of $\mathbf{B}_n$ is called a *closed* (resp. *open*) *hemisphere* of $\mathbf{S}_{n-1}$. By stereographic projection onto the diametral hyperplane, the closed (resp. open) hemisphere which does not contain the vertex of projection is mapped onto a *closed* (resp. *open*) ball of $n - 1$ dimensions, to which it is therefore *homeomorphic*.

If $n = 2$, we say " semicircle " instead of " hemisphere ".

3. REAL PROJECTIVE SPACES

In this section we shall need to invoke constantly the notions and results on *quotient spaces* (Chapter I, § 3), and in particular the following two properties, which for convenience we state as lemmas:

Let E be a topological space, R an equivalence relation on E, A a subset of E, R_A the equivalence relation induced on A by R and let f be the canonical mapping of E onto E/R. Then:

LEMMA 1. *If every open* (resp. *closed*) *set in* A *which is saturated with respect to* R_A *is the trace on* A *of an open* (resp. *closed*) *set in* E *which is saturated with respect to* R, *then the quotient space* A/R_A *is homeomorphic to the subspace* $f(A)$ *of* E/R. *This is so in particular if* A *is open or closed in* E *and saturated with respect to* R.

This follows from Proposition 10 of Chapter I, § 3, no. 6.

LEMMA 2. *If there is a continuous mapping* g *of* E *onto* A *such that, for all* $x \in E$, $g(x)$ *belongs to the equivalence class of* x, *then the quotient space* A/R_A *is homeomorphic to* E/R.

This is Corollary 2 of Proposition 10 of Chapter I, § 3, no. 6.

1. TOPOLOGY OF REAL PROJECTIVE SPACES

We recall the following definitions from algebra: given a division ring or a field K, the set of *lines* passing through 0 (i.e. vector subspaces of dimension 1) in the left vector space K_s^{n+1} over K is called *left projective space of* n *dimensions* over K, and is denoted by $\mathbf{P}_n$ (K).

If we make correspond to every line passing through 0 in K_s^{n+1} the same line *with the origin omitted*, we have a bijection of $\mathbf{P}_n(K)$ onto the *quotient* of K_{n+1}^* (the complement of $\{0\}$ in K_s^{n+1}) by the following equivalence relation $\Delta_n(K)$ between vectors x and y of K_{n+1}^*: " there exists $t \in K$ such that $t \neq 0$ and $y = tx$ ". In what follows we shall *identify* $P_n(K)$ with this quotient set. In the theory of projective spaces, we take the interval $[0, n]$ of $\mathbf{N}$ as index set for the coordinates of a point of K_{n+1}^*. The coordinates x_i $(0 \leqslant i \leqslant n)$ of any one of the points of K_{n+1}^* whose canonical image is $x \in \mathbf{P}_n(K)$ constitute what is called a *system of homogeneous coordinates* of the point x.

For each integer p such that $-1 \leqslant p \leqslant n$, the canonical image in $\mathbf{P}_n(K)$ of a vector subspace of $p+1$ dimensions (without the origin) of K_s^{n+1} is called a *projective linear variety* of p dimensions. A system of p points of $\mathbf{P}_n(K)$ is said to be *free* if it consists of the canonical images of p points of K_{n+1}^* which form a *free* system in the vector space K_s^{n+1}. The projective linear variety of $\mathbf{P}_n(K)$ generated by a free system of $p+1$ points (i.e. the smallest projective linear variety which contains these $p+1$ points) has dimension p.

When K is the field $\mathbf{R}$, the corresponding projective spaces can be topologized, and it is these we shall study.

DEFINITION 1. *The projective space* $\mathbf{P}_n(\mathbf{R})$ *endowed with the quotient topology of the topology of* $\mathbf{R}_{n+1}^*$ *by the equivalence relation* $\Delta_n(\mathbf{R})$ *is called real projective space of* n *dimensions.*

The projective space $\mathbf{P}_1(\mathbf{R})$ is called the *real projective line*, and $\mathbf{P}_2(\mathbf{R})$ is called the *real projective plane*.

Whenever there is no risk of confusion, we shall write $\mathbf{P}_n$ and Δ_n instead of $\mathbf{P}_n(\mathbf{R})$ and $\Delta_n(\mathbf{R})$.

PROPOSITION 1. *The projective space* $\mathbf{P}_n$ *is Hausdorff.*

We start by showing that the relation Δ_n is *open* (Chapter I, § 5, no. 2). Let A be an open set in $\mathbf{R}^*_{n+1}$; to saturate A with respect to Δ_n we have to take the union of the sets tA homothetic to A, as t runs through the set of real numbers $\neq 0$; since each of these sets is open, so is their union.

By Proposition 8 of Chapter I, § 8, no. 3, the proposition will be proved if we show that the subset M of $\mathbf{R}^*_{n+1} \times \mathbf{R}^*_{n+1}$ defined by the relation Δ_n is *closed*. Let then $(\boldsymbol{x}, \boldsymbol{y})$ be a point of $\mathbf{R}^*_{n+1} \times \mathbf{R}^*_{n+1}$ lying in the closure of M. If $\boldsymbol{x} = (x_i)$, there is an index i such that $x_i \neq 0$; hence there is a neighbourhood V of $(\boldsymbol{x}, \boldsymbol{y})$ such that for every point $(\boldsymbol{x}', \boldsymbol{y}') \in \mathrm{M} \cap \mathrm{V}$ the ith coordinate x'_i of $\boldsymbol{x}'$ is not 0. As $(\boldsymbol{x}', \boldsymbol{y}')$ tends to $(\boldsymbol{x}, \boldsymbol{y})$ while remaining in M, $y'_i x'^{-1}_i$ tends to $t = y_i x_i^{-1}$; since $\boldsymbol{y}' = (y'_i x'^{-1}_i)\boldsymbol{x}'$, we see by passing to the limit that $\boldsymbol{y} = \boldsymbol{x}$, which shows that $(\boldsymbol{x}, \boldsymbol{y}) \in \mathrm{M}$.

PROPOSITION 2. *The projective space* $\mathbf{P}_n$ *is compact and connected, and is homeomorphic to the quotient of the sphere* $\mathbf{S}_n$ *by the equivalence relation induced on the sphere by* Δ_n.

Let Δ'_n be the equivalence relation induced on $\mathbf{S}_n$ by Δ_n (the equivalence classes of Δ'_n are pairs of *diametrically opposite* points of $\mathbf{S}_n$). The mapping $\boldsymbol{x} \to \boldsymbol{x}/\|\boldsymbol{x}\|$ of $\mathbf{R}^*_{n+1}$ onto $\mathbf{S}_n$ is continuous, hence (Lemma 2) $\mathbf{P}_n$ is homeomorphic to $\mathbf{S}_n/\Delta'_n$. Since $\mathbf{S}_n$ is compact and connected, every Hausdorff quotient space of $\mathbf{S}_n$ is also compact and connected (Chapter I, § 9, no. 4, Theorem 2, Corollary 1; § 11, no. 3, Proposition 7).

PROPOSITION 3. *If* $n \geqslant 0$, *the projective space* $\mathbf{P}_n$ *is homeomorphic to the quotient space of the ball* $\mathbf{B}_n$ *obtained by identifying each point of* $\mathbf{S}_{n-1}$ *with its diametrically opposite point.*

Let H be the closed hemisphere of $\mathbf{S}_n$ defined by $x_0 \leqslant 0$. $\mathbf{P}_n$, which is homeomorphic to the quotient space of $\mathbf{S}_n$ by the relation Δ'_n, is also homeomorphic to the quotient of the subset H of $\mathbf{S}_n$ by the relation Δ''_n induced on H by Δ'_n. For every equivalence class of Δ'_n meets H in at least one point, and therefore (Lemma 1) it is enough to verify that if we saturate with respect to Δ'_n an open subset U of H which is saturated with respect to V''_n, we get an open subset V of $\mathbf{S}_n$. Now if $\boldsymbol{a} = (a_i) \in \mathrm{U}$ and if $a_0 < 0$, there is a neighbourhood W of $\boldsymbol{a}$ *in* $\mathbf{S}_n$ contained in U, and the union of W and $-$W is a neighbourhood of $\boldsymbol{a}$ saturated with respect to Δ'_n and contained in V. If on the other hand $a_0 = 0$, we have $-\boldsymbol{a} \in \mathrm{U}$, and there exists $r > 0$ such that the set of points $\boldsymbol{x} \in \mathrm{H}$ which satisfy one or the other of the relations

$\|x - a\| < r$, $\|x + a\| < r$ is contained in U; the set of points $x \in \mathbf{S}_n$ which satisfy one or the other of these relations is the neighbourhood of a which is saturated with respect to Δ'_n and is contained in V.

Observe that the quotient space H/Δ''_n is obtained by identifying, in H, each point of the intersection $\mathbf{S}_{n-1}$ of H and the hyperplane $x_0 = 0$ with its opposite point. To complete the proof it suffices to remark that the stereographic projection with vertex e_0 (§ 2, no. 4) is a homeomorphism of H onto $\mathbf{B}_n$ which leaves invariant the points of $\mathbf{S}_{n-1}$.

2. PROJECTIVE LINEAR VARIETIES

Every injective linear mapping f of $\mathbf{R}^{n+1}$ into $\mathbf{R}^{m+1}$ $(m \geqslant n)$ defines, by restriction to $\mathbf{R}^*_{n+1}$ and then by passage to the quotient with respect to the relations Δ_n and Δ_m (*Set Theory*, R, § 5, no 8), an injective mapping g of $\mathbf{P}_n$ into $\mathbf{P}_m$, called a *projective linear mapping*. If φ (resp. ψ) is the canonical mapping of $\mathbf{R}^*_{n+1}$ onto $\mathbf{P}_n$ (resp. $\mathbf{R}^*_{m+1}$ onto $\mathbf{P}_m$), we have $g \circ \varphi = \psi \circ f$, which shows that g is *continuous* on $\mathbf{P}_n$ (Chapter I, § 3, no. 4, Corollary to Proposition 6). In particular, every *projective linear transformation* of $\mathbf{P}_n$ (i.e. projective linear mapping of $\mathbf{P}_n$ *onto* itself) is a *homeomorphism* of $\mathbf{P}_n$ onto itself.

We recall also that, if V and V′ are two projective linear varieties of p dimensions in $\mathbf{P}_n$, there exists a projective linear transformation of $\mathbf{P}_n$ which transforms V into V′. In particular, if $p \geqslant 0$, there exists a projective linear transformation which transforms V into a *coordinate* projective linear variety, that is to say the canonical image of a coordinate variety W′ of $p + 1$ dimensions (without the point 0) of $\mathbf{R}^{n+1}$. If we identify W′ with $\mathbf{R}^*_{p+1}$, the relation induced by Δ_n on W′ is precisely Δ_p; since W′ is closed and saturated with respect to Δ_n, Lemma 1 shows that V′ is homeomorphic to $\mathbf{P}_p$ and is closed in $\mathbf{P}_n$; moreover, if $p < n$, its complement is dense in $\mathbf{P}_n$ (§ 1, no. 4, Corollary to Proposition 3). Hence:

PROPOSITION 4. *Every projective linear variety of* p *dimensions in a projective space* $\mathbf{P}_n$ *is closed in* $\mathbf{P}_n$ *and homeomorphic to* $\mathbf{P}_p$; *if* $p < n$ *its complement is dense in* $\mathbf{P}_n$.

This result is the origin of the names *projective line* and *projective plane* given to projective linear varieties of *one* and *two* dimensions in a projective space over an arbitrary division ring. We recall also that the projective linear varieties of $n - 1$ dimensions in $\mathbf{P}_n$ are called *projective hyperplanes*; every projective hyperplane is identical with the set of points whose

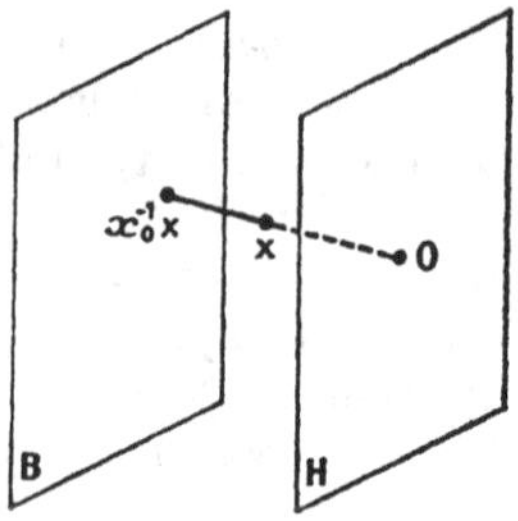

Figure 4.

homogeneous coordinates satisfy a relation of the form $\sum_{i=0}^{n} a_i x_i = 0$, where the a_i are not all zero (the " equation " of the hyperplane).

PROPOSITION 5. *In a projective space* $\mathbf{P}_n$ $(n \geqslant 0)$, *the complement of a projective hyperplane* H *is homeomorphic to* $\mathbf{R}^n$.

By making a projective linear transformation we may assume that H is the hyperplane whose equation is $x_0 = 0$. The set A of points $x = (x_i)$ of $\mathbf{R}^*_{n+1}$ such that $x_0 \neq 0$ is open and saturated with respect to Δ_n; its canonical image C in $\mathbf{P}_n$, which is the complement of H in $\mathbf{P}_n$, is therefore homeomorphic to the quotient of A by the equivalence relation Θ induced on A by Δ_n (Lemma 1). Let B be the hyperplane whose equation is $x_0 = 1$ in $\mathbf{R}^{n+1}$. To each point $x \in A$ let correspond the point $x_0^{-1}x$ where the line through 0 and x cuts B (Fig. 4); in this way we define a continuous mapping g of A onto B such that $g(x)$ is the only point of B congruent to $\mathbf{x}$ modulo Θ. It follows that B is homeomorphic to A/Θ (Lemma 2), hence to C; since B is homeomorphic to $\mathbf{R}^n$, the proof is complete.

COROLLARY. *Every point of* $\mathbf{P}_n$ *has an open neighbourhood homeomorphic to* $\mathbf{R}^n$.

It follows in particular that the real projective spaces are *locally connected* (this follows also from Chapter I, § 11, no. 6, Proposition 12).

3. EMBEDDING REAL NUMBER SPACE IN PROJECTIVE SPACE

Proposition 5 of no. 2 shows that if we are given a projective hyperplane H in $\mathbf{P}_n$ $(n \geqslant 0)$, there is a homeomorphism of $\mathbf{R}^n$ onto the complement $\complement H$ of this hyperplane. Once H has been chosen it is often convenient to *identify* $\mathbf{R}^n$ and $\complement H$ by means of the homeomorphism defined in Proposition 5; the projective hyperplane H is then said to be " at infinity ", and so are its points and subsets. Usually one takes H to be the " coordinate " hyperplane whose equation is $x_0 = 0$, and then the point $z = (z_i)$ of $\mathbf{R}^n$ is identified with the point of $\mathbf{P}_n$ whose homogeneous coordinates are $1, z_1, z_2, \ldots, z_n$.

Once this identification has been made, the *closure* in $\mathbf{P}_n$ of any *affine linear* variety V of p dimensions in $\mathbf{R}^n$ is a *projective linear* variety of p dimensions, not contained in the hyperplane at infinity, and identical with the projective linear variety *generated* by V. Conversely, every projective linear variety P of p dimensions which is not contained in the hyperplane at infinity has as its trace on $\mathbf{R}^n$ an affine linear variety of p dimensions, whose closure in $\mathbf{P}_n$ is P.

In the particular case $n = 1$, the hyperplane at infinity is a *point*; since $\mathbf{P}_1$ is compact, it follows from Alexandroff's theorem (Chapter I, § 9, no. 8, Theorem 4) that $\mathbf{P}_1$ is homeomorphic to the space $\tilde{\mathbf{R}}$ obtained by compactifying the locally compact space $\mathbf{R}$ by the adjunction of a point (the " point at infinity "). By Proposition 4 of § 2, no. 4, we see also that the *real projective line* $\mathbf{P}_1(\mathbf{R})$ *is homeomorphic to the circle* $\mathbf{S}_1$ *and to the torus* $\mathbf{T}$.

On the other hand, if $n > 1$, $\mathbf{P}_n(\mathbf{R})$ is not homeomorphic to $\mathbf{S}_n$, as we shall see later (cf. Exercise 4).

The " point at infinity " of the space $\tilde{\mathbf{R}}$ is denoted by ∞, *with no sign attached.* It is important not to confuse $\tilde{\mathbf{R}}$ with the *extended real line* $\overline{\mathbf{R}}$ defined in Chapter IV, § 4, which has *two* " points at infinity "; indeed $\tilde{\mathbf{R}}$ is homeomorphic to the quotient space of $\overline{\mathbf{R}}$ obtained by *identifying* the two points $+\infty$ and $-\infty$.

4. APPLICATION TO THE EXTENSION OF REAL-VALUED FUNCTIONS

Since $\mathbf{R}$ can be considered as a subset of $\tilde{\mathbf{R}}$, every mapping of a set E into $\mathbf{R}$ (i.e. every real-valued function on E) can be considered as a mapping of E into $\tilde{\mathbf{R}}$; in particular, if E is a subset of a topological space F and f is a mapping of E into $\mathbf{R}$, it may happen that at some points of the closure $\overline{E}$ of E, $f(x)$ tends to the limit ∞ as x tends to one of these points while remaining in E; we may then extend the function f by continuity by assigning it the value ∞ at these points (Chapter I, § 8, no. 5, Theorem 1).

Consider in particular the case where E is a subset of $\mathbf{R}^n$, the space $\mathbf{R}^n$ itself being considered as embedded in projective space $\mathbf{P}_n$; if we suppose that the hyperplane at infinity is $x_0 = 0$, a real-valued function f defined on E may be identified with the mapping

$$(x_0, x_1, x_2, \ldots, x_n) \to f\left(\frac{x_1}{x_0}, \frac{x_2}{x_0}, \ldots, \frac{x_n}{x_0}\right)$$

of E into $\tilde{\mathbf{R}}$; from what has been said in the previous paragraph it may be possible to extend this function, not only to some points of $\mathbf{R}^n$ in the closure of E, but also to some of the " points at infinity " of $\mathbf{P}_n$ in the closure of E.

Let us show that we obtain in this way, for example, the extension by continuity to the whole of $\tilde{\mathbf{R}}$ of a *rational function* of a real variable, already defined in algebra. Let us identify $\tilde{\mathbf{R}}$ and $\mathbf{P}_1$, every real number $x \in \mathbf{R}$ being identified with the point whose homogeneous coordinates are $(1, x)$, and the point ∞ with the point whose homogeneous coordinates are $(0,1)$. Let $u(x)/v(x)$ be a rational function, u and v being two coprime polynomials of degrees m and n respectively; if we suppose for example that $m \leqslant n$, and if we put $u_1(x,y) = x^n u(y/x)$, $v_1(x, y) = x^n v(y/x)$, the rational function u/v may be considered as the restriction, to the set of real numbers x such that $v(x) \neq 0$, of the mapping $(x,y) \to (v_1(x,y), u_1(x, y))$. In other words, we extend u/v by continuity, by giving it the value ∞ at those points $x \in \mathbf{R}$ where $v(x) = 0$, and by giving it at the point ∞ the value 0 if $m < n$, the value ∞ if $m > n$, and the value of the ratio of the leading coefficients if $m = n$.

In particular, the function $1/x$ can be extended to the point 0 by taking the value ∞ there, to ∞ by taking the value 0 there; this extended function is evidently a *homeomorphism* of $\tilde{\mathbf{R}}$ onto itself. The same is true of the *homographic function* $(ax + b)/(cx + d)$ when $ad - bc \neq 0$.

Likewise, if n is an integer > 0, the function x^n extends to the point ∞ by taking the value ∞ there.

On the other hand, it is in general impossible to extend by continuity a rational function of two real variables either to the space $\mathbf{P}_1 \times \mathbf{P}_1$ or to the space $\mathbf{P}_2$ (cf. Exercise 5).

5. SPACES OF PROJECTIVE LINEAR VARIETIES

Given a division ring K, the set $\mathbf{P}_{n,p}(\mathrm{K})$ of *projective linear varieties of* $p \geqslant 0$ *dimensions* of the left projective space $\mathbf{P}_n(\mathrm{K})$ is clearly in one-to-one correspondence with the set of vector subspaces of $p + 1$ dimensions of the left vector space K_s^{n+1}. Let $\mathrm{L}_{n+1,\, p+1}(\mathrm{K})$ denote the set of *free systems* $(\mathbf{x}_k)_{1 \leqslant k \leqslant p+1}$ of $p + 1$ vectors of K_s^{n+1}; then the set $\mathbf{P}_{n,p}(\mathrm{K})$

is again in one-to-one correspondence with the quotient of $L_{n+1,\,p+1}(K)$ by the equivalence relation $\Delta_{n,\,p}(K)$: " $(\boldsymbol{x}^k)$ and $(\boldsymbol{y}^k)$ generate the same vector subspace of $p+1$ dimensions of K_s^{n+1} ". In what follows we shall *identify* $\mathbf{P}_{n,\,p}(K)$ with this quotient set. On the other hand, if to each free system $(\boldsymbol{x}_k)$ of $p+1$ vectors of K_s^{n+1} we make correspond the *matrix* X of $p+1$ rows and $n+1$ columns for which $\boldsymbol{x}_k$ is the kth row $(1 \leqslant k \leqslant p+1)$, we have a one-to-one correspondence between $L_{n+1,\,p+1}(K)$ and the set of all matrices of $p+1$ rows and $n+1$ columns which are of *rank* $p+1$; we shall *identify* $L_{n+1,\,p+1}(K)$ with this set of matrices, and the relation $\Delta_{n,\,p}(K)$ between two matrices X, Y is then the following: " there exists a non-singular square matrix T of order $p+1$ such that $Y = T.X$ ".

In what follows we shall take K to be the field $\mathbf{R}$, and we shall omit the letter K in the notation above. We can define a topology on $\mathbf{P}_{n,p}$ by a process which generalizes the definition of the topology of real projective spaces. Namely, $L_{n+1,\,p+1}$ is contained in the space $\mathbf{M}_{p+1,\,n+1}$ of all matrices of $p+1$ rows and $n+1$ columns with real elements; we endow $L_{n+1,\,p+1}$ with the topology induced by the topology of this matrix space (§ 1, no. 6).

Definition 2. *The space* $\mathbf{P}_{n,\,p}$ *which is the quotient of the topological space* $L_{n+1,\,p+1}$ *by the equivalence relation* $\Delta_{n,\,p}$ *is called the space of projective linear varieties of* $p \geqslant 0$ *dimensions in real projective space* $\mathbf{P}_n$.

We shall use the following notation: given a matrix X of $p+1$ rows and $n+1$ columns, and any strictly increasing sequence

$$\sigma = (i_1, \ldots, i_{p+1})$$

of $p+1$ indices belonging to the interval $[0, n]$ of $\mathbf{N}$, we denote by X_σ the square submatrix of X formed by the columns whose indices are $i_1, i_2, \ldots, i_{p+1}$. We denote by A_σ the subset of $L_{p+1,\,p+1}$ consisting of these matrices X such that X_σ is non-singular. By Proposition 6 of § 1, no. 6, A_σ is a *dense open* set in $\mathbf{M}_{p+1,\,n+1}$, and the function $X \to X_\sigma^{-1}$ is *continuous* on A_σ.

A geometrical interpretation of the set A_σ is as follows: let E_σ be the vector subspace of $\mathbf{R}^{n+1}$ generated by the vectors $\boldsymbol{e}_i$ of the canonical basis such that $i \in \sigma$, and let E'_σ be the complementary subspace generated by the $\boldsymbol{e}_i$ such that $i \notin \sigma$; then to say that a matrix x belongs to A_σ means that the projections on E_σ of its $p+1$ rows of $\boldsymbol{x}_k$ form a free system, or again that the vector subspace generated by the $\boldsymbol{x}_k$ is a *complement* of E'_σ (or that its intersection with E'_σ consists only of 0).

Proposition 6. *The space* $\mathbf{P}_{n,\,p}$ *is Hausdorff.*

We show first that the relation $\Delta_{n,p}$ is *open*. If U is an open set in $L_{n+1,p+1}$, to saturate U with respect to $\Delta_{n,p}$ we have to take the union of the images of U under the mapping $X \to T.X$, where T runs through the set of non-singular square matrices of order $p+1$; since each of these mappings is bicontinuous, all these images are open sets and therefore so is their union.

By Proposition 8 of Chapter I, § 8, no. 3, the proof will be complete if we show that the subset N of $L_{n+1,p+1} \times L_{n+1,p+1}$, defined by $\Delta_{n,p}$, is *closed*. Let (X, Y) be a point of the product space which lies in the closure of N, and let σ be a sequence of indices such that X_σ is non-singular: since A_σ is open, there is a neighbourhood V of (X, Y) such that, for each pair $(X', Y') \in N \cap V$, the matrix X'_σ is non-singular; as (X', Y') tends to (X, Y) while remaining in N, the matrix $Y'_\sigma X'^{-1}_\sigma$ therefore tends to $T = Y_\sigma X_\sigma^{-1}$; since we have $Y' = (Y'_\sigma X'^{-1}_\sigma)X'$, we see by passing to the limit that $Y = T.X$, and the proof is complete.

PROPOSITION 7. *The space* $\mathbf{P}_{n,p}$ *is compact.*

It is enough to show that there is a compact subspace of $L_{n+1,p+1}$ which meets every equivalence class mod $\Delta_{n,p}$ in at least one point; for $\mathbf{P}_{n,p}$ is then the image of this subspace under the canonical mapping of $L_{n+1,p+1}$ onto $\mathbf{P}_{n,p}$, and is therefore compact (Chapter I, § 9, no. 4, Theorem 2).

Let $V_{n+1,p+1}$ be the subspace of $L_{n+1,p+1}$ whose elements are the systems $(\mathbf{x}_k)$ of $p+1$ vectors forming an *orthonormal Euclidean basis* of the vector subspace they generate that is to say such that $(x_h|x_k) = 0$ whenever $h \neq k$, $(x_h|x_k) = 1$ for $1 \leqslant h \leqslant p+1$. Every vector subspace of $p+1$ dimensions of $\mathbf{R}^{n+1}$ has such a basis and therefore every class mod. $\Delta_{n,p}$ meets $V_{n+1,p+1}$. On the other hand, the matrices $X = (x_{ij})$ of $V_{n+1,p+1}$ are defined by the relations

$$\sum_{j=0}^{n} x_{ij}^2 = 1 \qquad (1 \leqslant i \leqslant p+1),$$

$$\sum_{j=0}^{n} x_{ij}x_{kj} = 0 \qquad (i \neq k);$$

therefore they form a *closed* set in $M_{p+1,n+1}$; and since these relations imply that $|x_{ij}| \leqslant 1$ for each pair of indices (i, j), this set is *bounded* and hence *compact*.

PROPOSITION 8. $\mathbf{P}_{n,p}$ *is connected and locally connected, and every point has an open neighbourhood homeomorphic to* $\mathbf{R}^{(p+1)(n-p)}$.

For every (strictly increasing) sequence of indices σ, the set A_σ is open in $L_{n+1,p+1}$ and saturated with respect to $\Delta_{n,p}$; its canonical image

C_σ in $\mathbf{P}_{n,p}$ is therefore an open set homeomorphic to the quotient of A_σ by the equivalence relation Θ_σ induced on A_σ by $\Delta_{n,p}$ (Lemma 1).

Let B_σ be the subset of A_σ consisting of matrices X such that X_σ is the *unit matrix* of order $p + 1$; the elements of X other than those of X_σ are then *arbitrary*, and therefore B_σ is homeomorphic to the space $\mathbf{R}^{(p+1)(n-p)}$. To each matrix $X \in A_\sigma$ let us make correspond the matrix $Y = X_\sigma^{-1}X$, which belongs to B_σ; then we have defined a continuous mapping g of A_σ onto B_σ, such that $g(X)$ is the only matrix of B_σ congruent to X mod Θ_σ. It follows that B_σ is homeomorphic to A_σ/Θ_σ (Lemma 2), hence to C_σ.

The set C_σ is therefore connected. Since A_σ is dense in $L_{n+1,p+1}$, C_σ is dense in $\mathbf{P}_{n,p}$ and therefore $\mathbf{P}_{n,p}$ is also connected (Chapter I, § 11, no. 1, Proposition 1). On the other hand, every point of $\mathbf{P}_{n,p}$ belongs to C_σ for at least one sequence of indices σ, and therefore has an open neighbourhood homeomorphic to $\mathbf{R}^{(p+1)(n-p)}$.

The matrix $Y = g(X)$ can be interpreted as follows: let us suppose for the sake of simplicity that the sequence σ consists of the $p + 1$ indices $n - p$, $n - p + 1, \ldots, n$, and let a_{ij} $(1 \leqslant i \leqslant p + 1,\ 0 \leqslant j \leqslant n - p - 1)$ denote the elements of the first $n - p$ columns of y; then the vector subspace of $\mathbf{R}^{n+1}$ generated by the rows of x is that defined by the equations

$$x_j = \sum_{i=1}^{p+1} a_{ij}x_{n-p+i-1} \qquad (0 \leqslant j \leqslant n - p - 1).$$

6. GRASSMANNIANS

If K is a *field* and X is any matrix of $L_{n+1,p+1}(K)$, let $c_\sigma(X)$ denote the determinant of X_σ; in this way, to each matrix X of $L_{n+1,p+1}(K)$ correspond

$$h = \binom{n+1}{p+1}$$

determinants, not all zero (the components of the exterior product of the $p + 1$ rows of X). If we make correspond to X the point of the projective space $\mathbf{P}_{h-1}(K)$ whose homogeneous coordinates are the $c_\sigma(X,)$ we have defined a mapping of $L_{n+1,p+1}(K)$ into $\mathbf{P}_{h-1}(K)$, compatible with the relation $\Delta_{n,p}(K)$; passing to the quotient, we have therefore a mapping f of $\mathbf{P}_{n,p}(K)$ into $\mathbf{P}_{h-1}(K)$. The image $G_{n,p}(K)$ of $\mathbf{P}_{n,p}(K)$ under this mapping is called the *Grassmannian* of indices n, p. We recall also that the mapping f is *injective*, for if X is a matrix such that X_σ is non-singular, the matrix $Y = X_\sigma^{-1}X$ of B_σ which corresponds to the class of X mod. $\Delta_{n,p}(K)$ is the matrix $(d_{ij}/c_\sigma(X))$ $(1 \leqslant i \leqslant p + 1,\ 0 \leqslant j \leqslant n)$, where d_{ij} denotes the determinant of the matrix obtained from X_σ by replacing the ith column of X_σ by the jth column of X [which implies that d_{ij} is, up to sign, equal to one of the $c_\tau(X)$].

When K is the field **R**, this mapping f is evidently *continuous.* The inverse mapping g is also continuous; for the elements of a matrix belonging to B_σ are rational functions of the homogeneous coordinates of the point of the Grassmannian to which it corresponds; since $f(B_\sigma) = B'_\sigma$ is the set of points of $G_{n,p}$ whose homogeneous coordinate with index σ is not 0, it is an open set in $G_{n,p}$; hence g is continuous at every point of B'_σ, and since every point of $G_{n,p}$ belongs to at least one set B'_σ, g is continuous at every point. Thus:

PROPOSITION 9. *The Grassmannian* $G_{n,p}$ *is homeomorphic to the space* $\mathbf{P}_{n,p}$.

We recall finally that the Grassmannians $G_{n,p}(K)$ and $G_{n,n-p-1}(K)$ are subsets of the same projective space $\mathbf{P}_{n-1}(K)$ and can be transformed into the other by a projective linear transformation; it follows that $G_{n,p}$ and $G_{n,n-p-1}$ are *homeomorphic.*

EXERCISES

§ 1

1) Give a proof of the fact that $\mathbf{R}^n$ is equipotent with $\mathbf{R}$ by using Cantor's theorem (Chapter IV, § 8, no. 6, Theorem 1) and the relation $2^{\mathfrak{a}} . 2^{\mathfrak{a}} = 2^{\mathfrak{a}}$, valid for all infinite cardinals $\mathfrak{a}$.

¶ 2) There exists a *continuous* mapping of the interval $I = [0, 1]$ of $\mathbf{R}$ *onto* the square $I \times I$ of $\mathbf{R}^2$ (the " Peano curve ", cf. Exercise 8). (Show first, with the help of Exercise 11 of Chapter IV, § 8, that there exists a continuous mapping f of Cantor's triadic set K onto $I \times I$, and then extend f to I).

¶ 3) Let A and B be two countable dense subsets of $\mathbf{R}^2$. Show that there exists a homeomorphism of $\mathbf{R}^2$ onto itself which maps A onto B. (Show first that, by means of a rotation, we can reduce to the case where the projections pr_1 and pr_2 are injective mappings of A and B into $\mathbf{R}$. Then define, by a suitable inductive process, a bijection of A onto B which determines a monotone mapping of $\mathrm{pr}_1 A$ onto $\mathrm{pr}_1 B$ and a monotone mapping of $\mathrm{pr}_2 A$ onto $\mathrm{pr}_2 B$. Finally, using Exercise 11, Chapter IV, § 2, show that this mapping is a homeomorphism which can be extended to a homeomorphism of $\mathbf{R}^2$ onto itself.) Deduce that, if C is a subset of $\mathbf{R}^2$ whose complement is dense, C is homeomorphic to a subset of the complement of $\mathbf{Q}^2$ in $\mathbf{R}^2$. Generalize these results to $\mathbf{R}^n$, $n > 2$.

4) Every countable subspace of $\mathbf{R}^n$ with no isolated points is homeomorphic to the rational line $\mathbf{Q}$ (apply Exercise 13 of Chapter IV, § 8).

5) Every totally discontinuous compact subspace of $\mathbf{R}^n$ with no isolated points is homeomorphic to Cantor's triadic set (use Chapter IV, § 8, Exercises 11 and 12, and Chapter II, § 4, no. 4, Proposition 6).

6) A subset L of $\mathbf{R}^n$ is a *broken line* if there exists a finite sequence $(x_i)_{0 \leqslant i \leqslant p}$ of points of $\mathbf{R}^n$ such that, if S_i denotes the segment with end-points x_{i-1} and x_i $(1 \leqslant i \leqslant p)$, L is the union of the S_i, which are called the *sides* of L. (In general there is an infinite number of finite sequences of points of $\mathbf{R}^n$ which define the same broken line.) A broken line is also the image of a mapping u of $[0, 1]$ into $\mathbf{R}^n$ such that there exists a strictly increasing sequence $(t_j)_{0 \leqslant j \leqslant q}$ in $[0, 1]$ with $t_0 = 0$ and $t_q = 1$, with the property that u is an affine linear mapping

$$t \to a_j + tb_j \quad \text{for} \quad t_{j-1} \leqslant t \leqslant t_j \text{ and } 1 \leqslant j \leqslant q$$

(such a mapping u is said to be "*piecewise linear*"). Given a non-empty subset A of $\mathbf{R}^n$ we say that two points a, b of A *can be joined by a broken line in* A if there exists a broken line $L \subset A$ defined by a sequence $(x_i)_{0 \leqslant i \leqslant p}$ such that $x_0 = a$ and $x_p = b$. If any two points of A can be joined by a broken line in A, then A is connected. Conversely, if A is a *connected open subset* of $\mathbf{R}^n$, show that any two points of A can be joined by a broken line in A (consider the relation " x and y can be joined by a broken line in A " between points x, y of A; show that this is an equivalence relation whose classes are open sets); we can even assume always that this broken line is a union of segments each of which is parallel to a coordinate axis (same method). Deduce that, if A is a non-empty open set in $\mathbf{R}^n$, the component of a point $a \in A$ is the set of all points of A which can be joined to a by a broken line in A.

7) Let A be a non-empty connected open set in $\mathbf{R}^n$ $(n > 1)$ and let $(V_p)_{p \in \mathbf{N}}$ be a countable family of linear varieties in $\mathbf{R}^n$, each of which is of dimension $\leqslant n - 2$. Show that if B is the union of the V_p, then $A \cap \complement B$ is dense in A and connected. (To show that $A \cap \complement B$ is dense in A, use induction on n. Then show that if x, y are two distinct points of A, there exists for each $\varepsilon > 0$ a point $y' \in A$ such that $\|y' - y\| \leqslant \varepsilon$ and such that the closed segment with end-points x and y' meets B only at x. Finally use Exercise 6 to show that $A \cap \complement B$ is connected.)

8) Deduce from Exercise 7 that, if $n > 1$, a non-empty open subset of $\mathbf{R}^n$ is not homeomorphic to a subset of $\mathbf{R}$ (*).

9) A broken line L in $\mathbf{R}^n$ $(n > 1)$ (Exercise 6) is said to be *simple* if it is homeomorphic to the interval $I = [0, 1]$ of $\mathbf{R}$. It amounts to the same thing to say that there exists a piecewise linear homeomorphism u

(*) It can be shown that an open set in $\mathbf{R}^m$ is not homeomorphic to any subset of $\mathbf{R}^n$, if $n < m$.

of I onto L (Exercise 6). Show that if A is a connected open set in $\mathbf{R}^2$ and if $L \subset A$ is a simple broken line, then $A - L$ is connected (argue by induction on the number of segments which make up L, and use Exercise 6 of Chapter I, § 11).

¶ 10) *a*) Let $\mathfrak{M}$ be a finite set and let $R\{x, y\}$ be a symmetric relation between elements x, y of $\mathfrak{M}$. Suppose that there exist two distinct elements a, b in $\mathfrak{M}$ with the following properties: there exist $x \neq a$ and $y \neq b$ such that x (resp. y) is the only element $z \in \mathfrak{M}$ such that $R\{a, z\}$ (resp. $R\{b, z\}$) is true; and for each $t \in \mathfrak{M}$ other than a and b, the set of all $z \in \mathfrak{M}$ such that $R\{t, z\}$ is true is a set consisting of two elements, distinct from t. Show that under these conditions there exists a bijection $i \to x_i$ of an interval $[0, n]$ of $\mathbf{N}$ onto $\mathfrak{M}$ such that $x_0 = a$, $x_n = b$ and such that $R\{x_{i-1}, x_i\}$ is true for $1 \leqslant i \leqslant n$ (define x_i by induction on i).

b) In $\mathbf{R}^n$ $(n \geqslant 2)$, identified with $\mathbf{R}^{n-1} \times \mathbf{R}$, let L and L′ be two broken lines, defined respectively by two sequences $(\boldsymbol{x}_i)_{0 \leqslant i \leqslant p}$ and $(\boldsymbol{x}'_j)_{0 \leqslant j \leqslant q}$, with the following properties: (i) if we put $\boldsymbol{x}_i = (y_i, z_i)$, $\boldsymbol{x}'_j = (y'_j, z'_j)$, we have $z_0 = z'_0 = 0$, $z_p = z'_q = 1$, $0 < z_i < 1$ for $1 \leqslant i \leqslant p - 1$ and $0 < z'_j < 1$ for $1 \leqslant j \leqslant q - 1$; (ii) the $p + q - 2$ numbers z_i, z'_j $(1 \leqslant i \leqslant p - 1,\ 1 \leqslant j \leqslant q - 1)$ are all distinct; (iii) two distinct sides of L (resp. L′) have at most one point in common [which may or may not be one of the $\boldsymbol{x}_i$ (resp. $\boldsymbol{x}'_j$)]. Show that there exist two surjective piecewise linear mappings (Exercise 6) $\boldsymbol{u}: \mathrm{I} \to \mathrm{L}$, $\boldsymbol{u}': \mathrm{I} \to \mathrm{L}'$, where I is the interval $[0, 1]$ of $\mathbf{R}$, such that, if we put $\boldsymbol{u}(t) = (\boldsymbol{v}(t), \zeta(t))$, $\boldsymbol{u}'(t) = (\boldsymbol{v}'(t), \zeta'(t))$ $(t \in \mathrm{I})$, we have $\zeta(t) = \zeta'(t)$ for *all* $t \in \mathrm{I}$,

$$\zeta(0) = \zeta'(0) = 0 \quad \text{and} \quad \zeta(1) = \zeta'(1) = 1.$$

[Let $(a_k)_{1 \leqslant k \leqslant p+q-2}$ be the strictly increasing sequence formed by the z_i and the z'_j other than 0 and 1, and put $a_0 = 0$, $a_{p+q-1} = 1$; let B_i denote the set of all $\boldsymbol{x} = (y, z) \in \mathbf{R}^n$ such that

$$a_{i-1} \leqslant z \leqslant a_i \quad (1 \leqslant i \leqslant p + q - 1);$$

let $\mathfrak{M}$ denote the set of all pairs $\gamma = (\mathrm{C}, \mathrm{C}')$, where C (resp. C′) is the intersection of the *same* B_i with a side of L (resp. L′), such that neither C nor C′ consists of a single point; and let $R\{\alpha, \beta\}$ denote the following relation between elements α, β of $\mathfrak{M}$: "$\alpha \neq \beta$; if $\alpha = (\mathrm{C}_1, \mathrm{C}'_1)$, $\beta = (\mathrm{C}_2, \mathrm{C}'_2)$, then $\mathrm{C}_1 \cap \mathrm{C}_2$ and $\mathrm{C}'_1 \cap \mathrm{C}'_2$ are not empty, one of them consists of a single point, and if this point is not one of the $\boldsymbol{x}_i$ (resp. $\boldsymbol{x}'_j$) then C_1 and C_2 (resp. C'_1 and C'_2) are both contained in the same side of L (resp. L′)". Show that *a*) is applicable to the relation R.]

c) In $\mathbf{R}^n$, identified with $\mathbf{R}^{n-1} \times \mathbf{R}$, let K be a connected compact set such that $K \cap (\mathbf{R}^{n-1} \times \{0\})$ contains at least two distinct points. Show that if $K' = K - K$ (the set of all $x - y$, where $x \in K$ and $y \in K$), then $K' \cap (\mathbf{R}^{n-1} \times \{0\})$ contains a connected set which does not consist of a single point. [Use *b*), applying Chapter II, § 4, Proposition 6 and Exercise 15.]

11) Extend Exercises 14 and 15 of Chapter IV, § 8 to the spaces $\mathbf{R}^n$ $(n \geqslant 2)$.

¶ 12) *a*) Let f be a mapping of $\mathbf{R}$ into $\mathbf{R}$, and let $G \subset \mathbf{R}^2$ be the graph of f. Suppose that G is dense in $\mathbf{R}^2$ and meets every perfect set in $\mathbf{R}^2$ which is not contained in any countable union of lines $\{x_n\} \times \mathbf{R}$. Show that G is connected. (If this were false, show that there would exist two non-empty disjoint open sets A, B in $\mathbf{R}^2$ such that G was the union of $G \cap A$ and $G \cap B$, and there would exist at least one point of A and one point of B with the same abscissa, by using the fact that $\mathbf{R}$ is connected; then use Exercise 11.)

b) Deduce from *a*) that there is a *linear* mapping f of $\mathbf{R}$ into $\mathbf{R}$ whose graph G is *dense* in $\mathbf{R}^2$ and *connected*. [Using *a*) and Exercise 11, define by transfinite induction the values of f at the points of a Hamel base of $\mathbf{R}$, using the same method as in *Set Theory*, Chapter III, § 6, Exercise 24.] Deduce that the subgroup G of $\mathbf{R}^2$ satisfies conditions (R_{I}) and (R_{II}) of Chapter V, § 3, Exercise 4, but is not locally compact, and therefore has a topology strictly finer than the usual topology of $\mathbf{R}$; and that G is not locally connected.

13) Identifying $\mathbf{R}^n$ with $\mathbf{R}^{n-1} \times \mathbf{R}$, let f be a continuous mapping of an open subset A of $\mathbf{R}^{n-1}$ into $\mathbf{R}$, and let S be its graph. Show that the subspace S of $\mathbf{R}^n$ is homeomorphic to A, and that the complement of S in the " cylinder " $A \times \mathbf{R}$ is a dense open set in $A \times \mathbf{R}$, and has exactly two components if A is connected.

§ 2

* 1) Let I be the closed cube of $\mathbf{R}^n$ which is the product of n intervals identical with $[-\frac{1}{2}\pi, \frac{1}{2}\pi]$. To each point $x = (x_i)$ of I we make correspond the point $y = (y_j)$ of $\mathbf{R}^{n+1}$ such that

$$\begin{aligned}
y_1 &= \sin x_1 \\
y_2 &= \cos x_1 \sin x_2 \\
&\cdots\cdots\cdots \\
y_p &= \cos x_1 \cos x_2 \ldots \cos x_{p-1} \sin x_p \quad (2 \leqslant p \leqslant n-1) \\
&\cdots\cdots\cdots \\
y_n &= \cos x_1 \cos x_2 \ldots \cos x_{n-1} \sin 2x_n \\
y_{n+1} &= \cos x_1 \cos x_2 \ldots \cos x_{n-1} \cos 2x_n.
\end{aligned}$$

Show that the image of I under this mapping is the sphere $\mathbf{S}_n$, and that the restriction of this mapping to the interior of I is a homeomorphism onto a dense open set in $\mathbf{S}_n$. *

2) Define a homeomorphism of $\mathbf{S}_p \times \mathbf{S}_q$ onto a subset of $\mathbf{S}_{p+q+1}$ [note that the equation of $\mathbf{S}_{p+q+1}$ is

$$(x_1^2 + \cdots + x_{p+1}^2) + (x_{p+2}^2 + \cdots + x_{p+q+2}^2) = 1].$$

3) *a*) Show that, if f is a continuous mapping of $\mathbf{S}_1$ into $\mathbf{R}^n$, f can be extended to a continuous mapping of $\mathbf{B}_2$ into $\mathbf{R}^n$ [map the point $tx \in \mathbf{B}_2$ $(t \geqslant 0,\ x \in \mathrm{S}_1)$ to $tf(x)$].

b) Deduce that, if f is a continuous mapping of $\mathbf{S}_1$ into $\mathbf{S}_n$ such that $f(\mathbf{S}_1) \neq \mathbf{S}_n$, then f can be extended to a continuous mapping of $\mathbf{B}_2$ into $\mathbf{S}_n$ [use a stereographic projection whose vertex does not lie in $f(\mathbf{S}_1)$]. (*)

4) Show that there is no homeomorphism of $\mathbf{S}_1$ into $\mathbf{R}$. (Observe that the complement in $\mathbf{S}_1$ of any point of $\mathbf{S}_1$ is connected, and that every connected subset of $\mathbf{R}$ is an interval.)

Deduce that a homeomorphism of $\mathbf{S}_1$ onto a subspace of $\mathbf{S}_1$ is necessarily a homeomorphism of $\mathbf{S}_1$ *onto* $\mathbf{S}_1$ (use Proposition 4 of no. 4).

5) Show that, if $n > 1$, the sphere $\mathbf{S}_n$ is not homeomorphic to the circle $\mathbf{S}_1$ (cf. § 1, Exercise 8).

6) Identifying $\mathbf{S}_1$ with the torus $\mathbf{T}$, the quotient of $\mathbf{R}$ by the relation $x \equiv y \pmod 1$ (no. 4, Proposition 4, Corollary 2), let φ denote the canonical mapping of $\mathbf{R}$ onto $\mathbf{S}_1$. A continuous mapping f of a topological space E into $\mathbf{S}_1$ is *inessential* if there exists a continuous mapping g of E into $\mathbf{R}$ such that $f = \varphi \circ g$. (**) A mapping which is not inessential is said to be *essential*.

Show that the identity mapping of $\mathbf{S}_1$ onto $\mathbf{S}_1$ is essential (use Exercise 4).

7) Show that there exists an entourage U of the uniformity of $\mathbf{S}_1$ such that, if f is an inessential mapping of a topological space E into $\mathbf{S}_1$, every continuous mapping $f' :\ \mathrm{E} \to \mathbf{S}_1$ such that $(f(x), f'(x)) \in \mathrm{U}$ for all $x \in \mathrm{E}$ is also inessential.

8) Show that there exists no continuous mapping f of $\mathbf{B}_2$ onto $\mathbf{S}_1$ which extends the identity mapping of $\mathbf{S}_1$ (using Exercise 7, show that f,

(*) We shall see later that, for $n > 1$, the result remains true even if $f(\mathbf{S}_1) = \mathbf{S}_n$.

(**) We shall give later a general definition of an inessential mapping of a topological space into an arbitrary topological space, and we shall show there that, for mappings S into $\mathbf{S}_1$, this definition is equivalent to that given here.

restricted to the circle with centre 0 and radius $r \leqslant 1$, would always be an inessential mapping of this circle into $\mathbf{S}_1$, and hence obtain a contradiction when $r = 1$).

9) Let E be a topological space containing more than one point. Show that there exists a continuous mapping f of $\mathbf{S}_1$ into $\mathrm{F} = \mathrm{E} \times \mathbf{S}_1$ such that $f(\mathbf{S}_1) \neq \mathrm{F}$ and which does not extend to a continuous mapping of $\mathbf{B}_2$ into F (use Exercise 8). Deduce that, if $n > 1$, $\mathbf{S}_n$, $\mathbf{R}^n$ and $\mathbf{B}_n$ are not homeomorphic to any space of the form $\mathrm{E} \times \mathbf{S}_1$, where E is any topological space (see Exercise 3). In particular, if $n > 1$, $\mathbf{S}_n$ is not homeomorphic to $\mathbf{S}_1^n$, and for all n, $\mathbf{B}_n$ is not homeomorphic to $\mathbf{S}_1^n$.

Show likewise that $\mathbf{R}^2$ is not homeomorphic to the complement of a point in $\mathbf{R}^2$, and that $\mathbf{S}_2$ is not homeomorphic to $\mathbf{B}_2$ (if false, $\mathbf{R}^2$ would be homeomorphic to the product of $\mathbf{S}_1$ and the interval $]0, 1]$).

10) Let $\mathrm{H}_{n,p,q}$ denote the "quadric" in $\mathbf{R}^n$ defined by the equation

$$x_1^2 + x_1^2 + \cdots + x_p^2 - x_{p+1}^2 - \cdots - x_{p+q}^2 = 1 \qquad (p + q \leqslant n).$$

Show that $\mathrm{H}_{n,p,q}$ is homeomorphic to $\mathbf{S}_{p-1} \times \mathbf{R}^{n-p}$.

11) Let $\mathrm{C}_{n,p}$ be the "quadric cone" in $\mathbf{R}^n$ defined by the equation

$$x_1^2 + x_2^2 + \cdots + x_p^2 - x_{p+1}^2 - \cdots - x_n^2 = 0 \qquad (1 \leqslant p \leqslant n - 1).$$

Show that the complement of $\{0\}$ in $\mathrm{C}_{n,p}$ is homeomorphic to $\mathbf{S}_{p-1} \times \mathbf{S}_{n-p-1} \times \mathbf{R}$.

¶ 12) A subset E of $\mathbf{R}^n$ which contains the origin 0 is said to be *starlike* (with respect to 0) if, for all $x \in \mathrm{E}$ and all $t \in [0, 1]$, we have $tx \in \mathrm{E}$. The intersection of E with a closed ray of origin 0 is either the whole ray or a segment of which 0 is one end-point. The *shell* of E is the set K consisting of the non-zero end-points of these segments, together with 0 if there exists a ray whose intersection with E consists of 0 alone. In what follows we shall assume that $0 \notin \mathrm{K}$.

a) Show that the shell of E is contained in the frontier of E. Give an example in which these two sets are different.

b) Show that, if the shell K of E is compact, there is a homeomorphism of $\mathbf{R}^n$ onto itself which maps K onto $\mathbf{S}_{n-1}$, $\overline{\mathrm{E}}$ onto $\mathbf{B}_n$ and the interior of E onto the interior of $\mathbf{B}_n$ (map each point $x \in \mathrm{K}$ to its central projection on $\mathbf{S}_{n-1}$, and then extend this mapping to the whole of $\mathbf{R}^n$). Deduce that the frontier of E coincides with its shell, and that the interior of E is the set of points tx, where $x \in \mathrm{K}$ and $t \in [0, 1[$.

c) If E is unbounded and if its frontier coincides with its shell, then the interior of E is homeomorphic to $\mathbf{R}^n$, and its shell K is homeomorphic to an open subset of $\mathbf{S}_{n-1}$ [show that the image of E under the homeomorphism $x \to x/(1 + \|x\|)$ satisfies the conditions of *b*)].

d) Give an example of an unbounded starlike set E whose shell is closed but is not identical with the frontier of E.

13) Show that, in the space $\mathbf{S}_n$, the frontier of a non-empty open set whose exterior is non-empty has the power of the continuum (use Proposition 4 of no. 4).

§ 3

1) Let f be the mapping of $\mathbf{S}_2$ into $\mathbf{R}^4$ such that

$$f(x_1, x_2, x_3) = (x_1^2 - x_2^2, x_1x_2, x_1x_3, x_2x_3).$$

This function has the same value at diametrically opposite points of $\mathbf{S}_2$. Show that on passing to the quotient it defines a homeomorphism of $\mathbf{P}_2$ onto a subspace of $\mathbf{R}^4$.

Show likewise that the mapping g of $\mathbf{S}_3$ into $\mathbf{R}^6$ defined by

$$g(x_1, x_2, x_3, x_4) = (x_1^2 - x_2^2, x_1x_2, x_1x_3 + x_2x_4, x_3^2 - x_4^2, x_3x_4, x_1x_4 - x_2x_3)$$

defines, on passing to the quotient, a homeomorphism of $\mathbf{P}_3$ into $\mathbf{R}^6$.

2) Identify the projective space $\mathbf{P}_n$ with the quotient space of $\mathbf{B}_n$ defined in no. 1, Proposition 3, and let φ denote the canonical mapping of $\mathbf{B}_n$ onto $\mathbf{P}_n$. A continuous mapping f of a topological space E into $\mathbf{P}_n$ is said to be *inessential* if there exists a continuous mapping g of E into $\mathbf{B}_n$ such that $f = \varphi \circ g$, and *essential* in the converse case (cf. § 2, Exercise 6). Show that there exists an entourage U of the uniformity of $\mathbf{P}_n$ such that, if f is an inessential mapping of a topological space E into $\mathbf{P}_n$, every continuous mapping f' of E into $\mathbf{P}_n$ such that $(f(x), f'(x)) \in \mathrm{U}$ for all $x \in \mathrm{E}$ is also inessential.

3) If a continuous mapping of $\mathbf{S}_1$ into $\mathbf{P}_n$ is essential (Exercise 2) it cannot be extended to a continuous mapping of $\mathbf{B}_2$ into $\mathbf{P}_n$ (cf. § 2, Exercise 8).

4) If $n > 1$, there exists an essential mapping f of $\mathbf{S}_1$ into $\mathbf{P}_n$ such that $f(\mathbf{S}_1) \neq \mathbf{P}_n$ [take $f(\mathbf{S}_1)$ to be the image under φ of a diameter of $\mathbf{B}_n$; this image is a projective line]. Deduce that, if $n > 1$, $\mathbf{P}_n$ is homeomorphic to neither $\mathbf{S}_n$ nor $\mathbf{B}_n$ (use Exercise 3 of § 3 and Exercise 3 of § 2).

5) If we consider $\mathbf{R}^2$ as embedded in $\tilde{\mathbf{R}} \times \tilde{\mathbf{R}}$, and $\mathbf{R}$ as embedded in $\tilde{\mathbf{R}}$, show that the mapping $(x, y) \to x + y$ of $\mathbf{R}^2$ into $\mathbf{R}$ can be extended by continuity to the points (∞, a) and (a, ∞) of $\tilde{\mathbf{R}} \times \tilde{\mathbf{R}}$

for all finite values of a, but that it cannot be extended by continuity to (∞, ∞). If we consider $\mathbf{R}^2$ as embedded in $\mathbf{P}_2$ ($\mathbf{R}$ still considered as embedded in $\tilde{\mathbf{R}}$), $x + y$ can be extended by continuity to all points of the line at infinity except the point with homogeneous coordinates $(1, -1, 0)$.

State and prove the analogues of these results for the product xy.

6) If $n \geqslant 0$, let $\mathfrak{F}$ be the set of all closed subsets of $\mathbf{P}_n$ endowed with the uniformity induced by that of $\mathbf{P}_n$ by the procedure of Chapter II, § 2, Exercise 6 *b*). Show that the set $\mathrm{P}_{n,p}$ of projective linear varieties of $p \geqslant 0$ dimensions is closed in $\mathfrak{F}$, and that the topology induced on $\mathbf{P}_{n,p}$ by the topology of $\mathfrak{F}$ is the same as that defined in no. 5, Definition 2. [To define the entourages of the uniformity of $\mathbf{P}_n$ we may consider $\mathbf{P}_n$ as a quotient space of $\mathbf{S}_n$ (no. 1, Proposition 2) and take finite coverings of $\mathbf{S}_n$ by balls whose radii tend to 0.]

7) *a*) Show that, in the notation of no. 5, the space $\mathrm{L}_{n,p}$ is homeomorphic to the product of $\mathrm{V}_{n,p}$ and $\mathbf{R}^{p(p+1)/2}$. Note that every matrix $X \in \mathrm{L}_{n,p}$ can be expressed uniquely in the form $U.Y$, where $y \in \mathrm{V}_{n,p}$ and $U = (u_{ij})$ is such that $u_{ij} = 0$ whenever $i < j$, and $u_{ii} > 0$ for $1 \leqslant i \leqslant p$. Put $Y = f(X)$.

b) In the notation of no. 5, Proposition 8, show that f is a homeomorphism of B_σ onto $f(\mathrm{B}_\sigma)$ [note that $X \to f(X_\sigma^{-1} X)$ is a continuous mapping of A_σ onto $f(\mathrm{B}_\sigma)$ and that $f(X_\sigma^{-1}X)$ belongs to the same class as X mod $\Delta_{n,p}$; then apply Lemma 2]. Deduce that the intersection D_σ of A_σ and $\mathrm{V}_{n,p}$ is homeomorphic to the product of $\mathrm{V}_{p,p}$ and $\mathbf{R}^{p(n-p)}$ [every matrix belonging to D_σ is uniquely expressible as the product of a matrix of $\mathrm{V}_{p,p}$ and a matrix of $f(\mathrm{B}_\sigma)$].

8) Let g be the mapping such that, for each matrix $X \in \mathrm{L}_{n,p}$, $X'g(X)$ is the matrix formed by the first q rows of X $(q < p)$. Show that the restriction of g to $\mathrm{V}_{n,p}$ is a continuous open mapping of $\mathrm{V}_{n,p}$ onto $\mathrm{V}_{n,q}$ [use Exercise 7 *a*), by noting that if $X = U.Y$ then

$$g(X) = U'.g(Y),$$

where U' is the matrix obtained from U by removing the rows and columns with indices $> q$; on the other hand, observe that f is open].

9) Show that $\mathrm{V}_{n,p}$ is connected if $p < n$ (use Exercise 8 and Chapter I, § 11, no. 3, Proposition 7).

10) In a projective space $\mathbf{P}_n$, let $\mathrm{H}_{n,p}$ be the " quadric " defined by the equation

$$x_0^2 + x_1^2 + \cdots + x_{p-1}^2 - x_p^2 - \cdots - x_n^2 = 0 \qquad (1 \leqslant p \leqslant n).$$

Show that $H_{n,1}$ and $H_{n,n}$ are homeomorphic to S_{n-1}. If $2 \leqslant p \leqslant n-1$, $H_{n,p}$ is homeomorphic to the space obtained by identifying every point of the product $S_{p-1} \times S_{n-p}$ (considered as a subspace of $\mathbf{R}^p \times \mathbf{R}^{n-p+1}$) with its diametrically opposite point. Every point of $H_{n,p}$ has an open neighbourhood homeomorphic to $\mathbf{R}^{n-1}$.

Show that $H_{3,2}$ is homeomorphic to $S_1 \times S_1$ [identify $S_1 \times S_1$ with $\mathbf{T} \times \mathbf{T}$, a pair of diametrically opposite points of $S_1 \times S_1$ being identified with a pair of points (u, v) and $(1/2 + u, 1/2 + v)$ of $\mathbf{T} \times \mathbf{T}$; then consider the mapping $(u, v) \rightarrow (u + v, u - v)$ of $\mathbf{T} \times \mathbf{T}$ into itself].

11) In a projective space $\mathbf{P}_n$, let $C_{n,p}$ be the "quadric cone" defined by the equation

$$x_1^2 + x_2^2 + \cdots + x_p^2 - x_{p+1}^2 - \cdots - x_n^2 = 0 \qquad (1 \leqslant p \leqslant n-1).$$

Show that the complement of $\{0\}$ in $C_{n,p}$ is homeomorphic to $\mathbf{R} \times H_{n-1,p}$, with the notation of Exercise 10.

HISTORICAL NOTE

(Numbers in brackets refer to the bibliography at the end of this note.)

We have already had occasion to remark that the development of analytic geometry of the plane and in space led mathematicians to the notion of *n*-dimensional space, which provided them with an extremely convenient geometrical language for expressing simply and concisely algebraic theorems about equations in an arbitrary number of variables, and in particular all the general results of linear algebra. But although this language had become customary with many geometers by the middle of the 19th century, it remained purely a matter of convenience, and the absence of an " intuitive " representation of spaces of more than three dimensions appeared to forbid, in such spaces, the arguments " by continuity " which, founded exclusively on " intuition ", were permitted in the plane and in (threedimensional) space. In his researches on *analysis situs* and on the foundations of geometry, Riemann was the first to use considerations of this sort, by analogy with three-dimensional space (see the Historical Note on Chapter I) (*); following his example, many mathematicians were encouraged to use such arguments, with great success, particularly in the theory of algebraic functions of several complex variables. But in view of the very limited control over spatial intuition at that period, one might justifiably remain skeptical of the demonstrative value of such considerations, and not allow their use except on a purely heuristic basis, as tending to make plausible the truth of certain theorems. Thus Poincaré, in his memoir of 1887 on the residues of double integrals of two complex variables, avoided, as far as he could, all recourse to intuition in four-dimensional space: " *Comme cette langue hypergéométrique répugne encore à beaucoup de bons esprits*,

(*) See also the works of L. Schläfli, which date from the same period but remained unpublished until this century [3].

je n'en ferai qu'un usage peu fréquent "; the artifices which he used for this purpose enabled him to make topological arguments suffice when he was discussing three-dimensional space, he no longer hesitated to appeal for which to intuition ([1]).

Moreover, the discoveries of Cantor, and particularly the famous theorem which asserts that $\mathbf{R}$ and $\mathbf{R}^n$ are equipotent [which seemed to undermine the whole concept of dimension (*)], showed that, in order to put geometry and topology on a solid foundation, it was indispensable to free them entirely from all dependence on intuition. We have already noted (cf. Historical Note on Chapter I) that this need was the origin of the modern conception of general topology; but even before the creation of this latter theory there had already begun a rigorous study of the topology of real n-dimensional spaces and their immediate generalizations (" n-dimensional manifolds ") by methods which belong properly to that branch of topology known as " combinatorial topology ", or better " algebraic topology ". A volume in this series will be devoted to this branch of topology, and the reader will find information there on the historical stages of its development; in this chapter we have limited ourselves to establishing the most elementary topological properties of real number spaces and real projective spaces which, historically, have served as the starting point for the methods of algebraic topology.

(*) It is interesting to note that Dedekind, as soon as he knew of this result, understood the reason for its paradoxical appearance, and remarked to Cantor that it ought to be possible to prove the impossibility of a *bicontinuous one-to-one* correspondence between $\mathbf{R}^n$ and $\mathbf{R}^m$ when $m \neq n$ [2].

BIBLIOGRAPHY

[1] L. Schläfli, *Gesammelte mathematische Abhandlungen*, vol. 1, Basel (Birkhauser), 1950, pp. 169-387.

[2] H. Poincaré, *Œuvres*, vol. 3, p. 443 *et seq*. Paris (Gauthier-Villars) (1934) [*Acta Mathematica*, **9** (1887), p. 321].

[3] G. Cantor and R. Dedekind, Briefwechsel, *Act. Sci. et Ind.*, no. 518, Paris (Hermann) (1937).

CHAPTER VII

The additive groups $\mathbf{R}^n$

1. SUBGROUPS AND QUOTIENT GROUPS OF $\mathbf{R}^n$

Let us first introduce the following convention: if G is a topological group, we have defined (Chapter III, § 2) the product group G^n of n factors equal to G, for each integer $n > 0$. In this section, we shall extend this definition to the case $n = 0$ by the convention that G^0 denotes a group consisting of only one element. If H is any group, we shall identify $G^0 \times H$ with H.

On the set $\mathbf{R}^n$, we shall have to consider on the one hand its (additive) *topological group* structure and on the other hand its *vector space* structure over the field $\mathbf{R}$ (Chapter VI, § 1, no. 3). Given a subset A of $\mathbf{R}^n$, we may envisage the *subgroup* of $\mathbf{R}^n$ generated by A (the set of all linear combinations of points of A, with *integer* coefficients) and also the *vector subspace* generated by A (the set of all linear combinations of points of A, with *real* coefficients); these two notions must be carefully distinguished. In accordance with the definitions made in algebra, the *rank* of A is the *dimension* of the vector subspace V of $\mathbf{R}^n$ generated by A; to say that A is of rank p is therefore equivalent to saying that there are p points $x_i \in A$ which form a *free system* with respect to the field $\mathbf{R}$ (in other words, the relation $\sum_i t_i x_i = 0$, where the t_i are *real*, implies $t_i = 0$ for each i) and form a *basis* of V (which means that every point of V is a linear combination of the x_i with real coefficients).

In what follows we shall also have to make use of the notion of a system of points of $\mathbf{R}^n$ which are *free with respect to the field* $\mathbf{Q}$ *of rational numbers*; such a system is a finite subset (x_i) of $\mathbf{R}^n$ such that the relation $\sum_i r_i x_i = 0$, where the r_i are *rational numbers* (or *integers* — it comes to the same thing), implies that $r_i = 0$ for each i. This notion must be carefully distinguished from that of a free system with respect to $\mathbf{R}$: every system which is free with respect to $\mathbf{R}$ is free with respect to $\mathbf{Q}$, but the converse is false (for example, the numbers 1 and $\sqrt{2}$ in $\mathbf{R}$ form a free system with respect to $\mathbf{Q}$, but not a free system with respect to $\mathbf{R}$). Whenever we speak of a *free system* without qualification, we shall always mean a free system *with respect to* $\mathbf{R}$. It is therefore necessary to distinguish on $\mathbf{R}^n$ the vector space structure *with respect to* $\mathbf{R}$ from the vector space structure *with respect to* $\mathbf{Q}$; in particular, the vector subspace *with respect to* $\mathbf{Q}$ generated by a subset A of $\mathbf{R}^n$ is the set U of all linear combinations of A with *rational* coefficients; it is contained in the vector subspace V (with respect to $\mathbf{R}$) generated by A, but is in general distinct from V. The dimension of U (*with respect to* $\mathbf{Q}$) is called the *rational rank* of A; it is *at least equal* to the *rank* of A defined above (the dimension of V with respect to $\mathbf{R}$); it may be *infinite* if A is an infinite set, while the rank of any non-empty subset of $\mathbf{R}^n$ is always $\leqslant n$; in particular, the rational rank of any *uncountable* subset of $\mathbf{R}^n$ is always infinite, because any finite-dimensional vector space over $\mathbf{Q}$ is countable.

In this section we shall first determine the structure of the *closed subgroups* of the additive group $\mathbf{R}^n$.

1. DISCRETE SUBGROUPS OF $\mathbf{R}^n$

We have seen in Chapter V (§ 1, no. 1, Proposition 1) that the only closed subgroups of $\mathbf{R}$, other than $\mathbf{R}$ itself, are the *discrete* subgroups, generated by a *single* element. We shall begin by considering the *discrete* subgroups of $\mathbf{R}^n$.

First of all, the subgroup of $\mathbf{R}^n$ generated by p vectors $(p \leqslant n)$ of the canonical basis (Chapter VI, § 1, no. 3) of $\mathbf{R}^n$ is a discrete group isomorphic to the product $\mathbf{Z}^p$ of p groups which are equal to $\mathbf{Z}$. More generally, consider the subgroup G generated by p points $\boldsymbol{a}_i$ $(1 \leqslant i \leqslant p)$ which form a free system. There is a bijective linear mapping of $\mathbf{R}^n$ onto itself which maps $\boldsymbol{a}_i$ to $\boldsymbol{e}_i$ $(1 \leqslant i \leqslant p)$; since such a mapping is an automorphism of the topological group $\mathbf{R}^n$, G is isomorphic as a topological group to the subgroup generated by the $\boldsymbol{e}_i$ $(1 \leqslant i \leqslant p)$ and is therefore a *discrete* subgroup of rank p isomorphic to $\mathbf{Z}^p$.

The structure of the group $\mathbf{Z}^p$, and hence that of G, has been studied in algebra. We recall the main results of this study. The *bases* of G with respect to the ring $\mathbf{Z}$ are systems of p points

$$\boldsymbol{b}_i = \sum_{i=1}^{p} r_{ij}\boldsymbol{a}_j,$$

where the r_{ij} are integers such that the determinant $\det(r_{ij})$ is equal to $+1$ or -1. Every *subgroup* H of G is discrete and of rank

$q \leqslant p$; furthermore, if H is a given subgroup of rank q, there exists a free system of p points $\boldsymbol{b}_i$ $(1 \leqslant i \leqslant p)$ which generates G, and a system of q points $\boldsymbol{c}_i$ $(1 \leqslant i \leqslant q)$ which generates H, such that we have $\boldsymbol{c}_i = e_i\boldsymbol{b}_i$ for $1 \leqslant i \leqslant q$, where the e_i are integers (the *invariant factors* of H with respect to G) such that

$$e_{i+1} \equiv 0 \pmod{e_i} \qquad \text{for} \qquad 1 \leqslant i \leqslant q - 1.$$

The quotient group G/H is a discrete group isomorphic to $\mathbf{Z}^{p-q} \times \mathrm{F}$, where F is a *finite* abelian group, the direct product of q cyclic subgroups of respective orders $e_1, \ldots, e_q$.

We shall show now that the discrete subgroups of $\mathbf{R}^n$ which we have just been considering are the only ones that exist.

PROPOSITION 1. *Let* G *be a discrete subgroup of* $\mathbf{R}^n$ *of rank* p, *let* $(\boldsymbol{a}_i)_{1 \leqslant i \leqslant p}$ *be a free system of* p *points of* G, *and let* P *be the closed parallelotope with centre* 0 *and basis vectors* $\boldsymbol{a}_i$ (Chapter VI, § 1, no. 3). *Then the set* $\mathrm{G} \cap \mathrm{P}$ *is finite and generates* G, *and every point of* G *is a linear combination of the* $\boldsymbol{a}_i$ *with rational coefficients.*

$\mathrm{G} \cap \mathrm{P}$ is compact and discrete, hence *finite*. Let $\boldsymbol{x}$ be any point of G; it is equal to a linear combination $\sum_{i=1}^{p} t_i\boldsymbol{a}_i$ of the $\boldsymbol{a}_i$ with real coefficients. For each integer $m > 0$, consider the point

$$z_m = m\boldsymbol{x} - \sum_{i=1}^{p} [mt_i]\boldsymbol{a}_i = \sum_{i=1}^{p} (mt_i - [mt_i])\boldsymbol{a}_i \text{ (*)};$$

it belongs to G, and since $0 \leqslant mt_i - [mt_i] < 1$, it lies in P. Hence, first, $\boldsymbol{x} = z_1 + \sum_{i=1}^{p} [t_i]\boldsymbol{a}_i$, so that G is generated by $\mathrm{G} \cap \mathrm{P}$; and secondly, since $\mathrm{G} \cap \mathrm{P}$ is finite, there exist two distinct integers h, k such that $z_h = z_k$, so that $(h - k)t_i = [ht_i] - [kt_i]$ less $(1 \leqslant i \leqslant p)$, and therefore the t_i are rational numbers.

COROLLARY. *Let* $(\boldsymbol{a}_i)_{1 \leqslant i \leqslant p}$ *be a free system of* p *points of* $\mathbf{R}^n$ *and let*

$$\boldsymbol{b} = \sum_{i=1}^{p} t_i\boldsymbol{a}_i$$

be a linear combination of the $\boldsymbol{a}_i$, *with real coefficients. Then the subgroup* G *of* $\mathbf{R}^n$ *generated by the* $p + 1$ *points* $\boldsymbol{a}_1, \boldsymbol{a}_2, \ldots, \boldsymbol{a}_p$ *and* $\boldsymbol{b}$ *is discrete if and only if the numbers* t_i *are rational.*

(*) We recall (Chapter IV, § 8, no. 2) that, for each real number x, $[x]$ is the *integral part* of x, i.e., the largest rational integer $\leqslant x$.

Proposition 1 shows that the condition is necessary. Also it is sufficient, for if it is satisfied we can write $t_i = m_i/d$, where d and the m_i are integers $(1 \leqslant i \leqslant p)$; $\boldsymbol{b}$ is therefore a linear combination, with integer coefficients, of the p points $(1/d)\boldsymbol{a}_i$; hence G is a subgroup of the discrete group generated by these p points, and so is itself discrete.

The result of Proposition 1 can be expressed as follows: if q points $\boldsymbol{x}_i$ $(1 \leqslant i \leqslant q)$ of a *discrete* subgroup G of $\mathbf{R}^n$ form a system which is *dependent with respect to* $\mathbf{R}$, then they form a system which is *dependent with respect to* $\mathbf{Q}$. It follows immediately that the *rational rank* of a discrete subgroup of $\mathbf{R}^n$ is equal to its *rank*.

The Corollary to Proposition 1, applied to the case where the $\boldsymbol{a}_i$ are the n vectors $\boldsymbol{e}_i$ of the canonical basis of $\mathbf{R}^n$, gives us the following proposition:

PROPOSITION 2 (Kronecker). *Let* $\theta_1, \theta_2, \ldots, \theta_n$ *be* n *real numbers. In order that, for each* $\varepsilon > 0$, *there exist an integer* q *and* n *integers*

$$p_i \qquad (1 \leqslant i \leqslant n)$$

such that

$$|q\theta_i - p_i| \leqslant \varepsilon \qquad (1 \leqslant i \leqslant n),$$

where the left-hand side of at least one of these inequalities does not vanish, it is necessary and sufficient that at least one of the θ_i *be irrational.*

THEOREM 1. *Every discrete subgroup* G *of* $\mathbf{R}^n$, *of rank equal to* p, *is generated by a free system of* p *points.*

By virtue of the properties of groups isomorphic to $\mathbf{Z}^p$ recalled earlier, it is enough to show that G is a *subgroup* of a discrete group generated by a free system of p points. Now since G is of rank p there exists a free system of p points $\boldsymbol{a}_i$ $(1 \leqslant i \leqslant p)$ of G such that every $\boldsymbol{x} \in$ G is equal to a linear combination $\sum_{i=1}^{p} t_i\boldsymbol{a}_i$ of the $\boldsymbol{a}_i$ with real coefficients; since G is discrete, Proposition 1 shows that the t_i are *rational*. Furthermore, Proposition 1 shows that G is generated by a *finite* number of points; since these points are linear combinations of the $\boldsymbol{a}_i$ with rational coefficients, there exists an integer d such that they are linear combinations with *integer* coefficients of the p points $(1/d)\boldsymbol{a}_i = \boldsymbol{a}'_i$. It follows that G is a subgroup of the group generated by the $\boldsymbol{a}'_i$.

Theorem 1 can also be proved without recourse to the theory of invariant factors (cf. Exercise 1).

Discrete subgroups of $\mathbf{R}^n$ which are of rank n are also called *lattices* in $\mathbf{R}^n$.

2. CLOSED SUBGROUPS OF $\mathbf{R}^n$

We know already of two types of closed subgroups of $\mathbf{R}^n$; on the one hand, the *vector subspaces* of $\mathbf{R}^n$, which are isomorphic to the groups $\mathbf{R}^p$ $(p \leqslant n)$ (Chapter IV, § 1, no. 4, Proposition 2); on the other hand, the *discrete* subgroups (Chapter III, § 2, no. 1, Proposition 5) which are isomorphic to groups $\mathbf{Z}^q$ $(q \leqslant n)$, as we have just seen. We shall now determine the structure of an *arbitrary* closed subgroup of $\mathbf{R}^n$ by showing that such a subgroup is isomorphic to a *product* of the form

$$\mathbf{R}^p \times \mathbf{Z}^q \quad \text{where} \quad 0 \leqslant p + q \leqslant n.$$

The proof rests on the following proposition:

PROPOSITION 3. *Every non-discrete closed subgroup of* $\mathbf{R}^n$ *contains a line passing through* 0.

Let $(\boldsymbol{x}_p)_{p \in \mathbf{N}}$ be an infinite sequence of points of G such that $\boldsymbol{x}_p \neq 0$ and $\lim_{p \to \infty} \boldsymbol{x}_p = 0$; by hypothesis such a sequence exists. Let P be an open cube with centre 0, containing the $\boldsymbol{x}_p$. Let k_p denote the largest integer $h > 0$ such that $h\boldsymbol{x}_p \in \mathrm{P}$ (since P is a bounded box and $\boldsymbol{x}_p \neq 0$, the existence of k_p follows from Archimedes' axiom). The points $k_p \boldsymbol{x}_p$ lie in a compact set $\overline{\mathrm{P}}$, hence the sequence $(k_p \boldsymbol{x}_p)_{p \in \mathbf{N}}$ has a cluster point $\boldsymbol{a} \in \overline{\mathrm{P}}$. If $\|k_p \boldsymbol{x}_p - \boldsymbol{a}\| \leqslant \varepsilon$, we have

$$\|(k_p + 1)\boldsymbol{x}_p - \boldsymbol{a}\| \leqslant \varepsilon + \|\boldsymbol{x}_p\|,$$

and since $\lim_{p \to \infty} \boldsymbol{x}_p = 0$, it follows that $\boldsymbol{a}$ is also a cluster point of the sequence $((k_p + 1)\boldsymbol{x}_p)$, whose points belong to the closed set $\complement \mathrm{P}$, by definition of k_p; hence $\boldsymbol{a} \in \overline{\mathrm{P}} \cap \complement \mathrm{P}$ (the frontier of P, Fig. 5), which implies $\boldsymbol{a} \neq 0$. Moreover, since G is closed, we have $\boldsymbol{a} \in \mathrm{G}$. Let t be any real number; since $|tk_p - [tk_p]| < 1$, the relation $\|k_p \boldsymbol{x}_p - \boldsymbol{a}\| \leqslant \varepsilon$ implies that $\|[tk_p]\boldsymbol{x}_p - t\boldsymbol{a}\| \leqslant |t|\,\varepsilon + \|\boldsymbol{x}_p\|$; since $\lim_{p \to \infty} \boldsymbol{x}_p = 0$, $t\boldsymbol{a}$ is a

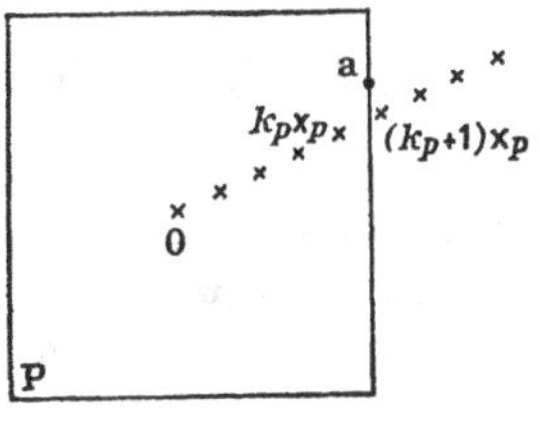

Figure 5.

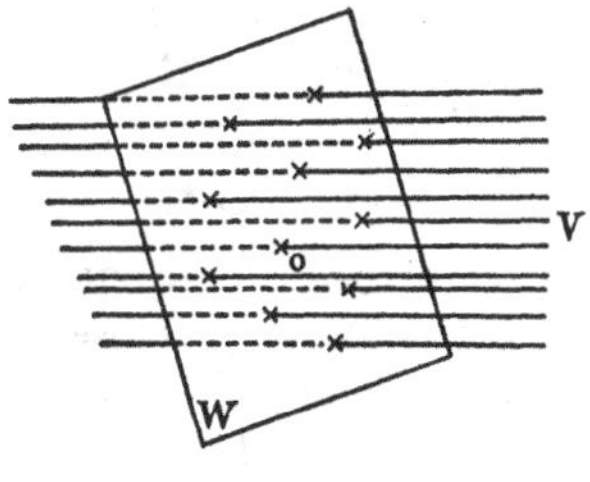

Figure 6.

cluster point of the sequence $([tk_p]\boldsymbol{x}_p)$; but the points of this sequence belong to G, and consequently $t\boldsymbol{a} \in G$, since G is closed. This completes the proof.

THEOREM 2. *Let* G *be a closed subgroup of* $\mathbf{R}^n$, *of rank* r $(0 \leqslant r \leqslant n)$. *Then there is a largest vector subspace* V *contained in* G; *for every vector subspace* W *complementary to* V, $W \cap G$ *is discrete and* G *is the direct sum of* V *and* $W \cap G$.

Let us first establish the existence of V by showing that the union of all the lines through 0 which lie in G is a vector subspace: indeed, the vector subspace generated by the union of these lines is the same as the subgroup they generate. The group G is the *direct sum* of V and $W \cap G$; for if $\boldsymbol{x} \in G$, we have $\boldsymbol{x} = \boldsymbol{y} + \boldsymbol{z}$ where $\boldsymbol{y} \in V$ and $\boldsymbol{z} \in W$; since $V \subset G$, $\boldsymbol{z} = \boldsymbol{x} - \boldsymbol{y} \in G$ and therefore $\boldsymbol{z} \in W \cap G$. It remains to show that $W \cap G$ is *discrete*; this follows from Proposition 3, since $W \cap G$ is a closed subgroup which contains no lines, by reason of the definition of V.

If $G \neq V$, we may say that G is the union of a countable infinity of linear varieties, *parallel* to V and passing through the points of the discrete group $W \cap G$ (Fig. 6).

If p is the dimension of the vector subspace V, we have $p \leqslant r$, and $W \cap G$ is a discrete group of rank $r - p$.

COROLLARY 1. *There exists a basis* $(\boldsymbol{a}_i)_{1 \leqslant i \leqslant n}$ *of* $\mathbf{R}^n$, *such that*

$$\boldsymbol{a}_i \in G \quad (1 \leqslant i \leqslant r), \qquad \boldsymbol{a}_i \in V \quad (1 \leqslant i \leqslant p)$$

and such that G *is the set of points* $\sum_{i=1}^{p} t_i \boldsymbol{a}_i + \sum_{j=p+1}^{r} n_j \boldsymbol{a}_j$, *where the* t_i *take all real values and the* n_j *take all integer values.*

This follows from Theorem 2, and Theorem 1 of no. 1 applied to the discrete group $W \cap G$.

COROLLARY 2. *There exists an automorphism of* $\mathbf{R}^n$ *which maps* G *onto the group* G′, *isomorphic to* $\mathbf{R}^p \times \mathbf{Z}^{r-p}$, *which is the direct sum of the vector subspace generated by* $\boldsymbol{e}_1, \boldsymbol{e}_2, \ldots, \boldsymbol{e}_p$ *and the* (discrete) *additive subgroup generated by* $\boldsymbol{e}_{p+1}, \boldsymbol{e}_{p+2}, \ldots, \boldsymbol{e}_r$.

This is an immediate consequence of Corollary 1. Corollary 2 of Theorem 2 shows that a closed subgroup G of $\mathbf{R}^n$ is completely determined up to isomorphism by two integers $\geqslant 0$: its *rank*, which we denote by $r(G)$, and the dimension of the largest vector subspace contained in G, which we call the *dimension* of G and denote by $d(G)$.

The only conditions which these integers have to satisfy are the inequalities $0 \leqslant d(G) \leqslant r(G) \leqslant n$.

3. ASSOCIATED SUBGROUPS

Let G be an arbitrary subgroup (closed or not) of $\mathbf{R}^n$. Consider the set G^* of points $\boldsymbol{u} = (u_i) \in \mathbf{R}^n$ such that, for *all* $\boldsymbol{x} = (x_i) \in G$, the number $(\boldsymbol{u}|\boldsymbol{x}) = \sum_{i=1}^{n} u_i x_i$ is an *integer*. It is immediately seen that G^* is a *subgroup* of $\mathbf{R}^n$; it is called the subgroup *associated* with G (*). If G and H are two subgroups of $\mathbf{R}^n$ such that $H \subset G$, it is clear that $G^* \subset H^*$.

PROPOSITION 4. *The subgroup* G^* *associated with a subgroup* G *of* $\mathbf{R}^n$ *is closed, and we have* $(\overline{G})^* = G^*$.

For each $\boldsymbol{x} \in G$, let $f_{\boldsymbol{x}}(\boldsymbol{u})$ denote $(\boldsymbol{u}|\boldsymbol{x})$; $f_{\boldsymbol{x}}$ is a linear form, therefore continuous. Since G^* is the intersection of the sets $\overset{-1}{f}_{\boldsymbol{x}}(\mathbf{Z})$ as $\boldsymbol{x}$ runs through G, and since each of these sets is closed, it follows that G^* is closed. On the other hand, if $\boldsymbol{u} \in G^*$, we have $(\boldsymbol{u}|\boldsymbol{x}) \in \mathbf{Z}$ for all $\boldsymbol{x} \in G$, and therefore, as $\mathbf{Z}$ is closed in $\mathbf{R}$, $(\boldsymbol{u}|\boldsymbol{y}) \in \mathbf{Z}$ for all $\boldsymbol{y} \in \overline{G}$; thus $\boldsymbol{u} \in (\overline{G})^*$. But we have $(\overline{G})^* \subset G^*$ (since $G \subset \overline{G}$), so that $(\overline{G})^* = G^*$.

Consider the structure of G^* when G is *closed*. By Corollary 1 to Theorem 2 of no. 2, there exists a base $(\boldsymbol{a}_i)_{1 \leqslant i \leqslant n}$ of $\mathbf{R}^n$ such that G coincides with the set of points

$$\boldsymbol{x} = \sum_{i=1}^{p} t_i \boldsymbol{a}_i + \sum_{j=p+1}^{p+q} n_j \boldsymbol{a}_j,$$

where the t_i take all real values and the n_j take all integer values. Hence $(\boldsymbol{u}|\boldsymbol{x})$ is an integer for *all* these points $\boldsymbol{x}$ if and only if $(\boldsymbol{u}|\boldsymbol{a}_i) = 0$ for $1 \leqslant i \leqslant p$ and $(\boldsymbol{u}|\boldsymbol{a}_i)$ is an integer for $p+1 \leqslant i \leqslant p+q$. Let us denote by $(\boldsymbol{a}'_i)_{1 \leqslant i \leqslant n}$ the basis of $\mathbf{R}^n$ such that $(\boldsymbol{a}'_i|\boldsymbol{a})_j = 0$ for $i \neq j$ and

$$(\boldsymbol{a}'_i|\boldsymbol{a}_i) = 1 \qquad \text{for} \qquad 1 \leqslant i \leqslant n$$

(*) This notion is a particular case, corresponding to the group $\mathbf{R}^n$, of a general notion in the theory of *duality* of locally compact abelian groups (see, e.g., A. WEIL, " L'integration dans les groupes topologiques et ses applications ", *Act. Sci. et Ind.* no. 869, Paris, Hermann, 1950, pp. 108-109). The reader will observe the close analogy which exists between the properties of associated subgroups in $\mathbf{R}^n$ and those of *orthogonal* vector subspaces of a vector space and its dual.

[the basis " dual " to $(\boldsymbol{a}_i)$]; if we put $\boldsymbol{u} = \sum_{i=1}^{n} u_i \boldsymbol{a}'_i$, it is clear that the points $\boldsymbol{u} \in G^*$ are characterized by the following conditions: $u_i = 0$ for $1 \leqslant i \leqslant p$, and $u_i \in \mathbf{Z}$ for

$$p + 1 \leqslant i \leqslant p + q;$$

hence G^* is the direct sum of a vector subspace W with a basis consisting of the $\boldsymbol{a}'_i$ such that $p + q + 1 \leqslant i \leqslant n$, and a discrete subgroup generated by the $\boldsymbol{a}'_i$ such that $p + 1 \leqslant i \leqslant p + q$. In other words:

PROPOSITION 5. *For every closed subgroup* G *of* $\mathbf{R}^n$,

$$r(G^*) = n - d(G) \quad \textit{and} \quad d(G^*) = n - r(G).$$

Let us apply the same reasoning to G^*. Observing that the basis dual to $(\boldsymbol{a}'_i)$ is $(\boldsymbol{a}_i)$, we see that:

PROPOSITION 6. *For every subgroup* G *of* $\mathbf{R}^n$, *we have* $(G^*)^* = \overline{G}$.

COROLLARY. *A point* $\boldsymbol{x}$ *lies in the closure of a subgroup* G *of* $\mathbf{R}^n$ *if and only if* $(\boldsymbol{u}|\boldsymbol{x})$ *is an integer for all* $\boldsymbol{u} \in \mathbf{R}^n$ *such that* $(\boldsymbol{u}|\boldsymbol{y})$ *is an integer for all* $\boldsymbol{y} \in G$.

Let us apply this characterization of points lying in the closure of a subgroup G to the case of the subgroup G generated by the n vectors $\boldsymbol{e}_j$ of the canonical basis $(1 \leqslant j \leqslant n)$ and by an arbitrary number m of points $\boldsymbol{a}_i$ $(1 \leqslant i \leqslant m)$ of $\mathbf{R}^n$. To say that $(\boldsymbol{u}|\boldsymbol{e}_j)$ is an integer for $1 \leqslant j \leqslant n$ means that the n coordinates of $\boldsymbol{u}$ are integers; therefore:

PROPOSITION 7 (Kronecker). *Let* $\boldsymbol{a}_i = (a_{ji})$ $(1 \leqslant i \leqslant m,\ 1 \leqslant j \leqslant n)$ *be* m *points of* $\mathbf{R}^n$ *and let* $\boldsymbol{b} = (b_j)$ $(1 \leqslant j \leqslant n)$ *be a point of* $\mathbf{R}^n$. *In order that, for each* $\varepsilon > 0$, *there exist* m *integers* g_i $(1 \leqslant i \leqslant m)$ *and* n *integers* p_j $(1 \leqslant j \leqslant n)$ *such that*

$$|q_1 a_{1j} + q_2 a_{2j} + \cdots + q_m a_{mj} - p_j - b_j| \leqslant \varepsilon \qquad (1 \leqslant j \leqslant n)$$

it is necessary and sufficient that, for each finite sequence (r_j) $(1 \leqslant j \leqslant n)$ *of* n *integers such that the* m *numbers* $\sum_{j=1}^{n} a_{ij}\, r_j$ $(1 \leqslant i \leqslant m)$ *are all integers, the number* $\sum_{j=1}^{n} b_j\, r_j$ *should also be an integer.*

COROLLARY 1. *In order that, for each* $\boldsymbol{x} = (x_j)$ $(1 \leqslant j \leqslant n)$ *and each* $\varepsilon > 0$, *there exist* m *integers* q_i $(1 \leqslant i \leqslant m)$ *and* n *integers*

p_j $(1 \leqslant j \leqslant n)$ *such that*

$$|q_1 a_{1j} + q_2 a_{2j} + \cdots + q_m a_{mj} - p_j - x_j| \leqslant \varepsilon \qquad (1 \leqslant j \leqslant n)$$

it is necessary and sufficient that there exist no finite sequence (r_j) *of* n *integers, not all zero, such that each of the* m *numbers* $\sum_{j=1}^{n} a_{ij} r_j$ *is an integer.*

For if G is dense in $\mathbf{R}^n$, that is if $\overline{G} = \mathbf{R}^n$, then $G^* = \{0\}$, and conversely.

In particular $(m = 1)$:

Corollary 2. *Let* $\theta_1, \theta_2, \ldots, \theta_n$ *be* n *real numbers. In order that, given any* n *real numbers* $x_1, x_2, \ldots, x_n$ *and a real number* $\varepsilon > 0$, *there should exist an integer* q *and* n *integers* p_j *such that*

$$|q\theta_j - p_j - x_j| \leqslant \varepsilon \qquad (1 \leqslant j \leqslant n)$$

it is necessary and sufficient that there exist no relation of the form $\sum_{j=1}^{n} r_j \theta_j = h$, *where the* r_j *are* n *integers not all zero, and* h *is an integer* [which implies, in particular, that the θ_j and the ratios θ_j/θ_k $(j \neq k)$ must be *irrational*].

We may interpret this result as follows: for each integer $q \in \mathbf{Z}$, let $\boldsymbol{x}_q$ denote the point with coordinates $q\theta_j - [q\theta_j]$ $(1 \leqslant j \leqslant n)$; then Corollary 2 gives a necessary and sufficient condition for the set of points $\boldsymbol{x}_q$ to be *dense* in the cube which is the product of n intervals $[0, 1]$ in the factors of $\mathbf{R}^n$.

Proposition 8. *If* G_1, G_2 *are any closed subgroups of* $\mathbf{R}^n$, *we have*

$$(G_1 = G_2)^* + G_1^* \cap G_2^* \qquad \textit{and} \qquad (G_1 \cap G_2)^* = \overline{G_1^* + G_2^*}.$$

The real number $(\boldsymbol{u}|\boldsymbol{x} + \boldsymbol{y})$ is an integer for all $\boldsymbol{x} \in G_1$ and all $\boldsymbol{y} \in G_2$ if and only if $(\boldsymbol{u}|\boldsymbol{x})$ is an integer for all $\boldsymbol{x} \in G_1$ and $(\boldsymbol{u}|\boldsymbol{y})$ is an integer for all $\boldsymbol{y} \in G_2$, because $(\boldsymbol{u}|\boldsymbol{x} + \boldsymbol{y}) = (\boldsymbol{u}|\boldsymbol{x}) + (\boldsymbol{u}|\boldsymbol{y})$; hence

$$(G_1 + G_2)^* = G_1^* \cap G_2^*$$

for any two subgroups G_1, G_2 of $\mathbf{R}^n$. If now we suppose that G_1 and G_2 are closed, we have $(G_1^* + G_2^*)^* = G_1 \cap G_2$ by Proposition 6; hence, taking the associated subgroups and applying Proposition 6 again, $(G_1 \cap G_2)^* = \overline{G_1^* + G_2^*}$.

Remark. Let G_1, G_2 be two *lattices* in $\mathbf{R}^n$ (no. 1) such that $G_2 \subset G_1$; then (Proposition 5) G_1^* and G_2^* are *lattices* in $\mathbf{R}^n$ such that $G_1^* \subset G_2^*$. We have seen in no. 1 that there is an integer $m > 0$ such that $mG_1 \subset G_2$;

consequently for $x \in G_1$ and $u \in G_2^*$ we have $m(u|x) \in \mathbf{Z}$ and therefore $(u|x) \in \mathbf{Q}$. If $x \in G_2$ and $u \in G_2^*$, or if $x \in G_1$ and $u \in G_1^*$, we have $(u|x) \in \mathbf{Z}$ by definition. Consequently on passing to the quotients the $\mathbf{Z}$-bilinear mapping $(x, u) \to (u|x)$ of $G_1 \times G_2^*$ into $\mathbf{Q}$ defines a $\mathbf{Z}$-bilinear mapping B of $(G_1/G_2) \times (G_2^*/G_1^*)$ into $\mathbf{Q}/\mathbf{Z}$. Moreover, it is clear that if $\bar{x}_0 \in G_1/G_2$ (resp. $\bar{u}_0 \in G_2^*/G_1^*$) is such that, for *each* $\bar{u} \in G_2^*/G_1^*$ (resp. for each $\bar{x} \in G_1/G_2$) we have

$$B(\bar{x}_0, u) = 0 \qquad [\text{resp. } B(\bar{x}, \bar{u}_0) = 0],$$

then necessarily $\bar{x}_0 = 0$ (resp. $\bar{u}_0 = 0$). It follows that there is a $\mathbf{Z}$-linear *bijection* h of G_2^*/G_1^* onto the *dual* of G_1/G_2, such that $\langle \bar{x}, h(\bar{u}) \rangle = B(\bar{x}, \bar{u})$ for $\bar{x} \in G_1/G_2$ and $\bar{u} \in G_2^*/G_1^*$; in particular, the finite groups G_1/G_2 and G_2^*/G_1^* are *isomorphic*.

4. HAUSDORFF QUOTIENT GROUPS OF $\mathbf{R}^n$

Every Hausdorff quotient group of $\mathbf{R}^n$ is of the form $\mathbf{R}^n/H$, where H is a closed subgroup of $\mathbf{R}^n$ (Chapter III, § 2, no. 6, Proposition 18). By Corollary 2, Theorem 2 of no. 3, there exists an automorphism f of $\mathbf{R}^n$ which transforms H into a subgroup H′, the direct sum of a vector subspace generated by p of the vectors e_i of the canonical basis and the discrete group generated by g of the $n-p$ remaining vectors

$$e_i, \quad 0 \leqslant p + q \leqslant n.$$

Passing to the quotients, f induces an isomorphism of $\mathbf{R}^n/H$ onto $\mathbf{R}^n/H'$ (Chapter III, § 2, no. 8, Remark 3); now, $\mathbf{R}^n/H'$ is isomorphic to $\mathbf{R}^{n-p-q} \times T^q$ (Chapter III, § 2, no. 9, Corollary to Proposition 26). Consequently:

PROPOSITION 9. *Every Hausdorff quotient group of* $\mathbf{R}^n$ *is isomorphic to a product group* $\mathbf{R}^h \times \mathbf{T}^k$ $(0 \leqslant h + k \leqslant n)$.

The product space $\mathbf{T}^n$ (and, by abuse of language, the topological group $\mathbf{T}^n$) is called the *n-dimensional torus*; by Proposition 4 of Chapter V, § 1, no. 2, $\mathbf{T}^n$ is compact, connected and locally connected.

Furthermore, if C denotes a closed cube of side 1 in $\mathbf{R}^n$, $\mathbf{T}^n$ is homeomorphic to the quotient space of C by the equivalence relation " $x_i \equiv y_i \pmod 1$ $(1 \leqslant i \leqslant n)$ " between the points $x = (x_i)$ and $y = (y_i)$ of C. Thus $\mathbf{T}^n$ is formed from the cube C by " identification of opposite faces ".

PROPOSITION 10. *The topological group* $\mathbf{T}^n$ *is locally isomorphic to* $\mathbf{R}^n$.

For $\mathbf{T}^n = (\mathbf{R}/\mathbf{Z})^n$ is isomorphic to $\mathbf{R}^n/\mathbf{Z}^n$ (Chapter III, § 2, no. 9, Corollary to Proposition 26) and $\mathbf{Z}^n$ is a discrete subgroup of $\mathbf{R}^n$ (Chapter III, § 2, no. 6, Proposition 19).

It follows that the groups $\mathbf{R}^p \times \mathbf{T}^{n-p}$ are all locally isomorphic to $\mathbf{R}^n$ $(0 \geqslant p \geqslant n)$; in § 2, no. 2, we shall see that they are the only *connected* groups which have this property.

5. SUBGROUPS AND QUOTIENT GROUPS OF $\mathbf{T}^n$

Let us identify $\mathbf{T}^n$ with $\mathbf{R}^n/\mathbf{Z}^n$, and let φ be the canonical homomorphism of $\mathbf{R}^n$ onto $\mathbf{R}^n/\mathbf{Z}^n$. Every subgroup of $\mathbf{T}^n$ is of the form $G = \varphi(H)$ where H is a subgroup of $\mathbf{R}^n$ which contains $\mathbf{Z}^n$ and is isomorphic to $H/\mathbf{Z}^n$ (Chapter III, § 2, no. 7, Proposition 20); for G to be *closed* in $\mathbf{T}^n$ it is necessary and sufficient for H to be closed in $\mathbf{R}^n$ (Chapter I, § 3, no. 4). Thus in order to find the closed subgroups of $\mathbf{T}^n$ we have to determine all the closed subgroups H of $\mathbf{R}^n$ which contain $\mathbf{Z}^n$; to do this we shall use Proposition 6 of no. 3, and we start by determining the subgroup H^* associated with H. Since $\mathbf{Z}^n$ is associated with itself, we have $H^* \subset \mathbf{Z}^n$; consequently (no. 1) there exists a basis $(\boldsymbol{a}_i)_{1 \leqslant i \leqslant n}$ of $\mathbf{R}^n$ which generates $\mathbf{Z}^n$, and a basis of H^* (with respect to the ring $\mathbf{Z}$) consisting of p points $\boldsymbol{b}_i$ $(1 \leqslant i \leqslant p)$ such that $\boldsymbol{b}_i = e_i \boldsymbol{a}_i$ $(1 \leqslant i \leqslant p)$, where the e_i are integers satisfying the congruences $e_{i+1} \equiv 0 \pmod{e_i}$ for $1 \leqslant i \leqslant p-1$. Let $(\boldsymbol{a}'_i)$ be the basis dual to $(\boldsymbol{a}_i)$; then $\boldsymbol{u} = \sum_{i=1}^{n} u_i \boldsymbol{a}_i$ belongs to $(H^*)^* = H$ if and only if $u_i e_i \in \mathbf{Z}$ for $1 \leqslant i \leqslant p$; in other words, H is the direct sum of the vector subspace V generated by $\boldsymbol{a}'_{p+1}, \ldots, \boldsymbol{a}_n$, and the discrete subgroup K generated by the p points

$$e_i^{-1}\boldsymbol{a}'_i \qquad (1 \leqslant i \leqslant p).$$

On the other hand, $\mathbf{Z}^n$ is the direct sum of $V \cap \mathbf{Z}^n$ and $K \cap \mathbf{Z}^n$, because the $\boldsymbol{a}'_i$ $(1 \leqslant i \leqslant n)$ generate $\mathbf{Z}^n$. The quotient group $H/\mathbf{Z}^n$ is therefore isomorphic to $(V/(V \cap \mathbf{Z}^n)) \times (K/(K \cap \mathbf{Z}^n))$ (Chapter III, § 2, no. 9, Corollary to Proposition 26); $V/(V \cap \mathbf{Z}^n)$ is isomorphic to $\mathbf{T}^{n-p}$, and $K/(K \cap \mathbf{Z}^n)$ is a *finite* group, the direct sum of p cyclic groups of orders e_i, respectively $(1 \leqslant i \leqslant p)$ (cf. no. 1).

Using the same notation, every Hausdorff quotient group of $\mathbf{T}^n$ is of the form $\mathbf{T}^n/\varphi(H)$ and is isomorphic to $\mathbf{R}^n/H$ (Chapter III, § 2, no. 7, Corollary to Proposition 22); if W is the vector subspace generated by K, $\mathbf{R}^n/H$ is isomorphic to W/K (Chapter III, § 2, no. 9, Corollary to Proposition 26), i.e., to $\mathbf{T}^p$. To sum up:

PROPOSITION 11. *Every closed subgroup of* $\mathbf{T}^n$ *is isomorphic to a group of the form* $\mathbf{T}^h \times F$ $(0 \leqslant h \leqslant n)$ *where* F *is a finite abelian group, such that the smallest number of cyclic subgroups of which* F *is a direct sum is* $\leqslant n - h$. *Every Hausdorff quotient group of* $\mathbf{T}^n$ *is isomorphic to a group of the form* $\mathbf{T}^h$ $(0 \leqslant h \leqslant n)$.

In particular $(n = 1)$:

COROLLARY. *Every closed subgroup of* $\mathbf{T}$, *other than* $\mathbf{T}$ *itself, is a finite cyclic group. Every Hausdorff quotient group of* $\mathbf{T}$, *other than* $\{0\}$, *is isomorphic to* $\mathbf{T}$.

6. PERIODIC FUNCTIONS

DEFINITION 1. *A function* f, *defined on* $\mathbf{R}^n$, *with values in an arbitrary set* E, *is said to be periodic if there exists a point* $\boldsymbol{a} \neq 0$ *in* $\mathbf{R}^n$ *such that*

$$f(\boldsymbol{x} + \boldsymbol{a}) = f(\boldsymbol{x}) \tag{1}$$

for all $\boldsymbol{x} \in \mathbf{R}^n$. *If* f *is periodic, every point* $\boldsymbol{a} \in \mathbf{R}^n$ *for which the relation* (1) *is an identity in* $\boldsymbol{x}$ *is called a period of* f.

The set G of all periods of a periodic function f is clearly a *subgroup* (which by hypothesis does not consist of 0 alone) of the additive group $\mathbf{R}^n$. If f is a *continuous* periodic mapping of $\mathbf{R}^n$ into a *Hausdorff* topological space E, its group of periods G is *closed*. For if $G_{\boldsymbol{x}}$ denotes the set of all $\boldsymbol{a} \in \mathbf{R}^n$ such that $f(\boldsymbol{x} + \boldsymbol{a}) = f(\boldsymbol{x})$ for a *given point* $\boldsymbol{x} \in \mathbf{R}^n$, then G is the intersection of the $G_{\boldsymbol{x}}$ as $\boldsymbol{x}$ runs through $\mathbf{R}^n$, and each $G_{\boldsymbol{x}}$ is *closed* (Chapter I, § 8, no. 1, Proposition 2). Let V be the largest vector subspace contained in G (no. 2, Theorem 2); the function f is *constant* on every coset mod V; if W is a vector subspace complementary to V, then f is determined by its restriction to W. In other words (W being a topological group isomorphic to $\mathbf{R}^p$ for some p), the study of continuous periodic functions on $\mathbf{R}^n$ is reduced to the study of such functions whose group of periods G is *discrete*; if this group is of rank g, the function f is said to be *q-ply periodic*, and every free system of q points which generate G is called a *principal system of periods* of f.

If $(\boldsymbol{a}_i)$ and $(\boldsymbol{b}_i)$ are two principal systems of periods of f, we have seen (no. 1) that each can be obtained from the other by a linear transformation with integer coefficients and determinant ± 1.

Let φ be the canonical mapping of $\mathbf{R}^n$ onto $\mathbf{R}^n/G$; to every mapping g of $\mathbf{R}^n/G$ into a set E corresponds the function $\dot{g} = g \circ \varphi$, which is a periodic mapping of $\mathbf{R}^n$ into E, having a group of periods which contains G; and conversely every mapping of $\mathbf{R}^n$ into E which has a group of periods containing G is of this form, since it is *compatible* with the relation $x \equiv y \pmod{G}$ (*Set Theory*, R, § 5, no. 7). In this way we define a *bijective* mapping $g \to \dot{g}$ of the set of all mappings of $\mathbf{R}^n/G$ into E onto the set of all mappings of $\mathbf{R}^n$ into E whose group of periods contains G. For $\dot{g}$ to be continuous (E being a topological space) it is necessary and sufficient that g should be continuous (Chapter I, § 3, no. 4, Proposition 6).

2. CONTINUOUS HOMOMORPHISMS OF $\mathbf{R}^n$ AND ITS QUOTIENT GROUPS

1. CONTINUOUS HOMOMORPHISMS OF THE GROUP $\mathbf{R}^m$ INTO THE GROUP $\mathbf{R}^n$

Every linear mapping (cf. Chapter VI, § 1, no. 3) of $\mathbf{R}^m$ into $\mathbf{R}^n$ is evidently a *continuous homomorphism* of the additive group $\mathbf{R}^m$ into the additive group $\mathbf{R}^n$. Conversely:

Proposition 1. *Every continuous homomorphism f of the additive group* $\mathbf{R}^m$ *into the additive group* $\mathbf{R}^n$ *is a linear mapping of* $\mathbf{R}^m$ *into* $\mathbf{R}^n$.

It is enough to show that $f(t\mathbf{x}) = tf(\mathbf{x})$ for all $\mathbf{x} \in \mathbf{R}^n$ and all $t \in \mathbf{R}$. The argument is the same as that of Proposition 5 of Chapter V, § 1, no. 3 if we replace x by $\mathbf{x}$ and $\mathbf{R}$ by $\mathbf{R}^m$.

2. LOCAL DEFINITION OF A CONTINUOUS HOMOMORPHISM OF $\mathbf{R}^n$ INTO A TOPOLOGICAL GROUP

Proposition 6 of Chapter V, § 1, no. 4 may be generalized to all the groups $\mathbf{R}^n$.

Proposition 2. *Let* A *be a parallelotope in* $\mathbf{R}^n$ *which contains* 0; *and let f be a continuous mapping of* A *into a topological group* G (written multiplicatively) *such that* $f(\mathbf{x}+\mathbf{y}) = f(\mathbf{x})f(\mathbf{y})$ *for each pair of points* $\mathbf{x}, \mathbf{y}$ *such that* $\mathbf{x} \in A$, $\mathbf{y} \in A$ *and* $\mathbf{x}+\mathbf{y} \in A$. *Then there exists a unique continuous homomorphism of* $\mathbf{R}^n$ *into* G *which extends f.*

By the same reasoning as in Proposition 6 of Chapter V, § 1, no. 4 we show that the homomorphism extending f, if it exists, is *unique*. Next, the subgroup G_1 of G, generated by $f(A)$, is *commutative*; for if $\mathbf{x}$

and y are any two points of A, then $\frac{1}{2}x$, $\frac{1}{2}y$ and $\frac{1}{2}(x+y)$ belong to A, therefore

$$f(\tfrac{1}{2}(x+y)) = f(\tfrac{1}{2}x)\ (f\tfrac{1}{2}y) = f(\tfrac{1}{2}y)f(\tfrac{1}{2}x),$$

which shows that $f(\frac{1}{2}x)$ and $f(\frac{1}{2}y)$ commute; therefore so do $f(x) = (f(\frac{1}{2}x))^2$ and $f(y) = (f(\frac{1}{2}y))^2$, which proves the assertion.

Let $a_1, a_2, \ldots, a_n$ be n non-zero vectors contained in A and proportional to the basis vectors of the parallelotope, and for each index i let D_i be the line through 0 and a_i, that is, the set of points ta_i as t runs through $\mathbf{R}$. Let A_i be the set of all $t \in \mathbf{R}$ such that $ta_i \in A$; then A_i is an interval of $\mathbf{R}$ containing $[0, 1]$, and the function

$$f_i(t) = f(ta_i)$$

is defined and continuous on A_i and satisfies the relation

$$f_i(t+t') = f_i(t)f_i(t')$$

whenever t, t' and $t+t'$ all belong to A_i. By Proposition 6 of Chapter V, § 1, no. 4, there exists a *continuous* homomorphism $\bar{f}_i$ of $\mathbf{R}$ into G which extends f_i. Since $\mathbf{R}^n$ is the direct sum of the subgroups D_i, we can define a homomorphism $\bar{f}$ of $\mathbf{R}^n$ into the abelian group G_1 by the rule $\bar{f}(x) = \prod_{i=1}^{n} \bar{f}_i(t_i)$ where $x = \sum_{i=1}^{n} t_i a_i$; $\bar{f}$ is an extension of f since, if $x \in A$, all the components x_i of x on the lines D_i also belong to A, by reason of the choice of the a_i; moreover, $\bar{f}$ is continuous on $\mathbf{R}^n$, because it is continuous on each of the lines D_i, and x_i is a linear function of x (and therefore continuous).

Corollary 1. *Let* V *be a neighbourhood of* 0 *in* $\mathbf{R}^n$ *and let* f *be a continuous mapping of* V *into a topological group* G *such that*

$$f(x+y) = f(x)f(y)$$

for each pair of points x, y *such that* $x \in V$, $y \in V$ *and* $x + y \in V$. *Then there exists a unique continuous homomorphism of* $\mathbf{R}^n$ *into* G *which agrees with* f *at all points of a neighbourhood* W *of* 0.

Take W to be an open box with centre 0, contained in V, and apply Proposition 2 to W.

We shall see later that this property of $\mathbf{R}^n$ extends to a larger class of topological groups, the "simply connected" groups.

COROLLARY 2. *Let f be a local isomorphism of $\mathbf{R}^n$ with a topological group* G. *Then there exists a unique strict morphism of $\mathbf{R}^n$ onto an open subgroup of* G *which agrees with f at all the points of some neighbourhood of* 0.

Let $\bar{f}$ be the continuous homomorphism of $\mathbf{R}^n$ into G which agrees with f at all points of a neighbourhood of 0; $\bar{f}(\mathbf{R}^n)$ contains, by hypothesis, a neighbourhood of the identity element of G and therefore (Chapter III, § 2, no. 1, Corollary to Proposition 4) is an open subgroup of G; moreover $\bar{f}$ is a strict morphism of $\mathbf{R}^n$ onto $\bar{f}(\mathbf{R}^n)$, by Chapter III, § 2, no. 8, Proposition 24.

THEOREM 1. *Every connected group* G *which is locally isomorphic to $\mathbf{R}^n$ is isomorphic to a group $\mathbf{R}^p \times \mathbf{T}^{n-p}$ $(0 \leqslant p \leqslant n)$.*

A suitably chosen local isomorphism f of $\mathbf{R}^n$ with G extends to a strict morphism of $\mathbf{R}^n$ onto an open subgroup of G (Proposition 2, Corollary 2) and therefore onto G itself, since G is connected. It follows that G is isomorphic to a quotient group $\mathbf{R}^n/\mathrm{H}$ of $\mathbf{R}^n$: H is *discrete*, otherwise there would exist points $x \neq 0$ of H arbitrarily close to 0 and such that $f(x) = f(0)$, contrary to the hypothesis that f is a local isomorphism. The theorem is therefore a consequence of Theorem 1 of § 1, no. 1.

3. CONTINUOUS HOMOMORPHISMS OF $\mathbf{R}^m$ INTO $\mathbf{T}^n$

PROPOSITION 3. *Every continuous homomorphism of $\mathbf{R}^m$ into $\mathbf{T}^n$ is of the form $x \to \varphi(u(x))$, where φ is the canonical homomorphism of $\mathbf{R}^n$ onto $\mathbf{T}^n$* (identified with $\mathbf{R}^n/\mathbf{Z}^n$) *and u is a linear mapping of $\mathbf{R}^m$ into $\mathbf{R}^n$.*

Let f be a continuous homomorphism of $\mathbf{R}^m$ into $\mathbf{T}^n$. We shall show that there is a linear mapping u of $\mathbf{R}^m$ into $\mathbf{R}^n$ such that the homomorphisms $x \to f(x)$ and $x \to \varphi(u(x))$ coincide at all points of a *neighbourhood of* 0 in $\mathbf{R}^m$; the proposition will then follow from Corollary 1 to Proposition 2 of no. 2. Now let V be a neighbourhood of 0 in $\mathbf{R}^n$ such that φ, restricted to V, is a local isomorphism of $\mathbf{R}^n$ with $\mathbf{T}^n$, and let ψ be the inverse of φ, defined on $\varphi(\mathrm{V})$. Since f is continuous, $\mathrm{V}' = \overset{-1}{f}(\varphi(\mathrm{V}))$ is a neighbourhood of 0 in $\mathbf{R}^m$; the mapping

$$x \to \psi(f(x)),$$

restricted to V′, is a continuous mapping of V′ into $\mathbf{R}^n$ such that $\psi(f(x+y)) = \psi(f(x)) + \psi(f(y))$ for each pair of points x, y in $\mathbf{R}^m$ such that $x \in \mathrm{V}'$, $y \in \mathrm{V}'$ and $x + y \in \mathrm{V}'$; hence (no. 2, Corollary 1 to Proposition 2) this mapping coincides with a well-determined continuous homomorphism u of $\mathbf{R}^m$ into $\mathbf{R}^n$ at all points of a neighbourhood W of 0 in $\mathbf{R}^m$. By Proposition 1 (no. 1) u is a linear mapping of $\mathbf{R}^m$ into $\mathbf{R}^n$; for all $x \in W$ we have therefore $f(x) = \varphi(u(x))$, which completes the proof.

Remark. The same argument shows, more generally, that if φ is a strict morphism of $\mathbf{R}^n$ into a group G, whose restriction to a suitable neighbourhood of 0 is a local isomorphism of $\mathbf{R}^n$ with G, then every continuous homomorphism of $\mathbf{R}^n$ into G is of the form $x \to \varphi(u(x))$, where u is a linear mapping of $\mathbf{R}^m$ into $\mathbf{R}^n$.

In the case $m = n = 1$, Proposition 3 gives:

PROPOSITION 4. *If φ is the canonical homomorphism of $\mathbf{R}$ onto $\mathbf{T}$, every continuous homomorphism of $\mathbf{R}$ into $\mathbf{T}$ is of the form $x \to \varphi(ax)$ where $a \in \mathbf{R}$; and it is a strict morphism of $\mathbf{R}$ onto $\mathbf{T}$* if $a \neq 0$.

4. AUTOMORPHISMS OF $\mathbf{T}^n$

Let H be a closed subgroup of $\mathbf{R}^n$ and let φ be the canonical homomorphism of $\mathbf{R}^n$ onto the quotient group $\mathbf{R}^n/H$. If f is a continuous homomorphism of $\mathbf{R}^n/H$ into a topological group G, then $\dot{f} = f \circ \varphi$ is a continuous homomorphism of $\mathbf{R}^n$ into G which is periodic and has a group of periods containing H; conversely every periodic continuous homomorphism of $\mathbf{R}^n$ into G, whose group of periods contains H, is of this form.

In the case where $H = \mathbf{Z}^n$, the quotient group $\mathbf{R}^n/\mathbf{Z}^n = \mathbf{T}^n$ is *compact*, and therefore every continuous homomorphism f of $\mathbf{T}^n$ into a topological group G is a *strict morphism* of $\mathbf{T}^n$ into G, provided that G is Hausdorff (Chapter III, § 2, no. 8, Remark 1); $\dot{f} = f \circ \varphi$ is a strict morphism of $\mathbf{R}^n$ into G; moreover, $f(\mathbf{T}^n) = \dot{f}(\mathbf{R}^n)$ is a *compact* subgroup of G, isomorphic to a group $\mathbf{T}^p$ $(0 \leqslant p \leqslant n)$.

In particular, we see that the only continuous homomorphism of $\mathbf{T}^n$ into a group $\mathbf{R}^m$ is the *zero* mapping, since $\{0\}$ is the only compact subgroup of $\mathbf{R}^m$.

Let us apply this to continuous homomorphisms of $\mathbf{T}^n$ into a group $\mathbf{T}^p$; if f is such a homomorphism, φ the canonical homomorphism of $\mathbf{R}^n$ onto $\mathbf{T}^n$, then $f \circ \varphi$ is a continuous homomorphism of $\mathbf{R}^n$ into $\mathbf{T}^p$; hence (no. 3, Proposition 3) if ψ denotes the canonical homomorphism of $\mathbf{R}^p$ onto $\mathbf{T}^p$, there exists a linear mapping u of $\mathbf{R}^n$ into $\mathbf{R}^p$ such that $f \circ \varphi = \psi \circ u$. If $x \in \mathbf{Z}^n$, $f(\varphi(x))$ is the identity element of $\mathbf{T}^p$, so that we must have $u(x) \in \mathbf{Z}^p$, i.e., $u(\mathbf{Z}^n) \subset \mathbf{Z}^p$. Conversely, if u is any linear mapping of $\mathbf{R}^n$ into $\mathbf{R}^p$ which satisfies this condition, then

$\psi \circ u$ is a periodic continuous homomorphism of $\mathbf{R}^n$ into $\mathbf{T}^p$, whose group of periods contains $\mathbf{Z}^n$, and therefore defines a continuous homomorphism of $\mathbf{T}^n$ into $\mathbf{T}^p$.

Consider under what conditions f is an *isomorphism* of $\mathbf{T}^n$ onto a subgroup of $\mathbf{T}^p$. First of all, u must be an *injective* mapping of $\mathbf{R}^n$ into $\mathbf{R}^p$, otherwise the vector subspace $\overset{-1}{u}(0)$ would contain points $x \neq 0$ arbitrarily close to the origin, and at such a point we should have

$$f(\varphi(x)) = f(\varphi(0)) \quad \text{and} \quad \varphi(x) \neq \varphi(0),$$

contrary to hypothesis. This condition implies that $p \geqslant n$. The image $u(\mathbf{Z}^n)$ is then a discrete subgroup of rank n of the group $\mathbf{Z}^p$; the *invariant factors* of $u(\mathbf{Z}^n)$ with respect to $\mathbf{Z}^p$ (§ 1, no. 1) must all be equal to 1, otherwise there would exist a point $x \in \mathbf{Z}^n$ and an integer $k > 1$ such that $u(k^{-1}x) \in \mathbf{Z}^n$ and $k^{-1} \notin \mathbf{Z}^n$, hence $f(\varphi(k^{-1}x)) = f(\varphi(0))$ and $\varphi(k^{-1}x) \neq \varphi(0)$, contrary to hypothesis. Conversely, if this condition is satisfied, $u(\mathbf{R}^n) \cap \mathbf{Z}^n = u(\mathbf{Z}^n)$, and f is an isomorphism of $\mathbf{T}^n$ onto $u(\mathbf{R}^n)/u(\mathbf{Z}^n)$.

If we apply this argument to the case $p = n$, we have the following proposition:

PROPOSITION 5. *Every isomorphism of the topological group* $\mathbf{T}^n$ *onto one of its subgroups is an automorphism of* $\mathbf{T}^n$ *which is obtained by passing to the quotient from a linear mapping* u *of* $\mathbf{R}^n$ *onto itself which, restricted to* $\mathbf{Z}^n$, *is an automorphism of* $\mathbf{Z}^n$.

Equivalently (§ 1, no. 1), if $u(e_i) = \sum_{j=1}^{n} a_{ij} e_j$, the a_{ij} must be *integers* such that $\det(a_{ij}) = \pm 1$. In particular, for $n = 1$:

PROPOSITION 6. *The only isomorphisms of the topological group* $\mathbf{T}$ *onto one of its subgroups are the identity mapping and the symmetry* $x \to -x$.

3. INFINITE SUMS IN THE GROUPS $\mathbf{R}^n$

1. SUMMABLE FAMILIES IN $\mathbf{R}^n$

Since every point of $\mathbf{R}^n$ has a *countable* fundamental system of neighbourhoods, a family (x_ι) of points of the additive group $\mathbf{R}^n$ is summable only if the set of indices ι such that $\mathbf{X}_\iota \neq 0$ is *countable* (Chapter III, § 5, no. 2, Corollary to Proposition 1); hence, essentially, the study of summable families in $\mathbf{R}^n$ is reduced to that of summable *sequences*. Nevertheless, for the same reasons as were given in Chapter IV, § 7, in connection with summable families in $\mathbf{R}$, we shall not impose any restriction, in what follows, on the cardinal of the index set.

PROPOSITION 1. *A family $(\boldsymbol{x}_\iota)_{\iota\in I}$ of points $\boldsymbol{x}_\iota = (x_{\iota,k})_{1\leqslant k\leqslant n}$ of $\mathbf{R}^n$ is summable if and only if each of the n families $(x_{\iota,k})_{\iota\in I}$ of real numbers is summable in $\mathbf{R}$.*

This follows from Chapter III, § 5, no. 4, Proposition 4.

The condition of Proposition 1 may be transformed as follows:

THEOREM 1. *A family $(\boldsymbol{x}_\iota)_{\iota\in I}$ of points of $\mathbf{R}^n$ is summable if and only if the family $(\|\boldsymbol{x}_\iota\|)$ of Euclidean norms of the $\boldsymbol{x}_\iota$ is summable in $\mathbf{R}$.*

This follows without trouble from Proposition 1, the condition for summability of a family of real numbers (Chapter IV, § 7, no. 2, Theorem 3), the inequalities

$$\sup_{1\leqslant k\leqslant n} |x_{\iota,k}| \leqslant \|\boldsymbol{x}_\iota\| \leqslant \sum_{i=1}^{n} |x_{\iota,i}|,$$

and the comparison principle (Chapter IV, § 7, no. 1, Theorem 2).

One can also proceed somewhat differently, by first establishing the following proposition:

PROPOSITION 2. *If $(\boldsymbol{x}_i)_{i\in I}$ is any finite family of points in $\mathbf{R}^n$, then*

$$\sum_{i\in I} \|\boldsymbol{x}_i\| \leqslant 2n.\sup_{J\subset I}\Big\|\sum_{i\in J} \boldsymbol{x}_i\Big\|. \tag{1}$$

For if $\boldsymbol{x}_i = (x_{ij})_{1\leqslant j\leqslant n}$, we have $\|\boldsymbol{x}_i\| \leqslant \sum_{j=1}^{n} |x_{ij}|$, hence

$$\sum_{i\in I} \|\boldsymbol{x}_i\| \leqslant \sum_{j=1}^{n}\Big(\sum_{i\in I} |x_{ij}|\Big).$$

Now $\sum_{i\in I} |x_{ij}| = \sum_{i\in I} x_{ij}^+ + \sum_{i\in I} x_{ij}^-$, and since for every subset J of I we have

$$-\sum_{i\in I} x_{ij}^- \leqslant -\sum_{i\in J} x_{ij}^- \leqslant \sum_{i\in J} x_{ij}^+ \leqslant \sum_{i\in I} x_{ij}^+,$$

it follows that

$$\sum_{i\in I} |x_{ij}| \leqslant 2.\sup_{J\subset I}\Big|\sum_{i\in J} x_{ij}\Big|.$$

But $\Big|\sum_{i\in J} x_{ij}\Big| \leqslant \Big\|\sum_{i\in J} \boldsymbol{x}_i\Big\|$, hence the inequality (1).

Now, Theorem 1 is equivalent to the following proposition (since $\mathbf{R}^n$ is a complete group): the family $(\boldsymbol{x}_\iota)$ satisfies Cauchy's criterion (Chapter III, § 5, no. 2, Theorem 1) if and only if the family $(\|\boldsymbol{x}_\iota\|)$ also satisfies Cauchy's criterion. Now the triangle inequality shows that this condition is sufficient, and the inequality (1) shows that it is necessary.

Furthermore we have the inequality

$$\left\|\sum_\iota \boldsymbol{x}_\iota\right\| \leqslant \sum_\iota \|\boldsymbol{x}_\iota\|, \tag{2}$$

which comes by passing to the limit from the analogous inequality for finite partial sums.

Corollary. *A family $(\boldsymbol{x}_\iota)$ of points of $\mathbf{R}^n$ is summable if and only if the set of finite partial sums of the family is bounded in $\mathbf{R}^n$.*

By Theorem 1 and the triangle inequality, this condition is necessary; it is sufficient by the inequality (1) and Theorem 1.

Proposition 3. *Let $(\boldsymbol{x}_\lambda)_{\lambda\in\mathrm{L}}$ be a summable family of points of $\mathbf{R}^m$, $(\boldsymbol{y}_\mu)_{\mu\in\mathrm{M}}$ a summable family of points of $\mathbf{R}^n$, and let f be a bilinear mapping of $\mathbf{R}^m \times \mathbf{R}^n$ into $\mathbf{R}^p$. Then the family $(f(\boldsymbol{x}_\lambda, \boldsymbol{y}_\mu))_{(\lambda,\mu)\in\mathrm{L}\times\mathrm{M}}$ is summable and we have*

$$\sum_{(\lambda,\mu)\in\mathrm{L}\times\mathrm{M}} f(\boldsymbol{x}_\lambda, \boldsymbol{y}_\mu) = f\Big(\sum_{\lambda\in\mu} \boldsymbol{x}_\lambda, \sum_{\mu\in\mathrm{M}} \boldsymbol{y}_\mu\Big). \tag{3}$$

To show that the family $(f(\boldsymbol{x}_\lambda, \boldsymbol{y}_\mu))$ is summable, it is sufficient by Proposition 1 to establish that each of the p families formed by the coordinates of the points $f(\boldsymbol{x}_\lambda, \boldsymbol{y}_\mu)$ in $\mathbf{R}^n$ is summable: in other words, we can restrict ourselves to the case where f is a bilinear *form*; but for such a form f we have

$$f(\boldsymbol{x}, \boldsymbol{y}) = \sum_{i,j} a_{ij} x_i y_j,$$

and therefore we are brought back to the case $f(\boldsymbol{x}, \boldsymbol{y}) = x_i y_j$, and in this case the result has already been proved (Chapter IV, § 7, no. 3, Proposition 1).

By specializing the function f we obtain in particular the following corollaries:

Corollary 1. *If $(a_\lambda)_{\lambda\in\mathrm{L}}$ is a summable family of real numbers and if $(\boldsymbol{x}_\mu)_{\mu\in\mathrm{M}}$ is a summable family of points of $\mathbf{R}^n$, then the family $(a_\lambda \boldsymbol{x}_\mu)_{(\lambda,\mu)\in\mathrm{L}\times\mathrm{M}}$ is summable and we have*

$$\sum_{(\lambda,\mu)\in\mathrm{L}\times\mathrm{M}} a_\lambda \boldsymbol{x}_\mu = \Big(\sum_{\lambda\in\mathrm{L}} a_\lambda\Big)\Big(\sum_{\mu\in\mathrm{M}} \boldsymbol{x}_\mu\Big). \tag{4}$$

COROLLARY 2. *If* $(\boldsymbol{x}_\lambda)_{\lambda\in L}$ *and* $(\boldsymbol{y}_\mu)_{\mu\in M}$ *are two summable families of points of* $\mathbf{R}^n$, *then the family* $(\boldsymbol{x}_\lambda|\boldsymbol{y}_\mu)$ (cf. Chapter VI, § 2, no. 2) *is summable in* $\mathbf{R}$, *and we have*

$$(5) \qquad \sum_{(\lambda,\mu)\in L\times M} (\boldsymbol{x}_\lambda|\boldsymbol{y}_\mu) = \Big(\sum_{\lambda\in L} \boldsymbol{x}_\lambda \Big| \sum_{\mu\in M} \boldsymbol{y}_\mu\Big).$$

2. SERIES IN $\mathbf{R}^n$

A series whose general term is $\boldsymbol{x}_m = (x_{mi})_{1\leqslant i\leqslant n}$ converges in $\mathbf{R}^n$ if and only if each of the n series $(x_{mi})_{m\in\mathbf{N}}$ converges in $\mathbf{R}$.

DEFINITION 1. *A series of points of* $\mathbf{R}^n$ *is said to be absolutely convergent if the series of Euclidean norms of its terms is convergent.*

PROPOSITION 4. *A series of points of* $\mathbf{R}^n$ *is commutatively convergent if and only if it is absolutely convergent.*

This is a consequence of Proposition 9 of Chapter III, § 5, no. 7 and Theorem 1 above.

The examples given in Chapter IV, § 7 show that a series in $\mathbf{R}^n$ can be *convergent* without being *absolutely convergent.*

EXERCISES

§ 1

1) Let G be a discrete subgroup of rank p in $\mathbf{R}^n$ and let $(\boldsymbol{a}_i)_{1 \leqslant i \leqslant p}$ be a free system of p points of G. The proof of Theorem 1 of no. 1 shows that G is a subgroup of the group generated by the p points $d^{-1}\boldsymbol{a}_i$, where d is some integer. Show, without using Theorem 1, that there exists a free system of p points $\boldsymbol{b}_i = \sum_{j=1}^{p} b_{ij}\boldsymbol{a}_j$ of G such that, if $\boldsymbol{x}_i = \sum_{j=1}^{p} x_{ij}\boldsymbol{a}_i$ $(1 \leqslant i \leqslant p)$ is any free system of p points of G, we have $|\det(x_{ij})| \geqslant |\det(b_{ij})| > 0$. Hence give a proof of Theorem 1 which does not rest on the theory of invariant factors, by showing that the $\boldsymbol{b}_i$ generate G. (Argue by contradiction: if $\boldsymbol{z} = \sum_{i=1}^{p} z_i\boldsymbol{b}_i \in \mathrm{G}$ and one of the z_i is not an integer, there exists a point $\boldsymbol{u} = \sum_{i=1}^{p} u_i\boldsymbol{b}_i$ of the group generated by $\boldsymbol{z}$ and the $\boldsymbol{b}_i$ such that $0 < u_i < 1$ for some index i, and hence obtain a contradiction.)

2) Let G be a discrete subgroup of $\mathbf{R}^n$. If G is the direct sum of two subgroups H and K, then the intersection of the vector subspaces generated by H and K consists only of 0, and the rank of G is therefore equal to the sum of the ranks of H and K (observe that, for every discrete subgroup G of $\mathbf{R}^n$, the vector subspace of $\mathbf{R}^n$ generated by G is canonically isomorphic to $\mathrm{G} \otimes_{\mathbf{Z}} \mathbf{R}$).

3) Let G be a discrete subgroup of $\mathbf{R}^n$. For a subgroup H of G to be a direct summand of G it is necessary and sufficient that $\mathrm{H} = \mathrm{V} \cap \mathrm{G}$ where V is a vector subspace of $\mathbf{R}^n$ (for necessity, use Exercise 2; for sufficiency, note that H is also the intersection of G and the vector subspace generated by H).

4) Let H and K be two discrete subgroups of $\mathbf{R}^n$ such that $H + K$ is a closed subgroup. Show that $H + K$ is then discrete and that $r(H) + r(K) = r(H \cap K) + r(H + K)$. (Let V be the vector subspace of $\mathbf{R}^n$ generated by $H \cap K$; split up H into the direct sum of $V \cap H$ and a discrete group H_1, and K into the direct sum of $V \cap K$ and a discrete group K_1, using Exercise 3; then show that the sum $H_1 + K_1$ is direct and use Exercise 2.)

5) Let G, G′ be two closed subgroups of $\mathbf{R}^n$ such that $G' \subset G$, and let V (resp. V′) be the largest vector subspace contained in G (resp. G′). Show that there is a vector subspace W complementary to V such that G is the direct sum of V and the discrete group $K = W \cap G$, and such that G′ is the direct sum of $V \cap G'$ and $K \cap G' = W \cap G'$. (If U is a vector subspace complementary to V′, use Exercise 3 to show that the discrete group $U \cap G'$ is the direct sum of $U \cap V \cap G'$ and a discrete group K′, and take W to be a vector subspace containing K′.) Deduce that:

a) The quotient group G/G' is isomorphic to a product group of the form

$$\mathbf{R}^p \times \mathbf{T}^q \times \mathbf{Z}^r \times F,$$

where F is a finite abelian group.

b) Every closed subgroup and every Hausdorff quotient group of a group of the form $\mathbf{R}^p \times \mathbf{T}^q \times \mathbf{Z}^r \times F$ (F finite and abelian) are of the same form.

¶ 6) Let H and K be two closed subgroups of $\mathbf{R}^n$ such that $H + K$ is a closed subgroup.

a) Show that, if V (resp. W) is the largest vector subspace contained in H (resp. K), then $V + W$ is the largest vector subspace contained in G, and therefore $d(H) + d(K) = d(H \cap K) + d(H + K)$ [note that the rational rank of $G/(V + W)$ is finite and consequently that G cannot contain a line which is not contained in $V + W$].

b) Let U be a subspace complementary to $V + W$, such that G is the direct sum of $V + W$ and $M = G \cap U$, and such that $H \cap K$ is the direct sum of $(V + W) \cap (H \cap K)$ and $L = H \cap K \cap U$ (Exercise 5). Let H′ (resp. K′) be the subgroup of M consisting of the components of the points of H (resp. K) in the decomposition of G into the direct sum of $V + W$ and M. Show that $M = H' + K'$ and that $H' \cap K' = L$.

c) If we put $H'' = H \cap (V + W)$ and $K'' = K \cap (V + W)$, show that $r(H'') + r(K'') = r(H'' \cap K'') + r(H'' + K'')$ (reduce to the case where $V \cap W = \{0\}$, and show that then $H'' \cap K''$ is the direct sum of $V \cap K''$ and $W \cap H''$).

d) Show that

$$r(H) + r(K) = r(H \cap K) + r(H + K)$$

[use *c*) and Exercise 4, noting that $r(H) = r(H') + r(H'')$ and $r(K) = r(K') + r(K'')$].

¶ 7) Let G be a closed subgroup of $\mathbf{R}^n$ and let H be a closed subgroup of G. Then G is the direct sum of H and another closed subgroup K if and only if H is the intersection of G and a vector subspace. (Use Exercise 5 for the necessity; to show sufficiency, apply Theorem 2 of no. 2 appropriately to G and H, and use Exercise 3).

8) *a*) Let H, K be two closed subgroups of $\mathbf{R}^n$. Show that if $r(H) + r(K)$ is equal to $r(H \cap K) + r(H + K)$, then $H + K$ is closed in $\mathbf{R}^n$. [Using Exercise 7 and the hypothesis given, reduce to the case where H, K and $H \cap K$ all generate the *same* vector subspace; if V (resp. W) denotes the largest subspace contained in H (resp. K), show with the help of Exercise 5 that V is generated by $V \cap K$, and W is generated by $W \cap H$; finally, split up $H \cap K$ into the direct sum of $H \cap K \cap (V + W)$ and a discrete group, by using Exercise 7.]

b) Deduce from *a*) and Exercise 6 that, if H and K are two closed subgroups of $\mathbf{R}^n$ such that $H + K$ is a closed subgroup, then $H^* + K^*$ is also a closed subgroup.

c) Deduce from *b*) that if H and K are two closed subgroups of $\mathbf{R}^n$ such that $d(H) + d(K) = d(H \cap K) + d(\overline{H + K})$, then the subgroup $H + K$ is closed.

9) Let G be a subgroup of $\mathbf{R}^n$, *not necessarily closed*, of rank p, and let V be the largest vector subspace contained in $\overline{G}$. If V has dimension $q \leqslant p$, show that G is the direct sum of $V \cap G$ and a discrete subgroup of rank $p - q$ contained in a vector subspace complementary to V [note that, for each $x \in \overline{G}$, $(x + V) \cap G$ is dense in the linear variety $x + V$].

10) Let a be a real number and n an integer > 0, and consider the numbers $x_k = ka - [ka] \in [0, 1[$ $(1 \leqslant k \leqslant n + 1)$. If I_h denotes the interval

$$\left[\frac{h-1}{n}, \frac{h}{n}\right[\qquad (1 \leqslant h \leqslant n),$$

show that there exist two distinct indices k, k' such that x_k and $x_{k'}$ belong to the same interval I_h (for a suitable value of h). Deduce that there exist two integers p, q such that $1 \leqslant p \leqslant n$ and $|pa - q| \leqslant 1/n$.

11) Let $\boldsymbol{a}_i = (a_{ij})$ $(1 \leqslant i \leqslant m,\ 1 \leqslant j \leqslant n)$ be m points of $\mathbf{R}^n$ and let q be an integer > 0. For each point $\boldsymbol{k} = (k_i)$ $(1 \leqslant i \leqslant m)$ of $\mathbf{R}^m$ with integer coordinates, *not all zero*, satisfying the inequalities $0 \leqslant k_i \leqslant q$, consider the point $\boldsymbol{x}_{\boldsymbol{k}} = (x_{j,\boldsymbol{k}})$ $(1 \leqslant j \leqslant n)$ of $\mathbf{R}^n$, all of whose coordinates belong to $[0, 1[$ and which is congruent mod $\mathbf{Z}^n$ to $\sum_{i=1}^{m} k_i \boldsymbol{a}_i$, i.e., the point with coordinates

$$x_{j,\boldsymbol{k}} = \sum_{i=1}^{m} k_i a_{ij} - \left[\sum_{i=1}^{m} k_i a_{ij}\right].$$

If p is the smallest integer such that $q + 1 \geqslant p^{n/m}$, show that there are two distinct points $\boldsymbol{k}, \boldsymbol{k}'$ such that $\boldsymbol{x}_{\boldsymbol{k}}$ and $\boldsymbol{x}_{\boldsymbol{k}'}$ both belong to the same cube of side $1/p$ (same method as Exercise 10). Deduce that there exist m integers p_i, not all zero, and n integers r_j $(1 \leqslant j \leqslant n)$ such that $0 \leqslant p_i \leqslant q$ for $1 \leqslant i \leqslant m$, and

$$\left|\sum_{i=1}^{m} p_i a_{ij} - r_j\right| \leqslant \frac{1}{p} \qquad (1 \leqslant j \leqslant n).$$

Use this result to give another proof of Proposition 2 of no. 1 (the "pigeon-hole principle").

¶ 12) For each pair of real numbers (θ, β) there exists an infinite number of triples (p, q, r) of integers such that $r > 0$, $|q| \leqslant \frac{1}{2}r$ and

$$-\frac{1}{r} < \theta q + p - \beta < \frac{1}{r}.$$

[If m, n are two integers such that $|n\theta - m| < 1/n$ (Exercise 10), take $r = n$ and choose p, q so that $pn + qm$ differs from $n\beta$ by less than $1/2$].

13) The *Farey series* of order n (n an integer > 0) is the set F_n of rational numbers p/q in their lowest terms such that $0 \leqslant p \leqslant q \leqslant n$, arranged in increasing order (i.e., the Farey series is a *sequence*).

a) Show that if two rational numbers $r = p/q$, $r' = p'/q'$ are such that $p'q - pq' = \pm 1$, then every pair of integers (p'', q'') can be expressed in the form $p'' = px + p'y$, $q'' = qx + q'y$ where x and y are integers. The fraction p''/q'' belongs to the closed interval with end-points r and r' if and only if x and y have the same sign.

b) Deduce from *a*) that if $r = p/q$ and $r' = p'/q'$ are two rational numbers in $[0, 1]$ such that $q > 0$ and $q' > 0$ and $qp' - pq' = \pm 1$, then r and r' are consecutive elements in the sequence F_n, where $n = \max(q, q')$. Furthermore, the smallest integer m such that the *open* interval with end-points r and r' contains a point of F_m is $m = q + q'$, and there is only one point of $F_{q+q'}$ in this interval, namely the fraction $(p + p')/(q + q')$.

c) Show that conversely, if r and r' are two consecutive elements of F_n, we have $p'q - pq' = \pm 1$ (use induction on n).

d) Deduce from *c*) that, for each real number $\theta \in [0, 1]$ and each integer $n \geqslant 1$, there is at least one fraction p/q in its lowest terms such that $1 \leqslant q \leqslant n$ and $|\theta - p/q| \leqslant 1/(n+1)q$ (cf. Exercise 10).

e) If p/q is a fraction in its lowest terms such that $|\theta - p/q| < 1/q^2$, then θ belongs to the open interval whose end-points are the two terms of the Farey series F_q consecutive to p/q.

14) Let $\varphi : \mathbf{R} \to \mathbf{T}$ be the canonical homomorphism; let θ be an element of infinite order in $\mathbf{T}$ and let θ_0 be an (irrational) real number such that $\varphi(\theta_0) = \theta$. For each integer $n > 0$, let S_n be the set of elements $k\theta$ of $\mathbf{T}$ for which $1 \leqslant k \leqslant n$, and for each interval I of $\mathbf{R}$ let $N(I, n)$ be the number of elements of the set $I \cap \overset{-1}{\varphi}(S_n)$.

a) If we take $I = [0, \theta_0[$, show that $N(I, n) = [n\theta_0]$ [note that $N(I, n)$ is then the number of pairs of integers (x, y) such that $1 \leqslant y \leqslant n$ and $x \leqslant y\theta_0 < x + \theta_0$].

b) More generally, if we take $I = [0, m\theta_0[$, m being an integer, we have $m.[(n - m)\theta_0] \leqslant N(I, n) \leqslant m.[n\theta_0]$ (same method).

c) Deduce that, if I is any interval, the number $N(I, n)/n$ tends to a limit equal to the length of I, as $n \to +\infty$. (Prove the result first when I is of the form $[m\theta_0 + a, m'\theta_0 + a']$, where m, m', a and a' are integers, by using *b*); then pass to the general case by approximating to the end-points of I by numbers of the form $m\theta_0 + a$.) ["*Equipartition of the sequence* $(k\theta)$ *mod* 1".]

* 15) Let I be a closed cube in $\mathbf{R}^n$ with side 2π. Map each point $x = (x_i)$ of I to the point $y = (y_j)$ of $\mathbf{R}^{n+1}$ such that

$$y_1 = \sin x_1$$
$$y_2 = (2 + \cos x_1) \sin x_2$$
$$\cdots\cdots\cdots\cdots\cdots\cdots$$
$$y_p = (2^{p-1} + \sum_{k=1}^{p-1} 2^{p-k-1} \cos x_{p-k} \cos x_{p-k+1} \ldots \cos x_{p-1}) \sin x_p \quad (2 \leqslant p \leqslant n)$$
$$\cdots\cdots\cdots\cdots\cdots\cdots$$
$$y_{n+1} = \sum_{k=1}^{n-1} 2^{p-k-1} \cos x_{n-k} \cos . x_{n-k+1} \ldots \cos x_n.$$

Show that the image of I under this mapping is homeomorphic to $\mathbf{T}^n$ (proof by induction on n: observe that y_p has the sign of $\sin x_p$ for $\leqslant p$ n). If $n=2$, the subset of $\mathbf{R}^3$ so defined is called a *torus of revolution.*$_*$

¶ 16) Let G be a subgroup of $\mathbf{R}^n$ which contains a compact *connected* subset K such that the affine linear variety generated by K is of dimension p. Show that G contains a vector subspace of dimension p. [Proof by induction on n, using Exercise 10 c) of Chapter VI, § 1.]

¶ 17) Let E be the product $(\mathbf{Q}_p)^n$ of n factors equal to the field $\mathbf{Q}_p$ of p-adic numbers (Chapter III, § 6, Exercise 23), endowed with the topological group structure which is the product of the additive topological structures of the factors $\mathbf{Q}_p$, and with its n-dimensional vector space structure over the field $\mathbf{Q}_p$. If v denotes the additive p-adic valuation on $\mathbf{Q}_p$, put $|x|_p = p^{-v_p(x)}$ for all $x \neq 0$ in $\mathbf{Q}_p$, and $|0|_p = 0$; and put $\|\boldsymbol{x}\| = \sup |x_i|_p$ for all $\boldsymbol{x} = (x_i)_{1 \leqslant i \leqslant n} \in \mathrm{E}$ (cf. Chapter IX, § 3, nos. 2 and 3). ^{i}

a) A subset A of E is relatively compact if and only if

$$\sup_{\boldsymbol{x} \in \mathrm{A}} \|\boldsymbol{x}\| < +\infty.$$

[Note that v_p is a continuous mapping of $\mathbf{Q}_p$ into $\mathbf{Z}$ (with the discrete topology).]

b) If G is a closed subgroup of E, then G is a topological module over the ring $\mathbf{Z}_p$ of p-adic integers (if $\boldsymbol{x} \in \mathrm{G}$, we have $n\boldsymbol{x} \in \mathrm{G}$ for all $n \in \mathbf{Z}$, and $\mathbf{Z}$ is dense in $\mathbf{Z}_p$).

c) If K is a compact subgroup of E, there exists a free system of $m \leqslant n$ points $\boldsymbol{a}_i$ $(1 \leqslant i \leqslant m)$ of E, such that K is the direct sum of the m groups $\mathbf{Z}_p.\boldsymbol{a}_i$. [Use a) to show that, if $(\boldsymbol{e}_j)_{1 \leqslant j \leqslant n}$ is the canonical base of E, there is an integer $k \in \mathbf{Z}$ such that K is contained in the direct sum of the n groups $\mathbf{Z}_p.p^k\boldsymbol{e}_j$; then apply the theory of modules over a principal ideal ring.]

d) If G is a closed (but not compact) subgroup of E, then G contains a vector subspace of dimension 1, $\mathbf{Q}_p.\boldsymbol{a}$ with $\boldsymbol{a} \neq 0$. [Let C be the compact subset of E consisting of all points $\boldsymbol{x} \in \mathrm{E}$ such that $\|\boldsymbol{x}\| = 1$; show that $\mathrm{G} \cap \mathrm{C}$ contains a sequence of points $(\boldsymbol{x}_r)_{r \in \mathbf{N}}$ such that $p^{-r}\boldsymbol{x}_r \in \mathrm{G}$, and take $\boldsymbol{a}$ to be a cluster point of this sequence.]

e) Let G be a closed subgroup of E, let V be the largest vector subspace contained in G and let W be a vector subspace complementary to V; show that $\mathrm{W} \cap \mathrm{G}$ is compact and that G is the direct sum of V and $\mathrm{W} \cap \mathrm{G}$ (argue as in Theorem 2 of no. 2).

f) Given a subgroup G of E, let G^* denote the set of all $\boldsymbol{u} = (u_i) \in \mathrm{E}$ such that

$$(\boldsymbol{u}|\boldsymbol{x}) = \sum_{i=1}^{n} u_i x_i$$

belongs to $\mathbf{Z}_p$ for *all* $x = (x_i) \in G$. Show that G^* is closed in E, that $(\overline{G})^* = G^*$ and that $(G^*)^* = \overline{G}$.

g) State and prove the analogues of Exercises 2 to 8 inclusive for closed subgroups of E. In particular, every closed subgroup and every Hausdorff quotient group of a group of the form

$$(\mathbf{Q}_p)^r \times (\mathbf{Q}_p/\mathbf{Z}_p)^s \times (\mathbf{Z}_p)^t \times F$$

(where F is the product of a finite number of cyclic groups of p-power order) is a group of the same form.

§ 2

1) Let φ be the canonical homomorphism of $\mathbf{R}^n$ onto $\mathbf{T}^n$ and let u be a linear mapping of $\mathbf{R}^m$ into $\mathbf{R}^n$. For $\varphi \circ u$ to be a strict morphism of $\mathbf{R}^m$ into $\mathbf{T}^n$ it is necessary and sufficient that either:

a) $\overset{-1}{u}(\mathbf{Z}^n)$ is of rank m; or

b) $u(\mathbf{R}^m) \cap \mathbf{Z}^n$ is of rank equal to the dimension of $u(\mathbf{R}^m)$.

2) Show that the only continuous homomorphism of the additive topological group $\mathbf{Q}_p$ of p-adic numbers (Chapter III, § 6, Exercises 23 *et seq.*) into the additive group $\mathbf{R}$ is the zero mapping.

¶ 3) Every p-adic number x (Exercise 2) can be written in the form $\sum_{-\infty}^{+\infty} \alpha_k p^k$ where the α_k are rational integers, those of index $k < 0$ being zero for all but a finite number of values of the index k. Moreover, if $\sum_{-\infty}^{+\infty} \alpha_k p^k = \sum_{-\infty}^{+\infty} \beta_k p^k$, the rational numbers $\sum_{-\infty}^{-1} \alpha_k p^k$ and $\sum_{-\infty}^{-1} \beta_k p^k$ are congruent mod 1. Let $\varphi_p(x)$ be the class mod 1 of the rational number $\sum_{-\infty}^{-1} \alpha_k p^k$.

a) Show that φ_p is a continuous homomorphism of the topological group $\mathbf{Q}_p$ into $\mathbf{T}$, and that for every continuous homomorphism f of $\mathbf{Q}_p$ into T there exists a unique $a \in \mathbf{Q}_p$ such that $f(x) = \varphi_p(ax)$ for all $x \in \mathbf{Q}_p$. [Note that the knowledge of the elements $f(p^k)$ for $k \leqslant 0$ determines f uniquely; show that each of these elements is the class mod 1 of a fraction whose denominator is a power of p; hence show that there exists $a \in \mathbf{Q}_p$ such that $f(p^k) = \varphi_p(ap^k)$ for all $k \leqslant 0$.]

b) Show that, if $\varphi : \mathbf{R} \rightarrow \mathbf{T}$ is the canonical homomorphism, then $\varphi(x) = \sum_p \varphi_p(x)$ for all $x \in \mathbf{Q}$, the sum being over all prime numbers p (observe that all but a finite number of terms of this sum will be zero).

¶ 4) *a*) Let G be a locally compact abelian group and let H be a subgroup of G isomorphic to $\mathbf{R}^p \times \mathbf{T}^q$ where p, q are integers $\geqslant 0$; suppose also that G/H is isomorphic to either **R** or **T**. Show that G is isomorphic to the product of H and a closed subgroup L isomorphic to G/H. [Apply Chapter V, § 3, Exercise 7 *d*), to obtain a continuous homomorphism f of **R** into G such that $f(\mathbf{R}) \not\subset H$; hence construct another continuous homomorphism $g: \mathbf{R} \to G$ such that $f(t) - g(t) \in H$ for all $t \in \mathbf{R}$ and $g(\mathbf{R}) \cap H = \{e\}$. Note that $f(H)$ is a proper closed subgroup of **R** and that, for every element $z \neq e$ in H, there exists a continuous homomorphism u of **R** into H such that $u(1) = z$.]

b) Let G be a Hausdorff abelian topological group, let H be a closed subgroup of G isomorphic to $\mathbf{R}^p \times \mathbf{T}^q$, and suppose that there exists a surjective continuous homomorphism $\varphi: \mathbf{R} \to G/H$. Show that there exists a continuous homomorphism $f: \mathbf{R} \to G$ such that $f(\mathbf{R}) \cap H = \{e\}$ and $G = H . f(\mathbf{R})$. [Reduce to case *a*) by considering the least upper bound on G of the given topology and the topology for which a fundamental system of neighbourhoods of 0 is formed by the $\overset{-1}{p}(\varphi(I))$, where $p: G \to G/H$ is the canonical mapping and I runs through a fundamental system of neighbourhoods of 0 in **R**; show that G is locally compact in this new topology (cf. Chapter III, § 4, Exercise 10 *e*)).]

c) Give an example where the restriction of p to $f(\mathbf{R})$ is not bicontinuous [cf. § 1, Exercise 2 *f*)].

5) Let G be a connected topological group and let H be a compact normal subgroup of G with no arbitrarily small subgroups (Chapter III, § 2, Exercise 30). Show that H is contained in the centre of G. (Observe that for s close to e in G, the image of H under $x \to sxs^{-1}$ is arbitrarily close to e.)

6) Let G be a topological group and let H be a closed normal subgroup of G, isomorphic to $\mathbf{R}^n$ ($n \geqslant 1$) and such that G/H is abelian.

a) Suppose, moreover, that H contains no connected subgroup normal in G except for H itself and $\{e\}$, and that H is not contained in the centre of G. Given $x_0 \in G$ which does not commute with every element of H, show that, for every $v \in H$, there exists a unique $u \in H$ such that $x_0^{-1}ux_0u^{-1} = v$. (Observe that in the additive notation in H the equation becomes $u - x_0^{-1}ux_0 = -v$ and that the continuous endomorphism $u \to u - x_0^{-1}ux_0$ of H is a linear mapping of $\mathbf{R}^n$ into itself. Show that its kernel is a normal subgroup of G.) Furthermore, if $f(u)$ is the unique $u \in H$ such that $x_0^{-1}ux_0u^{-1} = v$, then f is a bicontinuous automorphism of H.

b) Under the hypotheses of *a*), show that the continuous mapping $g: y \to (f(x_0^{-1}yx_0y^{-1}))^{-1}y$ of G into itself has as image the subgroup L of G consisting of all elements which commute with x_0 and that the relation $g(y) = g(y')$ is equivalent to $yy'^{-1} \in \mathrm{H}$. Deduce that there exists a continuous bijection $h: \mathrm{G/H} \to \mathrm{L}$ whose inverse is the restriction to L of the canonical mapping $p: \mathrm{G} \to \mathrm{G/H}$, and deduce that G is isomorphic to the topological semi-direct product of H and L.

¶ 7) Let G be a topological group, H a normal subgroup isomorphic to $\mathbf{R}^p \times \mathbf{T}^q$ ($p \geqslant 0$, $q \geqslant 0$), and suppose that there exists a surjective continuous homomorphism $\varphi: \mathbf{R} \to \mathrm{G/H}$. Show that there exists a continuous homomorphism $f: \mathbf{R} \to \mathrm{G}$ such that $f(\mathbf{R}) \cap \mathrm{H} = \{e\}$ and $\mathrm{G} = \mathrm{H}.f(\mathbf{R})$. [Using Exercise 4 *b*) and Chapter V, § 3, Exercise 8, reduce to the case where G is not commutative and then, using Exercise 5, to the case $q = 0$; now use induction on p, together with Exercise 6.]

§ 3

¶ 1) *a*) Let $(x_k)_{1 \leqslant k \leqslant m}$ be a finite sequence of real numbers such that $|x_k| \leqslant 1$ for $1 \leqslant k \leqslant m$ and $\sum_{k=1}^{m} x_k = 0$. Show that there exists a permutation σ of the interval $[1, m]$ of $\mathbf{N}$ such that

$$\left|\sum_{k=1}^{p} x_{\sigma(k)}\right| \leqslant 1$$

for *all* $p = 1, 2, \ldots, m$, and which preserves the order of the indices h for which $x_h > 0$ and the order of the indices h for which $x_h < 0$ [i.e., if $h < k$ and $x_h > 0$ and $x_k > 0$, then $\sigma(h) < \sigma(k)$, and similarly for indices h such that $x_h < 0$].

b) Let $(\boldsymbol{x}_k)_{1 \leqslant k \leqslant m}$ be a finite sequence of points $\boldsymbol{x}_k = (x_{ki})_{1 \leqslant i \leqslant n}$ of $\mathbf{R}^n$, such that $\|\boldsymbol{x}_k\| \leqslant 1$ for $1 \leqslant k \leqslant m$, and $\sum_{k=1}^{m} \boldsymbol{x}_k = 0$. Show that there exists a permutation σ of the interval $[1, m]$ of $\mathbf{N}$ such that

$$\left\|\sum_{k=1}^{p} \boldsymbol{x}_{\sigma(k)}\right\| \leqslant 5^{(n-1)/2} \quad \text{for all} \quad p = 1, 2, \ldots, m.$$

[Proof by induction on n, considering $\mathbf{R}^n$ as a product $\mathbf{R}^{n-1} \times \mathbf{R}$, and putting $\boldsymbol{x}_k = (\boldsymbol{x}'_k, x_{kn})$ with $\boldsymbol{x}'_k \in \mathbf{R}^{n-1}$. Take a subset H of the interval $[1, m]$ of $\mathbf{N}$ such that $\left\|\sum_{k\in H} x_k\right\|$ is a maximum; by means of a rotation, reduce to the case where $\sum_{k\in H} x'_k = 0$ and show that in this case we must have $x_{kn} \geqslant 0$ for $k \in H$ and $x_{kn} \leqslant 0$ for $k \notin H$; then use *a*) and the inductive hypothesis.]

c) Let $(x_k)_{1\leqslant k\leqslant m}$ be a finite sequence of points of $\mathbf{R}^n$ such that $\|x_k\| \leqslant 1$ for $1 \leqslant k \leqslant m$ and $\left\|\sum_{k=1}^{m} x_k\right\| = a > 0$. Show that there exists a permutation σ of the interval $[1, m]$ of $\mathbf{N}$ such that

$$\left\|\sum_{k=1}^{p} x_{\sigma(k)}\right\| \leqslant (a + 1)5^{(n-1)/2}$$

for *all* p such that $1 \leqslant p \leqslant m$ [reduce to case *b*)].

¶ 2) Let $(x_m)_{m\in\mathbf{N}}$ be an infinite sequence of points of $\mathbf{R}^n$ such that $\lim_{m\to\infty} x_m = 0$. For each finite subset H of $\mathbf{N}$, put $s_H = \sum_{m\in H} x_m$. Show that there are two possibilities: either

(i) $\lim_{\mathfrak{F}}\|S_H\| = +\infty$, where $\mathfrak{F}$ is the directed set of all finite subsets of $\mathbf{N}$; or else:

(ii) there exist permutations σ of $\mathbf{N}$ such that the series whose general term is $x_{\sigma(m)}$ is convergent in $\mathbf{R}^n$; in this case the set A of sums $\mathop{\mathsf{S}}_{m=0}^{\infty} x_{\sigma(m)}$ of these series, for all permutations σ with this property, is an *affine linear variety* in $\mathbf{R}^n$. [Show first, with the help of Exercise 1 *b*), that every cluster point (with respect to $\mathfrak{F}$) of the mapping $H \to S_H$ is the sum of a convergent series $(x_{\sigma(m)})$ for a suitable permutation σ. Use Chapter III, § 5, Exercise 3 to prove that the set A, if not empty, is a coset of a closed subgroup of $\mathbf{R}^n$. Show finally that A is *connected* by means of Exercise 1 *c*) and Chapter II, § 4, Exercise 15.]

HISTORICAL NOTE

(Numbers in brackets refer to the bibliography at the end of this note.)

The main results of the theory of subgroups and quotient groups of the additive groups $\mathbf{R}^n$ have been known since the end of the last century. Many questions of arithmetic and analysis led mathematicians to investigate the structure of subgroups of $\mathbf{R}^n$ generated by a *finite* number of points. Thus Lagrange, while developing the theory of "continued fractions", showed in passing that, for every real number θ, there exist integers $m, n,$ not both zero, such that $m - n\theta$ is arbitrarily small ([1], vol. 7, p. 27). In 1835 Jacobi, motivated by his researches on periodic analytic functions of several complex variables, showed that if x, y, z are three vectors of $\mathbf{R}^2$, there exist integers $m, n, p,$ not all zero, which make the vector $mx + ny + pz$ arbitrarily small ([2], vol. 2, p. 25). A little later Dirichlet, in the course of his work on the theory of algebraic numbers, discovered his famous "pigeon-hole principle" (*Schubfachprinzip*) (cf. § 1, Exercises 10 and 11), by means of which he showed that p forms $\alpha_{i1}m_1 + \alpha_{i2}m_2 + \cdots + \alpha_{in}m_n - q_i$ $(1 \leqslant i \leqslant p)$, where the α_{ij} are arbitrary real numbers and the m_j and the q_i are integers (not all zero), can simultaneously be made arbitrarily small ([3], vol. 1, p. 635). By an entirely different method Hermite arrived at the same result in 1850, for forms of the particular type $m\theta_i - q_i$ $(1 \leqslant i \leqslant p)$ ([4], vol. 1, p. 105). Finally, in 1884, Kronecker proved the general result stated in Proposition 7 of § 1 ([5], vol. 3_1, p. 47).

Of course these results were independent of the general theory of locally compact abelian groups, which is of recent origin (see the Historical Note to Chapter III); but this latter theory, particularly the theory of duality (*), has shed a new light on these old results, mainly by

(*) See for example A. Weil [8].

bringing out the fundamental concept of associated subgroups. The exposition in the text is based on these ideas (*).

The point of view we have taken in this chapter is purely *qualitative*: that is to say, we have established the *existence* of linear combinations of p points, with integer coefficients, which approximate arbitrarily closely to a given point (which may possibly have to satisfy certain conditions); but one may also ask whether there exist relations between the accuracy of the approximation and the magnitude of the coefficients in the linear combinations which give the approximation; this is the point of view of the *quantitative* theory of "Diophantine approximation", and the point of view of all the authors quoted in the first paragraph. Over the last hundred years these questions have been the object of many and diverse investigations, rich in applications to the theory of numbers; to trace their development in this context would take us far outside our present scope, and we shall therefore do no more than refer the reader who wishes to go into these theories to the fundamental writings of Minkowski [6] and H. Weyl [7], which are the origin of an abundant literature (**).

(*) An analogous exposition had already been sketched by Marcel Riesz [9].
(**) For a recent bibliography of the subject, see e.g., J. Koksma [10].

BIBLIOGRAPHY

[1] J. L. Lagrange, *Œuvres*, vol. 7, Paris (Gauthier-Villars), 1877.

[2] C. G. J. Jacobi, *Gesammelte Werke*, vol. 2, Berlin (G. Reimer), 1882.

[2 *bis*] C. G. J. Jacobi, *Über die vierfach periodischen Funktionen zweier Variabeln* [Ostwald's Klassiker, no. 64, Leipzig (Engelmann), 1895].

[3] P. G. Lejeune-Dirichlet, *Werke*, vol. 1, Berlin (G. Reimer), 1889.

[4] C. Hermite, *Œuvres*, vol. 1, Paris (Gauthier-Villars), 1905.

[5] L. Kronecker, *Werke*, vol. 3_1, Leipzig (Teubner), 1899.

[6] H. Minkowski, *Gesammelte Abhandlungen*, 2 vols., Leipzig-Berlin (Teubner), 1911.

[7] H. Weyl, "Über die Gleichverteilung von Zahlen mod. Eins", *Math. Ann.*, vol. 77 (1916), p. 313 [= *Selecta*, Basel-Stuttgart (Birkhäuser), 1956, p. 111].

[8] A. Weil, L'intégration dans les groupes topologiques et ses applications. *Act. Sci. et Ind.*, no. 869, Paris (Hermann) 1940, pp. 108-109 [2nd edition, *ibid.*, no. 1145 (1953)].

[9] M. Riesz, Modules reciproques [*Proc. International Congress of Mathematicians*, Oslo (1936), vol. 2, p. 36].

[10] J. Koksma, *Diophantische Approximationen*, Berlin (Springer), 1936.

CHAPTER VIII

Complex numbers

1. COMPLEX NUMBERS, QUATERNIONS

1. DEFINITION OF COMPLEX NUMBERS

The polynomial $X^2 + 1$ has no root in $\mathbf{R}$, because $x^2 + 1 \geqslant 1$ for all $x \in \mathbf{R}$; it is therefore irreducible over $\mathbf{R}$. [This is a particular case of the analogous result which applies to any *ordered field.*]

DEFINITION 1. *The field* $\mathbf{R}[X]/(X^2 + 1)$ *is called the field of complex numbers and is denoted by* $\mathbf{C}$. *The canonical image of* X *in* $\mathbf{C}$ *is denoted by* i, *so that* $\mathbf{C}$ *is obtained from the field* $\mathbf{R}$ *by algebraic adjunction of the root* i *of the polynomial* $X^2 + 1$. *The elements of* $\mathbf{C}$ *are called complex numbers.*

From an algebraic point of view the importance of the field $\mathbf{C}$ is due to the following fundamental theorem:

THEOREM 1. (d'Alembert-Gauss). *The field* $\mathbf{C}$ *of complex numbers is algebraically closed.*

For the proof it is enough to establish that (i) every element $\geqslant 0$ in $\mathbf{R}$ has a *square root,* and (ii) every polynomial of *odd* degree with coefficients in $\mathbf{R}$ has *at least one root* in $\mathbf{R}$. The first of these assertions has already been proved in Chapter IV, § 3, no. 3. As to the second, if $f(X) = a_0X^n + a_1X^{n-1} + \cdots + a_n$ is a polynomial of odd degree n $(a_0 \neq 0)$ with real coefficients, we may write $f(x) = a_0x^ng(x)$ for $x \neq 0$, where

$$g(x) = 1 + \frac{a_1}{a_0x} + \cdots + \frac{a_n}{a_0x^n}$$

tends to $+1$ as x tends to $+\infty$ or $-\infty$. Hence there is a number $a > 0$ such that $f(a)$ has the sign of a_0 and $f(-a)$ the sign of $-a_0$;

and so by Bolzano's theorem (Chapter IV, § 6, no. 1, Theorem 2), f has at least one root in $[-a, a]$.

Remarks. 1) Theorem 1 can be proved without invoking the theory of ordered fields by using properties of the *topology* of the field $\mathbf{C}$, which will be defined below (no. 2); see § 2, Exercise 2 and also the part of this series devoted to algebraic topology, where the theorem of d'Alembert-Gauss will appear as a consequence of results on the *degree of a mapping*.

2) Since $\mathbf{C}$ is of degree 2 over $\mathbf{R}$, it follows that $\mathbf{C}$ is, up to isomorphism, the *only* algebraic extension of $\mathbf{R}$ other than $\mathbf{R}$ itself, and that there is no field contained in $\mathbf{C}$ which contains $\mathbf{R}$, other than $\mathbf{R}$ and $\mathbf{C}$.

We know that $\mathbf{R}$ may be identified with a subfield of $\mathbf{C}$, and that every $z \in \mathbf{C}$ can be written uniquely in the form $x + iy$, where x and y are real; x is called the *real part* of z and is denoted by $\mathfrak{R}(z)$; y the *imaginary part* of z, denoted by $\mathfrak{I}(z)$. Complex numbers of the form iy (y real) are called *pure imaginary*. The relation $x + iy = 0$ (x, y real) is equivalent to $x = 0$ and $y = 0$.

Since $i^2 = -1$, the elements of $\mathbf{C}$ (when given by their real and imaginary parts) satisfy the following rules of calculation:

$$(1) \qquad (x + iy) + (x' + iy') = (x + x') + i(y + y')$$

$$(2) \qquad (x + iy)(x' + iy') = (xx' - yy') + i(xy' + x'y).$$

In particular, $(x + iy)(x - iy) = x^2 + y^2 \in \mathbf{R}$, so that, if $x + iy \neq 0$,

$$(3) \qquad \frac{1}{x + iy} = \frac{x}{x^2 + y^2} - i\frac{y}{x^2 + y^2}.$$

The second root of the polynomial $X^2 + 1$ in $\mathbf{C}$ is $-i$; consequently the only *automorphism* of $\mathbf{C}$, other than the identity mapping, which leaves all real numbers invariant, is that which maps $z = x + iy$ to $x - iy$; the latter is denoted by $\bar{z}$ and (in agreement with the general definitions) is called the complex number *conjugate* to z. We have

$$\mathfrak{R}(z) = \frac{1}{2}(z + \bar{z}) \quad \text{and} \quad \mathfrak{I}(z) = \frac{1}{2i}(z - \bar{z}).$$

By reason of this automorphism, if $f(z)$ is a polynomial with *real* coefficients, we have $f(\bar{z}) = \overline{f(z)}$ for all $z \in \mathbf{C}$.

The real number $z\bar{z} = x^2 + y^2$ is called the *algebraic norm* of z, or simply the *norm* whenever there is no risk of confusion; it is a real number $\geqslant 0$, which vanishes only if $z = 0$. The real number $\sqrt{z\bar{z}} = \sqrt{x^2 + y^2} \geqslant 0$ reduces to the absolute value of z when z is real, and we still call it the *absolute value* of z, and denote it by $|z|$, when z is any complex number. The relation $|z| = 0$ is equivalent to $z = 0$. If z and z'

are two complex numbers, the conjugate of zz' is $\overline{zz}'$, hence $|zz'|^2 = zz'\overline{zz}' = |z|^2|z'|^2$ and therefore $|zz'| = |z|.|z'|$: *the absolute value of a product is the product of the absolute values of the factors.* In particular, if $z \neq 0$ and $z' = 1/z$, we have $|1/z| = |1/z|$.

Finally, for all complex numbers z, z', we have the *triangle inequality*

$$|z + z'| \leqslant |z| + |z'|. \tag{4}$$

2. THE TOPOLOGY OF C

The mapping $(x, y) \rightarrow x + iy$ of the real plane $\mathbf{R}^2$ onto $\mathbf{C}$ is *bijective*; by means of this bijection we can *transport* to $\mathbf{C}$ the topology of $\mathbf{R}^2$ (cf. Chapter VI, § 1, no. 5). The topology thus defined on $\mathbf{C}$ is *compatible* with the field structure of $\mathbf{C}$ (Chapter III, § 6, no. 7), because it is compatible with the ring structure of $\mathbf{C}$ (Chapter VI, § 1, no. 5) and, by (3), $1/z$ is continuous on the complement $\mathbf{C}^*$ of 0 in $\mathbf{C}$.

If we endow the set $\mathbf{C}$ with this topology and with the field structure defined earlier (no. 1, Definition 1), we have defined on $\mathbf{C}$ the structure of a *topological field* (Chapter III, § 6, no. 7); whenever we speak of the topology of $\mathbf{C}$ it will always be the above topology that is meant.

In the future we shall generally *identify* the sets $\mathbf{C}$ and $\mathbf{R}^2$ considered as topological spaces; the subfield $\mathbf{R}$ of $\mathbf{C}$ is then identified with the abscissa of $\mathbf{R}^2$, which for this reason is called the *real axis*; likewise the ordinate is called the *imaginary axis* (note that this is *not* a subfield of $\mathbf{C}$). The ray with origin 0 and direction ratios $(1, 0)$ (identified with $\mathbf{R}_+$) is called the *positive real semi-axis*; the opposite ray, with the same origin and direction ratios $(-1, 0)$, is called the *negative real semi-axis*.

For the purposes of graphical illustration we use the representation of $\mathbf{R}^2$ (well known in elementary analytic geometry) by the points of a plane in which have been drawn two perpendicular coordinate axes, representing respectively the real axis and the imaginary axis of $\mathbf{C}$ (Fig. 7).

As in every topological field, every *rational function* of n complex variables with complex coefficients is *continuous* at every point of $\mathbf{C}^n$ at which the denominator does not vanish.

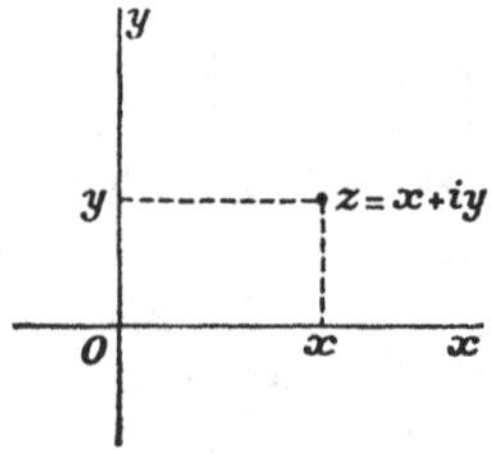

Figure 7.

The permutation $z \to \bar{z}$ of $\mathbf{C}$ is continuous, and is therefore an *automorphism* of the topological field $\mathbf{C}$.

In fact it is the *only* automorphism of the topological field $\mathbf{C}$ other than the identity automorphism (see Exercise 4).

The functions $\mathfrak{R}(z)$, $\mathfrak{J}(z)$ are just the *projections* of $\mathbf{R}^2$ onto its factors, and are therefore *continuous*; the same is true of the *absolute value* $|z|$, since it is the *Euclidean norm* (Chapter VI, § 2, no. 1) of the point (x, y) in $\mathbf{R}^2$.

The properties of the absolute value lead to another proof of the fact that the topology of $\mathbf{C}$ is compatible with its field structure (cf. Chapter IX, § 3, no. 2); the continuity of $z + z'$ follows from the triangle inequality $|z + z'| \leqslant |z| + |z'|$; that of zz' follows from the relation

$$\begin{aligned} |zz' - z_0z_0'| &= |z_0(z' - z_0') + (z - z_0)z_0' + (z - z_0)(z' - z_0')| \\ &\leqslant |z_0|.|z' - z_0'| + |z_0'|.|z - z_0| + |z - z_0|.|z' - z_0'|; \end{aligned}$$

lastly, the continuity of z^{-1} follows from the relation

$$|z_0^{-1} - z^{-1}| = |z|^{-1}.|z - z_0|.|z_0|^{-1}.$$

3. THE MULTIPLICATIVE GROUP $\mathbf{C}^*$

We know from Chapter III, § 6, no. 7 that the topology induced on the multiplicative group $\mathbf{C}^*$ of non-zero complex numbers is compatible with the group structure of $\mathbf{C}^*$. Since $\mathbf{C}^*$ is *open* in $\mathbf{C}$, it follows that $\mathbf{C}^*$ is a *locally compact* topological group (Chapter I, § 9, no. 7, Proposition 13) and therefore *complete* (with respect to the multiplicative uniformity; cf. Chapter III, § 6, no. 8, Proposition 8). The multiplicative group $\mathbf{R}_+^*$ of real numbers > 0 is a *closed subgroup* of $\mathbf{C}^*$. Another subgroup is the set $\mathbf{U}$ of complex numbers of *absolute value* 1, which is identified with the *unit circle* $\mathbf{S}_1$ of $\mathbf{R}^2$, and is therefore a *compact* group. Moreover:

PROPOSITION 1. *The topological group $\mathbf{C}^*$ is isomorphic to the product of the topological groups $\mathbf{R}^*$ and $\mathbf{U}$.*

For the mapping $z \to \left(|z|, \dfrac{z}{|z|}\right)$ is a *homeomorphism* of $\mathbf{C}^*$ onto $\mathbf{R}_+^* \times \mathbf{U}$ (Chapter VI, § 2, no. 3, Proposition 3); and it follows immediately that it is an isomorphism of the group structures.

The topological group $\mathbf{R}_+^*$ is already known to be isomorphic to the additive group $\mathbf{R}$ (Chapter V, § 4, Theorem 1); the study of the topological group $\mathbf{C}^*$ is therefore reduced to that of $\mathbf{U}$, which we shall consider in § 2.

4. THE DIVISION RING OF QUATERNIONS

It follows from Theorem 1 that the field $\mathbf{R}$ is a *maximal* ordered field, and therefore the only *non-commutative* division ring of *finite* rank over $\mathbf{R}$ is (up to isomorphism) the division ring of quaternions over $\mathbf{R}$; it is denoted by $\mathbf{H}$ and is called the *division ring of real quaternions* (or simply the *division ring of quaternions*, when there is no fear of confusion). Since $\mathbf{H}$ is of rank 4 over the field $\mathbf{R}$, we can define a topology on $\mathbf{H}$ homeomorphic to that of $\mathbf{R}^4$ (Chapter VI, § 1, no. 5). To be precise, we shall usually *identify* $\mathbf{H}$ with $\mathbf{R}^4$, the elements $1, i, j, k$ of the canonical basis of $\mathbf{H}$ being identified respectively with the vectors e_0, e_1, e_2, e_3 of the canonical basis of $\mathbf{R}^4$.

We recall that the multiplication table of the canonical basis of $\mathbf{H}$ is given by the formulae

$$i^2 = j^2 = k^2 = -1, \qquad ij = -ji = k,$$
$$jk = -kj = i, \qquad ki = -ik = j.$$

The topology of $\mathbf{H}$ is compatible not only with the ring structure of $\mathbf{H}$ (Chapter VI, § 1, no. 5) but also with its *division ring* structure; for if x is a non-zero quaternion, the coordinates of x^{-1} are rational functions of those of x, whose denominators do not vanish. The division ring $\mathbf{H}$, endowed with this topology, is therefore a *non-commutative topological division ring*. The quaternions $a + bi$ (a, b real) form a (topological) *subfield* of $\mathbf{H}$, isomorphic to the field $\mathbf{C}$, with which it is often identified.

We have thus a third example of a *locally compact*, *connected* topological division ring, the others being $\mathbf{R}$ and $\mathbf{C}$. In fact these are the *only* topological division rings with these two properties.

We know from algebra that the reduced norm of a quaternion $x = x_0 + x_1 i + x_2 j + x_3 k$ is

$$\mathrm{N}(x) = x_0^2 + x_1^2 + x_2^2 + x_3^2 = \|x\|^2$$

(it is therefore the square of the *Euclidean norm* of x). Since

$$\mathrm{N}(xy) = \mathrm{N}(x)\mathrm{N}(y),$$

it follows that the set of all quaternions of norm 1, which is identical with the sphere $\mathbf{S}_3$, forms a compact subgroup of the multiplicative group $\mathbf{H}^*$ of non-zero quaternions.

PROPOSITION 2. *The multiplicative group* $\mathbf{H}^*$ *of non-zero quaternions is a topological group isomorphic to the product of its subgroups* $\mathbf{R}_+^*$ *and* $\mathbf{S}_3$.

Every quaternion $x \neq 0$ can be written as $x = \|x\|.z$ where z is a quaternion with norm 1; since $\|xx'\| = \|x\|.\|x'\|$, the mapping $x \to (\|x\|, x/\|x\|)$ of $\mathbf{H}^*$ onto $\mathbf{R}_+^* \times \mathbf{S}_3$ is an isomorphism of the group structures, and from Chapter VI, § 2, no. 3, Proposition 3 it is a homeomorphism of $\mathbf{H}^*$ onto $\mathbf{R}_+^* \times \mathbf{S}_3$.

Remarks. 1) With the use of the relations $\|x + y\| \leqslant \|x\| + \|y\|$ and $\|xy\| = \|x\|.\|y\|$ it can be proved directly, as with the field of complex numbers in no. 2, that the topology of $\mathbf{R}^4$ is compatible with the division ring structure of H (cf. Chapter IX, § 3, no. 2).

2) From what has been proved it follows that the spheres $\mathbf{S}_1$ and $\mathbf{S}_3$ can carry a group structure compatible with their topology. We shall see later that, for each integer n other than 1 and 3, there exists *no* group structure on $\mathbf{S}_n$ compatible with the topology of $\mathbf{S}_n$.

3) Every point of the group $\mathbf{S}_3$ has a neighbourhood homeomorphic to $\mathbf{R}^3$ (Chapter VI, § 2, no. 4, Proposition 5) but $\mathbf{S}_3$ is not locally isomorphic to $\mathbf{R}^3$; for if it were it would be abelian, since it is connected (Chapter VII, § 2, no. 2, Theorem 1), and this is not the case, since i and j belong to $\mathbf{S}_3$ and $ij \neq ji$ (cf. Chapter V, § 3).

2. ANGULAR MEASURE, TRIGONOMETRIC FUNCTIONS

1. THE MULTIPLICATIVE GROUP U

THEOREM 1. *The (multiplicative) topological group* $\mathbf{U}$ *of complex numbers of absolute value* 1 *is isomorphic to the (additive) topological group* $\mathbf{T}$ *of real numbers mod* 1.

$\mathbf{U} = \mathbf{S}_1$ is *compact* and *connected*, and has a neighbourhood of the identity element $+1$ homeomorphic to an open interval of $\mathbf{R}$ (Chapter VI, § 2, no. 4, Proposition 5); the theorem is therefore a consequence of the topological characterization of $\mathbf{T}$ given in Chapter V, § 3 (Theorem 2).

COROLLARY. *The multiplicative group* $\mathbf{C}^*$ *of non-zero complex numbers is isomorphic to the group* $\mathbf{R} \times \mathbf{T}$ (cf. § 1. no. 3, Proposition 1).

Remark. The isomorphism of the groups $\mathbf{C}^*$ and $\mathbf{R} \times \mathbf{T}$ implies the existence of roots of every "binomial equation" $z^n = a$ in the field $\mathbf{C}$. Using this fact and the local compactness of $\mathbf{C}$ we can obtain another proof of the theorem of d'Alembert-Gauss (Exercise 2).

There are only *two* distinct isomorphisms of the group **T** onto the group **U**; for if g, g' are two isomorphisms of **T** onto **U**, and if h' is the inverse of g', then $h' \circ g$ is an automorphism of **T**, and therefore (Chapter VII, § 2, no. 4, Proposition 6) we have identically either $g'(x) = g(x)$ or $g'(x) = g(-x)$. We may always assume that g is such that i is the image under g of the class mod 1 of the point $\frac{1}{4}$; then, if φ denotes the canonical homomorphism of **R** onto **T**, every strict morphism of the additive group **R** onto the multiplicative group **U** is of the form $x \rightarrow g(\varphi(x/a))$, where a is a real number $\neq 0$ (Chapter VII, § 2, no. 3, Proposition 4); note that the interval $]-\frac{1}{2}|a|, \frac{1}{2}|a|[$ is the largest symmetric open interval of **R** which is mapped one-to-one onto its image by this strict morphism, and that we have $g(\varphi(\frac{1}{4})) = i$. We shall denote the homomorphism $x \rightarrow g(\varphi(x))$ by $x \rightarrow e(x)$; every strict morphism of **R** onto **U** is therefore of the form $x \rightarrow e(x/a)$, where $a \neq 0$. The function $e(x)$ is continuous on **R**, complex valued, and satisfies the identities

$$|e(x)| = 1, \tag{1}$$

$$e(x+y) = e(x)e(y), \tag{2}$$

together with the relations

$$(3) \quad e(0) = 1, \quad e(\tfrac{1}{4}) = i, \quad e(\tfrac{1}{2}) = -1, \quad e(\tfrac{3}{4}) = -i, \quad e(1) = 1.$$

From (1) and (2) it follows that

$$e(-x) = \frac{1}{e(x)} = \overline{e(x)}, \tag{4}$$

and from (2) and (3) that

$$e(x + \tfrac{1}{4}) = ie(x), \quad e(x + \tfrac{1}{2}) = -e(x),$$
$$e(x + \tfrac{3}{4}) = -ie(x), \quad e(x+1) = e(x).$$

Thus the function $e(x)$ is *periodic* and has 1 as a principal period.

Remark. The mapping $x + iy \rightarrow e^x e(y)$ is a strict morphism of the additive group **C** onto the multiplicative group **C***, and its restriction to a suitable neighbourhood of 0 is a local isomorphism of **C** with **C***. Consequently (Chapter VII, § 2, no. 3) every strict morphism of **C** onto **C*** is of the form $x + iy \rightarrow e^{\alpha x + \beta y} e(\gamma x + \delta y)$, where $\alpha, \beta, \gamma, \delta$ are any real numbers such that $\alpha\delta - \beta\gamma \neq 0$. We shall see later that there is just *one* of these homomorphisms, denoted by $z \to e^z$, such that

$$\lim_{z \to 0} \frac{e^z - 1}{z} = 1;$$

and the restriction of this homomorphism to the real axis is the same as e^x (whence the notation).

2. ANGLES

Since the field $\mathbf{R}$ is *ordered*, we may *orient* the real number plane $\mathbf{R}^2$ by taking $e_1 \wedge e_2$ as positive bivector (e_1, e_2 being the vectors of the canonical basis). In the oriented real number plane $\mathbf{R}^2$ (identified with $\mathbf{C}$ in what follows) we can then define the *angle* $(\widehat{\Delta_1, \Delta_2})$ of an arbitrary pair of rays (Δ_1, Δ_2) with origin o (*). The set $\mathfrak{A}$ of all angles has the structure of an abelian group (written additively) defined by

$$(\widehat{\Delta_1, \Delta_3}) = (\widehat{\Delta_1, \Delta_2}) + (\widehat{\Delta_2, \Delta_3}),$$

so that, in particular, $(\widehat{\Delta_1, \Delta_1}) = 0$ and $(\widehat{\Delta_2, \Delta_1}) = -(\widehat{\Delta_1, \Delta_2})$.

The *flat angle* ϖ is the solution $\neq 0$ of the equation $2\theta = 0$ in $\mathfrak{A}$; it is the angle which the negative real semi-axis makes with the positive real semi-axis.

If z is any non-zero complex number, the *amplitude* (or *argument*) of z, denoted by $\mathrm{Am}(z)$, is the angle which the ray through z with origin o makes with the positive real semi-axis. The mapping $z \to \mathrm{Am}\,(z)$ is a *homomorphism* of the multiplicative group $\mathbf{C}^*$ onto the additive group $\mathfrak{A}$, and therefore we have

$$\mathrm{Am}\,(zz') = \mathrm{Am}\,(z) + \mathrm{Am}(z') \quad \text{and} \quad \mathrm{Am}\,(\bar{z}) = \mathrm{Am}\,(z^{-1}) = -\mathrm{Am}(z).$$

The angle $\delta = \mathrm{Am}\,(i)$ is called the *positive right angle*; it is one of the solutions in A of the equation $2\theta = \varpi$, the other one being $-\delta = \delta + \varpi$.

The homomorphism $z \to \mathrm{Am}\,(z)$, restricted to the subgroup $\mathbf{U}$ of $\mathbf{C}^*$, is an isomorphism of the group structure of $\mathbf{U}$ *onto* that of $\mathfrak{A}$ (**); if we

---

(*) We know from algebra that an *equivalence relation* is defined on the set of all pairs (Δ_1, Δ_2) of rays with origin o by considering two pairs (Δ_1, Δ_2) and (Δ_1', Δ_2') as equivalent if there exists a *rotation* which transforms Δ_1 into Δ_1' and Δ_2 into Δ_2' simultaneously; the *angle* of the pair (Δ_1, Δ_2), or the *angle* Δ_2 *makes with* Δ_1, is then by definition the equivalence class of the pair (Δ_1, Δ_2).

(**) This is because every ray with origin o meets the circle $\mathbf{S}_1$, since the field $\mathbf{R}$ is *Pythagorean*.

use this isomorphism to *transport* the topology of $\mathbf{U}$ to the group $\mathfrak{A}$, the latter group becomes a compact topological group, and the mapping $z \rightarrow \mathrm{Am}(z)$ of $\mathbf{C}^*$ onto $\mathfrak{A}$ is a *strict morphism* of the topological group $\mathbf{C}^*$ onto the topological group $\mathfrak{A}$.

Let us denote by $\theta \rightarrow f(\theta)$ the isomorphism of $\mathfrak{A}$ onto $\mathbf{U}$ which is the inverse of the isomorphism $z \rightarrow \mathrm{Am}(z)$ of $\mathbf{U}$ onto $\mathfrak{A}$. By definition $\mathfrak{R}(f(\theta))$ is denoted by $\cos\theta$ and is called the *cosine* of the angle θ; $\mathfrak{I}(f(\theta))$ is denoted by $\sin\theta$ and is called the *sine* of the angle θ. These functions are continuous on the topological group $\mathfrak{A}$, and satisfy the following relations (*loc. cit.*), which are immediate consequences of the definitions above:

$$\cos 0 = 1, \qquad \sin 0 = 0, \qquad \cos \varpi = -1, \qquad \sin \varpi = 0,$$

$$\cos(-\theta) = \cos\theta, \qquad \sin(-\theta) = -\sin\theta,$$

$$\cos(\theta + \theta') = \cos\theta\cos\theta' - \sin\theta\sin\theta',$$

$$\sin(\theta + \theta') = \sin\theta\cos\theta' + \sin\theta'\cos\theta,$$

$$\cos^2\theta + \sin^2\theta = 1.$$

By definition, the *tangent* of an angle $\theta \in \mathfrak{A}$ is defined, whenever $\cos\theta \neq 0$, to be $\sin\theta/\cos\theta$ (*loc. cit.*) and is denoted by $\tan\theta$; it is a continuous function, which extends by continuity to $\tilde{\mathbf{R}}$ (Chapter VI, § 3, no. 4) by taking the value ∞ for the angles δ and $-\delta$. We have $\tan(\theta + \varpi) = \tan\theta$. The *cotangent* of θ, denoted by $\cot\theta$, is the element of $\tilde{\mathbf{R}}$ equal to $1/\tan\theta$.

Note that, if $\mathrm{Am}(z) = \theta$, we have $z = |z|(\cos\theta + i\sin\theta)$; this expression is called the *trigonometric form* of the complex number $z \neq 0$.

3. ANGULAR MEASURE

By Theorem 1 of no. 1, the topological group $\mathfrak{A}$ of angles is *isomorphic* to $\mathbf{T}$. Every strict morphism of $\mathbf{R}$ onto $\mathfrak{A}$ can be obtained by composing the isomorphism $z \rightarrow \mathrm{Am}(z)$ of $\mathbf{U}$ onto $\mathfrak{A}$ with a strict morphism of $\mathbf{R}$ onto $\mathbf{U}$; if we put $\mathfrak{z}(x) = \mathrm{Am}(\boldsymbol{e}(x))$, every strict morphism of $\mathbf{R}$ onto $\mathfrak{A}$ is therefore of the form $x \rightarrow \mathfrak{z}(x/a)$ $(a \neq 0)$. Given a real number $a > 0$, fixed once and for all, every angle θ corresponds, by the homomorphism $x \rightarrow \mathfrak{z}(x/a)$, to a *class* of real numbers mod a (i.e., an element of $\mathbf{Z}/a\mathbf{Z}$) which is called the *measure* of θ relative to the *base a*; by abuse of language, every real number in this class is also called *a measure* of θ; the angle $\mathfrak{z}(x/a)$ is called *the angle with measure x* (relative to the base a). If x is a measure of θ and x' a measure of θ' (relative to the same base) then $x + x'$ is a measure of $\theta + \theta'$, and $-x$ is a measure of $-\theta$. The *principal measure* of an angle (relative to the base a) is that one of its measures which lies in the interval $[0, a[$.

Choice of a base a. We restrict ourselves always to bases $a > 1$. To each $a > 1$ corresponds an angle $\omega = \mathfrak{I}(1/a)$ whose principal measure is 1, and which is called the *unit of angular measure* relative to the base a; conversely, to each angle $\omega \neq 0$ there corresponds a unique $a > 1$ such that $\mathfrak{I}(1/a) = \omega$, so that knowledge of the unit of angular measure determines the base $a > 1$ entirely.

In numerical calculations one usually takes either $a = 360$ or $a = 400$; the corresponding unit of angular measure is called the *degree* $(a = 360)$ or the *grade* $(a = 400)$.

In analysis, and indeed in all branches of mathematics where numerical calculation is not involved, the base a defined by the condition

$$\lim_{x \to 0} \frac{e(x/a) - 1}{x} = i$$

is universally used; this base is denoted by 2π. The corresponding unit of angular measure is called a *radian*, and the measure is called *radian* measure; with the definition of e^z for complex z mentioned earlier, we have $e(x) = e^{2\pi i x}$ for all $x \in \mathbf{R}$.

Once the base a has been chosen, when one speaks of an *angle* one usually means a *measure* of this angle relative to the base a; this abuse of language has no drawback provided (as is always the case when numerical calculations are not involved) the base a remains fixed throughout, and provided that one remembers that two real numbers which are congruent mod a correspond to the *same* angle.

For example, what is usually understood by the *amplitude* of a complex number $z \neq 0$ is a *radian measure* of this angle, determined by conventions which will depend on the question under consideration; once these conventions have been made, the measure of the amplitude thus chosen is denoted by $\mathrm{Am}(z)$.

4. TRIGONOMETRIC FUNCTIONS

If we compose the functions $\cos\theta$, $\sin\theta$, $\tan\theta$, $\cot\theta$ (defined on $\mathfrak{A}$) with the homomorphism $x \to \mathfrak{I}(x/a)$ of $\mathbf{R}$ onto $\mathfrak{A}$, the functions

$$\cos\mathfrak{I}\left(\frac{x}{a}\right), \quad \sin\mathfrak{I}\left(\frac{x}{a}\right), \quad \tan\mathfrak{I}\left(\frac{x}{a}\right), \quad \cot\mathfrak{I}\left(\frac{x}{a}\right)$$

so obtained are called respectively the *cosine*, *sine*, *tangent* and *cotangent* of the *number* x relative to the base a, and are written $\cos_a x$, $\sin_a x$, $\tan_a x$, $\cot_a x$. The mapping $x \to \cos_a x + i \sin_a x$ is the composition of $\theta \to \cos\theta + i\sin\theta$ and $x \to \vartheta(x/a)$, so that, from the definition of $\cos\theta$ and $\sin\theta$ in no. 2, we have the identity

$$e\left(\frac{x}{a}\right) = \cos_a x + i \sin_a x, \tag{5}$$

which is equivalent to

$$\cos_a x = \mathfrak{R}\left(e\left(\frac{x}{a}\right)\right), \qquad \sin_a x = \mathfrak{J}\left(e\left(\frac{x}{a}\right)\right),$$

and also, by (4), equivalent to

$$\cos_a x = \frac{1}{2}\left(e\left(\frac{x}{a}\right) + e\left(-\frac{x}{a}\right)\right), \qquad \sin_a x = \frac{1}{2i}\left(e\left(\frac{x}{a}\right) - e\left(-\frac{x}{a}\right)\right).$$

Hence the identities

$$\cos_b x = \cos_a\left(\frac{ax}{b}\right), \qquad \sin_b x = \sin_a\left(\frac{ax}{b}\right). \tag{6}$$

The only trigonometric functions which arise in branches of mathematics where numerical calculation is not involved are those relative to the base 2π referred to above; these functions are denoted simply by $\cos x$, $\sin x$, $\tan x$, $\cot x$ in place of $\cos_{2\pi} x$, $\sin_{2\pi} x$, $\tan_{2\pi} x$, $\cot_{2\pi} x$. For the purposes of numerical calculation, there are tables of the trigonometric functions corresponding to the bases $a = 360$ and $a = 400$; and the formulae (6) allow us to deduce the values of the trigonometric functions relative to any other base.

The relations recalled earlier between the cosines and sines of *angles* evidently give rise to the same relations between the cosines and the sines of the *numbers* which measure these angles; in particular, we have

$$\begin{gathered}\cos_a(x+y) = \cos_a x \cos_a y - \sin_a x \sin_a y,\\ \sin_a(x+y) = \sin_a x \cos_a y + \sin_a y \cos_a x,\\ \cos_a(-x) = \cos_a x, \qquad \sin_a(-x) = -\sin_a x,\\ \cos_a^2 x + \sin_a^2 x = 1.\end{gathered}$$

The functions $\cos_a x$ and $\sin_a x$ are continuous on $\mathbf{R}$, and are periodic with period a; moreover, a is a *principal period* of these functions, for the relation $\cos_a x = \cos_a y$ implies that either $\sin_a x = \sin_a y$ or

$\sin_a x = -\sin_a y$, i.e.,

$$e\left(\frac{x}{a}\right) = e\left(\frac{y}{a}\right) \quad \text{or} \quad e\left(\frac{x}{a}\right) = e\left(-\frac{y}{a}\right),$$

so that either

$$x \equiv y \pmod{a} \quad \text{or} \quad x \equiv -y \pmod{a};$$

and similarly

$$\sin_a x = \sin_a y$$

is equivalent to either $x \equiv y \pmod{a}$ or $x + y \equiv \frac{1}{2} a \pmod{a}$.

It follows from this that $\cos_a x$ never takes the same value twice in the interval $[0, \frac{1}{2} a]$; hence, when restricted to this interval, it is a *bijective* mapping of this interval onto the interval $[-1, +1]$. Since $\cos_a 0 = 1$ and $\cos_a(\frac{1}{2} a) = -1$, $x \to \cos_a x$ is a *strictly decreasing* mapping of $[0, \frac{1}{2} a]$ *onto* $[-1, 1]$ (Chapter IV, § 2, no. 6, Theorem 5 and Remark). We have $\cos_a x = 0$ for $x = a/4$, $\cos_a x > 0$ for $0 \leqslant x < a/4$, $\cos_a x < 0$ for $a/4 < x \leqslant a/2$. Since $\cos_a(-x) = \cos_a x$ we can deduce how $\cos_a x$ varies in the interval $[-\frac{1}{2} a, 0]$, and hence throughout the whole of $\mathbf{R}$ by periodicity (Fig. 8). Since $\sin_a x = -\cos_a(x + a/4)$ we can also deduce how $\sin_a x$ varies in $\mathbf{R}$ (Fig. 8).

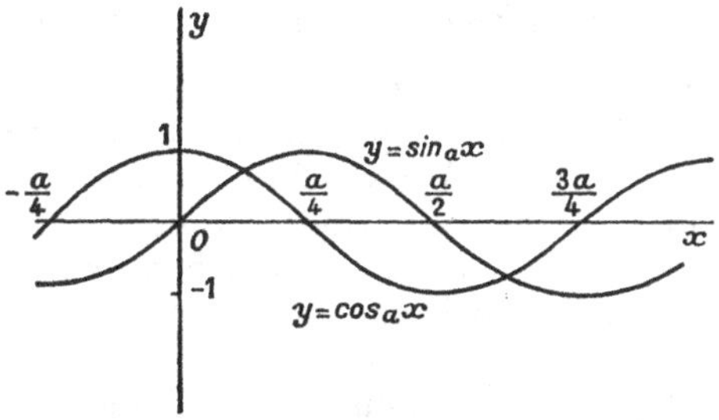

Figure 8.

The function $\tan_a x$ is a continuous mapping of $\mathbf{R}$ onto $\tilde{\mathbf{R}}$; it takes the value ∞ for the values $\frac{1}{4} a + \frac{1}{2} ka$ $(k \in \mathbf{Z})$. Since $\frac{1}{2} a$ is a period of $\tan_2 x$, it is a *principal period.* In the interval $[0, \frac{1}{4} a]$, $\sin_a x$ increases from 0 to 1, $\cos_a x$ decreases from 1 to 0, and therefore $\tan_a x$ is *strictly increasing* in $[0, \frac{1}{4} a[$ and maps this interval onto $[0, +\infty[$; it follows that $\tan_a x$ is strictly increasing in the interval

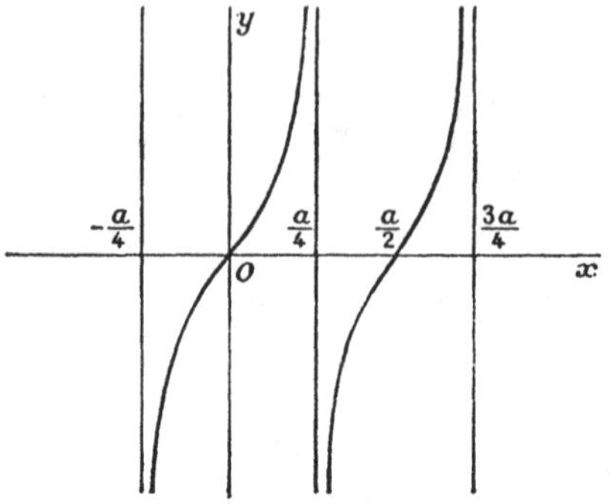

Figure 9.

$]-\frac{1}{4}a, +\frac{1}{4}a[$. and is a homeomorphism of this interval onto $\mathbf{R}$ (Fig. 9).

5. ANGULAR SECTORS

Given two *distinct* closed rays Δ_1, Δ_2 with origin o, let x be the principal measure of the angle $(\widehat{\Delta_1, \Delta_2})$ (relative to a base a, chosen once and for all). The union of the closed (resp. open) rays Δ with origin o which are such that the principal measure y of the angle $(\widehat{\Delta_1, \Delta})$ satisfies $0 \leqslant y \leqslant x$ (resp. $0 < y < x$) is the *closed* (resp. open) *angular sector* S with origin Δ_1 and extremity Δ_2, as defined in algebra. For by means of a rotation we can always reduce to the case where S does not contain the ray through the point -1. If α and β are then the angles which Δ_1 and Δ_2 respectively make with the positive real semi-axis, then the closed angular sector S is the union of the closed half-lines Δ which make an angle θ with the positive real semi-axis such that $\tan \frac{1}{2}\alpha \leqslant \tan \frac{1}{2}\theta \leqslant \tan \frac{1}{2}\beta$. Now if u, v, t are the measures of α, β, θ respectively which lie in the interval $]-\frac{1}{2}a, +\frac{1}{2}a[$, these inequalities are equivalent to $\tan_a \frac{1}{2}u \leqslant \tan_a \frac{1}{2}t \leqslant \tan_a \frac{1}{2}v$; and since $\tan_a x$ is an increasing function in the interval $]-\frac{1}{4}a, +\frac{1}{4}a[$, they are also equivalent to $u \leqslant t \leqslant v$, or to $0 \leqslant t - u \leqslant v - u$; since $x = v - u$, $y = t - u$, the result is proved for closed angular sectors, and the proof for open ones is similar.

A closed angular sector is a *closed* set in $\mathbf{R}^2$, and the *open* angular sector with the same origin and the same extremity is its *interior* in $\mathbf{R}^2$

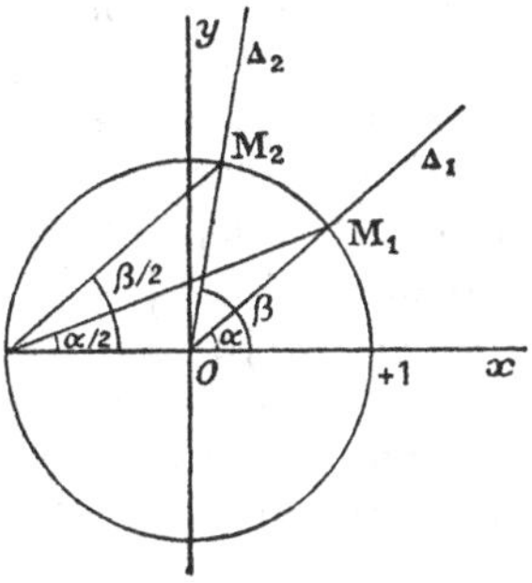

Figure 10.

(Chapter VI, § 2, no. 3, Proposition 3). The angle $(\widehat{\Delta_1, \Delta_2})$, with principal measure x, is called the *angle* of the sector S; S is said to be *salient* if $x < \frac{1}{2} a$, *flat* (or a *closed half-plane*) if $x = \frac{1}{2} a$; *re-entrant* if $x > \frac{1}{2} a$. A salient angular sector is *acute* if $x < \frac{1}{4} a$, *right* if $x = \frac{1}{4} a$, *obtuse* if $x > \frac{1}{4} a$. The *bisector* of the sector S is the ray Δ which makes an angle $y = \frac{1}{2} x$ with Δ_1.

Two distinct closed rays Δ_1, Δ_2 determine two closed angular sectors; their union is the real plane $\mathbf{R}^2$, and their intersection is $\Delta_1 \cup \Delta_2$.

6. CROSSES

We have also defined in algebra the *cross* of a pair of *lines* in a two-dimensional vector space over a maximal ordered field (*). This definition applies in particular to the real plane $\mathbf{R}^2$. The set $\mathfrak{A}_0$ of all crosses has the structure of an abelian group (written additively) defined by

$$(\widehat{D_1, D_3}) = (\widehat{D_1, D_2}) + (\widehat{D_2, D_3})$$

so that, in particular, $(\widehat{D_1, D_1}) = 0$ and $(\widehat{D_2, D_1}) = -(\widehat{D_1, D_2})$.

The *right cross* δ_0 is the solution $\neq 0$ of the equation $2\theta = 0$ in $\mathfrak{A}_0$; it is the cross which the imaginary axis makes with the real axis.

(*) We recall that an equivalence relation is defined on the set of all non-isotropic pairs of lines (D_1, D_2) by considering two pairs of lines (D_1, D_2) and (D_1', D_2') to be equivalent if there exists a *direct similitude* which transforms D_1 into D_1' and D_2 into D_2' simultaneously; the *cross* of the pair (D_1, D_2) is then the equivalence class of the pair.

We define a *canonical homomorphism* φ of the group $\mathfrak{A}$ of angles onto the group $\mathfrak{A}_0$ of crosses by making correspond to the angle which a ray Δ makes with the positive real semi-axis, the cross which the line D containing Δ makes with the real axis. A cross θ_0 is the image under φ of two angles θ and $\theta + \varpi$; thus $\mathfrak{A}_0$ is isomorphic to the quotient of $\mathfrak{A}$ by the subgroup $\{0, \varpi\}$. If we *transport* to $\mathfrak{A}_0$ the topology of the quotient group $\mathfrak{A}/\{0, \varpi\}$ by means of the bijective homomorphism associated with φ, then $\mathfrak{A}_0$ becomes a compact topological group and φ a strict morphism of $\mathfrak{A}$ onto $\mathfrak{A}_0$.

If we compose the homomorphism φ of $\mathfrak{A}$ onto $\mathfrak{A}_0$ with the homomorphism $x \rightarrow \vartheta(x/a)$ of $\mathbf{R}$ onto $\mathfrak{A}$, we have a homomorphism $x \rightarrow \vartheta_0(x/a)$ of $\mathbf{R}$ onto $\mathfrak{A}_0$; every cross $\theta_0 \in \mathfrak{A}_0$ corresponds, under this homomorphism, to a *class* of real numbers mod $\frac{1}{2}a$, which is called the *measure* of the cross θ_0 (relative to the base a); by abuse of language, every number in this class is called *a measure* of θ_0, and that one which belongs to the interval $[0, \frac{1}{2}a[$ is called the *principal measure* of θ_0; the cross $\vartheta_0(x/a)$ is the *cross of measure* x. Every measure of θ_0 is also a measure of one of the two angles θ, $\theta + \varpi$ whose image under the homomorphism φ is θ_0.

Here again, once the base a has been chosen, when we speak of a *cross* we generally mean, by abuse of language, a *measure* of this cross relative to the base a.

Remark. We can define a homomorphism of $\mathbf{C}^*$ onto $\mathfrak{A}_0$ by mapping each complex number $z \neq 0$ to the cross which the line through 0 and z makes with the real axis. Clearly this homomorphism is the composition of φ and the homomorphism $z \rightarrow \mathrm{Am}(z)$ of $\mathbf{C}^*$ onto $\mathfrak{A}$; it is therefore a *strict morphism* of the topological group $\mathbf{C}^*$ onto the topological group $\mathfrak{A}_0$, and the associated bijective homomorphism is an *isomorphism* of the quotient group $\mathbf{C}^*/\mathbf{R}^*$ onto $\mathfrak{A}_0$.

We know that if D denotes a line making a cross θ_0 with the real axis and if (a, b) is a pair of direction ratios of D, then the *tangent* of the cross θ_0 (denoted by $\tan\theta_0$) is the element b/a of $\tilde{\mathbf{R}}$ ($=\infty$ if $a=0$), which is also called the *slope* of the line D. If θ and $\theta+\varpi$ are the two angles whose image under φ is θ_0, then we have $\tan\theta_0 = \tan\theta = \tan(\theta+\varpi)$. The mapping $\theta_0 \rightarrow \tan\theta_0$ is a *homeomorphism* of $\mathfrak{A}_0$ onto $\tilde{\mathbf{R}}$, for the topological space $\mathbf{C}^*/\mathbf{R}^*$ is just the *real projective line* $\mathbf{P}_1$, and from Chapter VI, § 3, no. 3, the mapping of a line (considered as a point of $\mathbf{P}_1$) to its slope is a homeomorphism of $\mathbf{P}_1$ onto $\tilde{\mathbf{R}}$. If now we transport to $\tilde{\mathbf{R}}$ the group structure of $\mathfrak{A}_0$ by means of the mapping $\theta_0 \rightarrow \tan\theta_0$, we have defined on $\tilde{\mathbf{R}}$ the structure of an *abelian topological group*, in which the product of two elements t_1, t_2 is $\dfrac{t_1 + t_2}{1 - t_1 t_2}$ whenever t_1, t_2 belong to $\mathbf{R}$ and $t_1 t_2 \neq 1$; for pairs (t_1, t_2) which do not satisfy these conditions, the product of t_1 and t_2 is obtained by extending the function $\frac{x+y}{1-xy}$ by continuity to $\tilde{\mathbf{R}} \times \tilde{\mathbf{R}}$, and is still denoted by $\frac{t_1+t_2}{1-t_1t_2}$.

3. INFINITE SUMS AND PRODUCTS OF COMPLEX NUMBERS

1. INFINITE SUMS OF COMPLEX NUMBERS

Since the additive group of the field $\mathbf{C}$ is the same as the additive group $\mathbf{R}^2$, it is not necessary to make a special study of summable families and series in $\mathbf{C}$, since this is included in the general theory of Chapter VII, § 3; we leave to the reader the exercise of translating the results of this theory into the language of the theory of complex numbers. We state only the following proposition, which is a corollary to Proposition 3 of Chapter VII, § 3, no. 1:

PROPOSITION 1. *If $(u_\lambda)_{\lambda \in \mathbf{L}}$ and $(v_\mu)_{\mu \in \mathbf{M}}$ are two summable families of complex numbers, then the family $(u_\lambda v_\mu)_{(\lambda, \mu) \in \mathbf{L} \times \mathbf{M}}$ is summable and we have*

$$\sum_{(\lambda,\mu) \in \mathbf{L} \times \mathbf{M}} u_\lambda v_\mu = \Big(\sum_{\lambda \in \mathbf{L}} u_\lambda\Big)\Big(\sum_{\mu \in \mathbf{M}} v_\mu\Big). \tag{1}$$

We leave it to the reader to state the corresponding result for quaternions.

2. MULTIPLIABLE FAMILIES IN $\mathbf{C}^*$

In the multiplicative group $\mathbf{C}^*$ of non-zero complex numbers a family $(z_\iota)_{\iota \in \mathbf{I}}$ cannot be multipliable unless $\lim z_\iota = 1$ with respect to the filter of complements of finite subsets of $\mathbf{I}$ (Chapter III, § 5, no. 2, Proposition 1); furthermore, since every point of $\mathbf{C}^*$ has a countable fundamental system of neighbourhoods, the set of indices ι such that $z_\iota \neq 1$ is countable if the family (z_ι) is multipliable (Chapter III, § 5, no. 2, Corollary to Proposition 1).

PROPOSITION 2. *A family (z_ι) of complex numbers $z_\iota = r_\iota(\cos\theta_\iota + i \sin\theta_\iota)$ is multipliable if and only if the family (r_ι) of absolute values of the z_ι is multipliable in $\mathbf{R}_+^*$ and the family (θ_ι) of amplitudes of the z_ι is summable in the group of angles $\mathfrak{A}$.*

In view of the structure of the group $\mathbf{C}^*$ (§ 1, no. 3, Proposition 1), the proposition is an immediate consequence of Proposition 4 of Chapter III, § 5, no. 4.

If we map each angle θ to that one of its measures (to any given base a) which belongs to the interval $]-\frac{1}{2}a, \frac{1}{2}a]$, we have a *local isomorphism* of $\mathfrak{A}$ with $\mathbf{R}$ (§ 2, no. 2); since $\lim \theta_\iota = 0$ with respect to the filter of complements of finite subsets of I, we may replace the condition (in the statement of Proposition 2) that the family θ_ι should be summable in $\mathfrak{A}$ by the condition that the family (t_ι) of *measures* of the angles θ_ι which belong to $]-\frac{1}{2}a, \frac{1}{2}a]$ should be summable in $\mathbf{R}$.

The following theorem gives another criterion for a family of complex numbers, put in the form $(1 + u_\iota)$, to be multipliable in $\mathbf{C}^*$. (It generalizes Theorem 4 of Chapter IV, § 7, no. 4; see also Chapter IX, Appendix, no. 2, Proposition 1):

THEOREM 1. *The family* $(1 + u_\iota)_{\iota \in I}$ *is multipliable in* $\mathbf{C}^*$ *if and only if the family* $(|u_\iota|)$ *is summable in* $\mathbf{R}$.

For each finite subset J of I, put

$$p_J = \prod_{\iota \in J} (1 + a_\iota), \qquad s_J = \sum_{\iota \in J} a_\iota, \qquad \sigma_J = \sum_{\iota \in J} |a_\iota|.$$

LEMMA 1. *For each finite subset* J *of* I, *let* $\varphi(J) = \sup_{L \subset J} (p_L - 1)$. *Then for each subset* L *of* J *we have*

$$|p_L - 1 - s_L| \leqslant \varphi(J)\sigma_L. \tag{2}$$

This is clear if L is empty. We proceed by induction on card (L). Let $L = K \cup \{\lambda\}$, where $\lambda \notin K$; then $p_L = p_K(1 + a_\lambda)$ and $s_L = s_K + a_\lambda$, so that $p_L - 1 - s_L = (p_K - 1 - s_K) + (p_K - 1)a_\lambda$; hence, by the inductive hypothesis and the definition of $\varphi(J)$, we have

$$|p_L - 1 - s_L| \leqslant \varphi(J)\sigma_K + \varphi(J)|a_\lambda| = \varphi(J)\sigma_L,$$

which proves the lemma.

LEMMA 2. *If* J *is a finite subset of* I *such that* $\varphi(J) < 1/4$, *then*

$$|\sigma_J| \leqslant 4\varphi(J)/(1 - 4\varphi(J)).$$

For since $\sigma_L \leqslant \sigma_J$ for all subsets L of J, it follows from (2) that $|s_L| \leqslant \varphi(J)\sigma_J + |p_L - 1| \leqslant (1 + \sigma_J)\varphi(J)$; but by virtue of Chapter VII, § 3, no. 1, Proposition 2, we have $|\sigma_J| \leqslant 4 \sup_{L \subset J} |s_L|$, hence $\sigma_J \leqslant 4\varphi(J)(1 + \sigma_J)$, and the result follows.

Now let us show that the condition stated in Theorem 1 is *sufficient.* The hypothesis that the family $(|u_\iota|)$ is *summable* in $\mathbf{R}$ implies that the family $(1 + |u_\iota|)$ is *multipliable* in $\mathbf{R}_+^*$ (Chapter IV, § 7, no. 4, Theorem 4); hence, for each $\varepsilon > 0$, there is a finite subset J_0 of I such that, for each finite subset L of I which does not meet J_0, we have $\prod_{\iota \in L} (1 + |u_\iota|) - 1 \leqslant \varepsilon$. But we can write $\prod_{\iota \in L} (1 + u_\iota) - 1$ in the form $\sum_{M} \left(\prod_{\iota \in M} u_\iota \right)$ where M runs through all non-empty subsets of L; and since $\left| \prod_{\iota \in M} u_\iota \right| = \prod_{\iota \in M} |u_\iota|$, we have

$$\left| \prod_{\iota \in L} (1 + u_\iota) - 1 \right| \leqslant \sum_{M} \left(\prod_{\iota \in M} |u_\iota| \right) = \prod_{\iota \in L} (1 + |u_\iota|) - 1 \leqslant \varepsilon.$$

This proves our assertion, by virtue of Cauchy's criterion, since $\mathbf{C}^*$ is a complete group.

We still have to show that the condition of Theorem 1 is *necessary.* If $(1 + u_\iota)_{\iota \in I}$ is a multipliable family in $\mathbf{C}^*$, there exists a finite subset J of I such that, for every finite subset H of I which does not meet J, we have $\left| \prod_{\iota \in H} (1 + u_\iota) - 1 \right| \leqslant 1/8$. By Lemma 2, it follows that $\sum_{\iota \in H} |u_\iota| \leqslant 1$ for every finite subset H of I which does not meet J, and hence the family $(|u_\iota|)$ is summable in $\mathbf{R}$ (Chapter IV, § 7, no. 1, Theorem 1).

The proof above applies also, *mutatis mutandis*, to (ordered) infinite products in certain non-commutative division rings and algebras (see Exercise 6, and Chapter IX, Appendix).

3. INFINITE PRODUCTS OF COMPLEX NUMBERS

For an infinite product of non-zero complex numbers with general factor $z_n = r_n (\cos \theta_n + i \sin \theta_n)$ to be convergent in $\mathbf{C}^*$, it is necessary and sufficient, from the structure of the group $\mathbf{C}^*$, that the product with general factor r_n converge in $\mathbf{R}_+^*$ and the series with general term t_n (the measure of θ_n which lies in $]-\frac{1}{2}a, \frac{1}{2}a]$) converge in $\mathbf{R}$.

DEFINITION 1. *An infinite product of complex numbers, with general factor* $1 + u_n$, *is said to be absolutely convergent if the product with general factor* $1 + |u_n|$ *is convergent* (or, equivalently, if the series with general term $|u_n|$ is convergent).

PROPOSITION 3. *An infinite product of complex numbers is commutatively convergent if and only if it is absolutely convergent.*

This follows from Proposition 9 of Chapter III, § 5, no. 7, and Theorem 1 of no. 2 above.

Remarks. 1) The product with general factor $|1 + u_n|$ can be convergent, and indeed absolutely convergent in $\mathbf{R}_+^*$, without the convergence of the product with general factor $1 + |u_n|$ (see Exercise 4); of course, this cannot happen if all the u_n are real and > 0 from a certain index onwards.

2) As already remarked for products of factors > 0, the *convergence* of the series with general term u_n *is neither necessary nor sufficient* for the convergence of the product with general factor $1 + u_n$.

4. COMPLEX NUMBER SPACES AND PROJECTIVE SPACES

1. THE VECTOR SPACE $\mathbf{C}^n$

Since the topological space $\mathbf{C}$ can be identified with $\mathbf{R}^2$, the topological product $\mathbf{C}^n$ of n spaces identical with $\mathbf{C}$ can be identified with $\mathbf{R}^{2n}$ *quâ topological space*; likewise, the topological group structure of $\mathbf{C}^n$, the product of the additive (topological) group structures of the n factors, can be identified with that of the additive group $\mathbf{R}^{2n}$. But since $\mathbf{C}$ is a field, we may define on $\mathbf{C}^n$ the structure of an *n-dimensional vector space over* $\mathbf{C}$, the product az of a complex number a and a point $z = (z_i)$ of $\mathbf{C}^n$ being the point (az_i); this vector space structure should be carefully distinguished from the structure of a *2n-dimensional vector space over* $\mathbf{R}$, defined on $\mathbf{R}^{2n}$ (Chapter VI, § 1, no. 3). We shall reserve the notation $\mathbf{C}^n$ for the topological space which is the product of n spaces identical with $\mathbf{C}$, endowed in addition with the vector space structure over $\mathbf{C}$ which has just been defined; $\mathbf{C}^n$ is called *complex number space of* n *dimensions*. Note that the mapping $(t, z) \rightarrow tz$ is continuous on $\mathbf{C} \times \mathbf{C}^n$.

An affine linear mapping of $\mathbf{C}^n$ into $\mathbf{C}^m$ is also an affine linear mapping of $\mathbf{R}^{2n}$ into $\mathbf{R}^{2m}$, but the converse is false.

For example, the mapping $z \rightarrow \bar{z}$ is a linear mapping of the vector space $\mathbf{R}^2$ onto itself but is not a linear mapping of the vector space $\mathbf{C}$ onto itself.

Every affine linear mapping of $\mathbf{C}^n$ into $\mathbf{C}^m$ is therefore *uniformly* continuous; in particular, every affine linear mapping of $\mathbf{C}^n$ *onto* itself is a *homeomorphism*.

Every *affine linear variety of* p *dimensions* ($p \leqslant n$) in the vector space $\mathbf{C}^n$ is also an *affine linear variety of* $2p$ *dimensions* in the vector space $\mathbf{R}^{2n}$; here again, the converse is false. To avoid all confusion, (affine) linear varieties of p dimensions in $\mathbf{C}^n$ will be called *complex linear varieties of* p *dimensions* (the linear varieties of $\mathbf{R}^{2n}$ being called *real linear varieties* when necessary to prevent misunderstanding). In particular, complex linear varieties of one dimension (resp. two dimensions) are called *complex lines* (resp. *complex planes*), and complex linear varieties of $n - 1$ dimensions are called *complex hyperplanes*.

It is often convenient to regard real number space $\mathbf{R}^n$ as *embedded* in complex number space $\mathbf{C}^n$, by identifying $\mathbf{R}^n$ with the subset of $\mathbf{C}^n$ defined by the relations $\mathfrak{I}(z_k) = 0$ $(1 \leqslant k \leqslant n)$. The topological group structure induced on this subset by the topological group structure of $\mathbf{C}^n$ coincides with that of $\mathbf{R}^n$.

Note that $\mathbf{R}^n$, thus embedded in $\mathbf{C}^n$, is not a complex linear variety in $\mathbf{C}^n$.

A system of p vectors of $\mathbf{R}^n$ which is *free* over $\mathbf{R}$ is also *free* over $\mathbf{C}$. Every *real* linear variety V of p dimensions in $\mathbf{R}^n$ *generates* a *complex* linear variety V′ of p dimensions in $\mathbf{C}^n$ such that V is the *trace* of V′ on $\mathbf{R}^n$; if V is defined by a system of $n - p$ linear equations $f_k(x) = a_k$, where the f_k are linear forms on $\mathbf{R}^n$ (with real coefficients, and linearly independent over $\mathbf{R}$) and the a_k are real numbers, then the *same* equations define V′, but now the coordinates of x take complex values.

Conversely, if a complex linear variety of p dimensions has a non-empty intersection with $\mathbf{R}^n$, this intersection is a real linear variety, but its dimension may be $< p$.

2. TOPOLOGY OF VECTOR SPACES AND ALGEBRAS OVER THE FIELD C

All the definitions and all the results of nos. 5 and 6 of Chapter VI, § 1, relative to topologies on vector spaces and algebras over the field $\mathbf{R}$, and in particular spaces and algebras of matrices with elements in $\mathbf{R}$, remain valid with no modifications when we replace $\mathbf{R}$ by $\mathbf{C}$ throughout.

3. COMPLEX PROJECTIVE SPACES

With the notation recalled in Chapter VI, § 3, no. 1, we make the following definition, analogous to the definition of real projective spaces:

DEFINITION 1. *The projective space* $\mathbf{P}_n(\mathbf{C})$, *endowed with the topology which is the quotient of that of* $\mathbf{C}^*_{n+1}$ *by the equivalence relation* $\Delta_n(\mathbf{C})$, *is called complex projective space of* n *dimensions.*

The projective space $\mathbf{P}_1(\mathbf{C})$ is called the *complex projective line*, and $\mathbf{P}_2(\mathbf{C})$ is called the *complex projective plane*.

Most of the arguments relating to real projective spaces extend with very slight modifications to complex projective spaces.

In the first place we see that the topological space $\mathbf{P}_n(\mathbf{C})$ is *Hausdorff* by the argument of Proposition 1 of Chapter VI, § 3, no. 1, which applies word for word simply by replacing $\mathbf{R}$ by $\mathbf{C}$. Again, the proof of Proposition 2 of Chapter VI, § 3, no. 1 shows that $\mathbf{P}_n(\mathbf{C})$ is *compact and connected*, and homeomorphic to the quotient of the sphere $\mathbf{S}_{2n+1}$ (considered as embedded in the space $\mathbf{C}^*_{n+1}$, identified with $\mathbf{R}^*_{2n+2}$) by the equivalence relation induced on this sphere by $\Delta_n(\mathbf{C})$; the only point of difference is that now, if $n \geqslant 0$, the equivalence classes for this relation are homeomorphic to the circle $\mathbf{S}_1$.

For this reason, Proposition 3 of Chapter VI, § 3, no. 1 has no analogue for complex projective spaces.

Next one shows, as in Chapter VI, § 3, no. 2, that every p-dimensional projective linear variety in the space $\mathbf{P}_n(\mathbf{C})$ is a closed set, homeomorphic to $\mathbf{P}_p(\mathbf{C})$, and that its complement is dense in $\mathbf{P}_n(\mathbf{C})$ if $p < n$. The proof of Proposition 5 of Chapter VI, § 3, no. 2 can be transposed as it stands, simply by substituting $\mathbf{C}$ for $\mathbf{R}$, and shows that (if $n \geqslant 0$) the complement of a projective hyperplane in $\mathbf{P}_n(\mathbf{C})$ is homeomorphic to $\mathbf{C}^n$, and therefore that every point has a neighbourhood homeomorphic to $\mathbf{C}^n$. This result allows us to *embed* complex number space $\mathbf{C}^n$ in complex projective space $\mathbf{P}_n(\mathbf{C})$, by identifying $\mathbf{C}^n$ with the complement of a projective hyperplane, called the " hyperplane at infinity " (usually the hyperplane whose equation is $x_0 = 0$). In the particular case $n = 1$, the hyperplane at infinity is a *point*, and Alexandroff's theorem shows that $\mathbf{P}_1(\mathbf{C})$ is homeomorphic to the space $\tilde{\mathbf{C}}$ obtained by compactifying the locally compact space $\mathbf{C}$ by adjoining a " point at infinity ", denoted by ∞. Proposition 4 of Chapter VI, § 2, no. 4 then shows that *the complex projective line* $\mathbf{P}_1(\mathbf{C})$ *is homeomorphic to the sphere* $\mathbf{S}_2$.

We leave to the reader the task of enunciating the results analogous to those of Chapter VI, § 3, no. 4, for functions which take their values in $\mathbf{C}$.

Consider the space $\mathbf{R}^{n+1}$ as *embedded* in $\mathbf{C}^{n+1}$ (no. 1). Let f be the canonical mapping of $\mathbf{C}^*_{n+1}$ onto its quotient space $\mathbf{P}_n(\mathbf{C})$. The subspace $f(\mathbf{R}^*_{n+1})$ consists of the points of $\mathbf{P}_n(\mathbf{C})$ which have at least one system of *real* homogeneous coordinates; let us show that $f(\mathbf{R}^*_{n+1})$ is homeomorphic to real projective space $\mathbf{P}_n(\mathbf{R})$, which will allow us to consider the space $\mathbf{P}_n(\mathbf{R})$ as *embedded* in $\mathbf{P}_n(\mathbf{C})$. Now the relation induced by $\Delta_n(\mathbf{C})$ on $\mathbf{R}^*_{n+1}$ is $\Delta_n(\mathbf{R})$; the canonical mapping φ of

$$\mathbf{R}^*_{n+1}/\Delta_n(\mathbf{R}) = \mathbf{P}_n(\mathbf{R})$$

onto $f(\mathbf{R}^*_{n+1})$ is *continuous* (Chapter I, § 3, no. 6, Proposition 10); since

$P_n(\mathbf{R})$ is compact, φ must be a homeomorphism (Chapter I, § 9, no. 4, Theorem 2, Corollary 2).

We can also prove that φ is bicontinuous without using the compactness of $P_n(\mathbf{R})$, by invoking the criterion of Proposition 10 of Chapter I, § 3, no. 6 (see Exercise 3).

Since every vector subspace of $p+1$ dimensions of $\mathbf{R}^{n+1}$ generates a complex vector subspace of $p+1$ dimensions in $\mathbf{C}^{n+1}$, we see that every projective linear variety V of p dimensions in $P_n(\mathbf{R})$ (V is called a *real* projective linear variety) generates a projective linear variety V′ of p dimensions in $P_n(\mathbf{C})$ (V′ is called a *complex* projective linear variety), such that V is the trace of V′ on $P_n(\mathbf{R})$. Moreover, every system of (homogeneous) equations of V is also a system of (homogeneous) equations of V′ when we allow the variables to take complex values.

4. SPACES OF COMPLEX PROJECTIVE LINEAR VARIETIES

With the notation recalled in Chapter VI, § 3, no. 5, we define similarly the spaces of projective linear varieties in a complex projective space:

Definition 2. *The quotient space* $P_{n,p}(\mathbf{C})$ *of the topological space* $L_{n+1,\,p+1}(\mathbf{C})$ *by the equivalence relation* $\Delta_{n,p}(\mathbf{C})$ *is called the space of projective linear varieties of* $p \geqslant 0$ *dimensions in the projective space* $P_n(\mathbf{C})$.

By the argument of Proposition 6 of Chapter VI, § 3, no. 5, we see first of all that $P_{n,p}(\mathbf{C})$ is *Hausdorff*. Next we prove that it is *compact* by replacing the subspace $V_{n+1,\,p+1}$ (in the proof of Proposition 7 of Chapter VI, § 3, no. 5) by the subspace $W_{n+1,\,p+1}$ of $L_{n+1,\,p+1}(\mathbf{C})$ consisting of systems of $p+1$ vectors which form an *orthonormal Hermitian basis* of the vector subspace they generate; that is to say, $W_{n+1,\,p+1}$ consists of the matrices $X = (x_{ij})$ which satisfy the conditions

$$\sum_{j=0}^{n} x_{ij}\bar{x}_{ij} = 1 \qquad (1 \leqslant i \leqslant p+1),$$

$$\sum_{j=0}^{n} x_{ij}\bar{x}_{kj} = 0 \qquad (i \neq k).$$

The proof of Proposition 8 of Chapter VI, § 3, no. 5 extends unaltered for the space $P_{n,p}(\mathbf{C})$ and shows that this space is *connected* and *locally connected* and that each of its points has a neighbourhood homeomorphic to $\mathbf{C}^{(p+1)(n-p)}$. Finally the proof of Proposition 9 of Chapter VI, § 3, no. 6 applies without change, and therefore the Grassmannian $G_{n,p}(\mathbf{C})$ is homeomorphic to $\mathbf{P}_{n,p}(\mathbf{C})$.

Remark. Most of the properties common to the real and complex number spaces (resp. projective spaces) are also valid for number spaces (resp. projective spaces) defined in the same way over the *division ring of quaternions* $\mathbf{H}$; indeed, they are capable of extension to many other topological fields and division rings (cf. Exercises 2 and 6).

EXERCISES

§ 1

1) Let $f(z) = z^n + a_1 z^{n-1} + \cdots + a_{n-1}z + a_n$ be a polynomial of degree n with complex coefficients and put $f(z) = \prod_{i=1}^{n} (z - z_i)$. Let $r_0 = \sup_i |z_i|$.

a) Show that if the real number $r > 0$ is such that

$$r^n \leqslant |a_1|r^{n-1} + |a_2|r^{n-2} + \cdots + |a_{n-1}|r + |a_n|,$$

then $r_0 \leqslant r$, and deduce that

$$r_0 \leqslant \sup\left(1, \sum_{k=1}^{n} |a_k|\right).$$

b) Let $(\lambda_i)_{1 \leqslant i \leqslant n}$ be a finite sequence of n real numbers > 0 such that $\sum_{i=1}^{n} \lambda_i^{-1} = 1$. Show that $r_0 \leqslant \sup_k (\lambda_k |a_k|)^{1/k}$ [use *a*)].

c) Deduce from *a*) that, if the coefficients a_i are all non-zero, we have

$$r_0 \leqslant \sup\left(2|a_1|, 2\left|\frac{a_2}{a_1}\right|, \ldots, 2\left|\frac{a_{n-1}}{a_{n-2}}\right|, \left|\frac{a_n}{a_{n-1}}\right|\right).$$

d) Deduce from *a*) that

$$r_0 \leqslant |a_1 - 1| + |a_2 - a_1| + \cdots + |a_{n-1} - a_n| + |a_n|$$

[consider the polynomial $(z - 1)f(z)$]. Hence show that, if the a_i are all real and > 0, we have

$$r_0 \leqslant \sup\left(a_1, \frac{a_2}{a_1}, \ldots, \frac{a_{n-1}}{a_{n-2}}, \frac{a_n}{a_{n-1}}\right).$$

2) An *algebraic number* is a complex number which is algebraic over the field $\mathbf{Q}$ of rational numbers. Show that the ordered field B of all real algebraic numbers is a maximal ordered field, but that it is not complete in the topology induced by that of $\mathbf{R}$ [this topology is the same as $\mathcal{T}_0(\mathrm{B})$; cf. Chapter IV, § 3, Exercise 2 *b*)].

¶ 3) *a*) Let K be a maximal ordered field, endowed with the topology $\mathcal{T}_0(\mathrm{K})$, which is compatible with its field structure (Chapter IV, § 3, Exercise 2). Let

$$f(\mathrm{X}) = a_0\mathrm{X}^n + a_1\mathrm{X}^{n-1} + \cdots + a_n$$

be a polynomial in K[X] which has a *simple* root α in K. Show that, for every element $\varepsilon > 0$ in K, there exists $\eta > 0$ in K such that every polynomial $g(\mathrm{X}) = b_0\mathrm{X}^n + b_1\mathrm{X}^{n-1} + \cdots + b_n$, for which $|a_i - b_i| \leqslant \eta$ for all i, has exactly one simple root β such that $|\beta - \alpha| \leqslant \varepsilon$. (Expand f and g in powers of $\mathrm{X} - \alpha$.) Give an upper bound for η in terms of the a_i, α and ε.

b) Deduce that, if K is a maximal ordered field, its *completion* $\hat{\mathrm{K}}$ with respect to $\mathcal{T}_0(\mathrm{K})$ (which is a field with a natural order structure: Chapter IV, § 4, Exercise 2) is a maximal ordered field.

c) Let K_0 be an ordered field and let $\mathrm{S} = \mathrm{K}_0((\mathrm{X}))$ be the field of formal power series in one indeterminate over K_0; linearly order S by taking the elements $\geqslant 0$ in S to be 0 and the formal power series whose term of lowest degree has coefficient > 0. Show that S, endowed with the topology $\mathcal{T}_0(\mathrm{S})$, is *complete*, but that S is not a maximal ordered field (show that the polynomials $\mathrm{Y}^p - \mathrm{X}$ of S[Y] are irreducible for each integer $p > 1$).

d) In the example of *c*), take $\mathrm{K}_0 = \mathbf{R}$. Let K be a maximal ordered algebraic extension of S; show that K is not complete in the topology $\mathcal{T}_0(\mathrm{K})$. (Embed S in the field E of formal power series with well-ordered exponents, ordered in the same manner as S, and embed K in a maximal ordered extension Ω of E. Observe that E is complete, and give an example of a Cauchy sequence in K which converges to an element $f \in \mathrm{E}$ which is not algebraic over S; choose f so that there is an infinite number of S-isomorphisms of E into an algebraic closure of Ω, such that the images of f under these isomorphisms are all distinct.]

4) Show that every continuous homomorphism (necessarily injective) f of the topological field $\mathbf{C}$ onto a subfield of $\mathbf{C}$ is either the identity automorphism or the automorphism $z \to \bar{z}$. [Note that we must have $f(x) = x$ for all $x \in \mathbf{Q}$, and deduce that $f(x) = x$ for all $x \in \mathbf{R}$.] Show that there exist an infinite number of discontinuous isomorphisms of $\mathbf{C}$ onto subfields of $\mathbf{C}$ other than $\mathbf{C}$ itself.

¶ 5) *a*) Let K be a *non-commutative* division subring of the division ring of quaternions **H**; show that the centre Z of K is a subfield of the centre **R** of **H**. (Observe that every element of K commutes with every element of the field L generated by $Z \cup \mathbf{R}$; if $Z \not\subset \mathbf{R}$, the field L would be a maximal subfield of **H**, and K would be contained in L, contrary to hypothesis.)

b) Deduce that every isomorphism f (not necessarily continuous) of **H** onto a division subring of **H** is an inner automorphism $x \to axa^{-1}$ of **H**. [The restriction of f to **R** is an isomorphism of **R** onto a subfield of **R**, by virtue of *a*); use Exercise 3 of Chapter IV, § 3.]

§ 2

1) Let a be a non-zero complex number, and n an integer > 0. For each real number $r > 0$ such that $r^n \leqslant |a|$, show that there exists $z \in \mathbf{C}$ such that $|z| = r$ and $|a + z^n| = |a| - r^n$. Deduce that, if f is a polynomial of degree > 0, with complex coefficients, we cannot have $|f(z_0)| \leqslant |f(z)|$ at all points z of a neighbourhood of a point z_0, provided that $f(z_0) \neq 0$.

2) Show that, if f is any polynomial with complex coefficients, not identically zero, there exists a real number $r > 0$ such that

$$|f(z)| > |f(0)|$$

whenever $|z| \geqslant r$. Hence, with the help of Exercise 1 and Weierstrass' theorem (Chapter IV, § 6, no. 1, Theorem 1), give another proof of the fact that **C** is algebraically closed (consider the function f on the compact set of points z such that $|z| \leqslant r$).

3) Show, without using Theorem 1, that the mapping $z \to z/\bar{z}$ is a strict morphism of the topological group $\mathbf{C}^*$ onto the topological group **U**. Deduce that the mapping $t \to (1 + it)/(1 - it)$ $[(1 + it)/(1 - it) = -1$ if $t = \infty]$ is an isomorphism of the topological group $\tilde{\mathbf{R}}$ (no. 6) onto the topological group **U**, and hence give another proof of Theorem 1.

¶ 4) Let K be the smallest Pythagorean subfield of **R** and let $K' = K(i)$. Let G be the multiplicative group of elements of K′ of absolute value 1 (a subgroup of **U**). Show that G is not isomorphic to the additive group of numbers of K mod 1. (Observe that in the latter group there exist elements of any prime order p; on the other hand, if p is a prime such that $p - 1$ is not a power of 2, show that G contains no pth root of unity other than 1, by noting that the degree over **Q** of every element of K′ is a power of 2.)

§ 3

¶ 1) For every finite sequence $s = (z_k)_{k\in I}$ of complex numbers, put

$$\sum_{k\in I} |z_k| = \rho_s . \sup_{J\subset I}\left|\sum_{k\in J} z_k\right|$$

(cf. Chapter VII, § 3, no. 1, Proposition 2).

a) * If $z_k = r_k(\cos\varphi_k + i\sin\varphi_k)$, where $r_k = |z_k|$ and φ_k is the principal radian measure of the amplitude of z_k, put

$$f(\varphi) = \sum_{k\in I} r_k(\cos(\varphi - \varphi_k))^+$$

for every $\varphi\in[0, 2\pi[$. Show that the least upper bound of $f(\varphi)$ in the interval $[0, 2\pi[$ is equal to $\sup_{J\subset I}\left|\sum_{k\in J} z_k\right|$. By considering the integral $\int_0^{2\pi} f(\varphi)\,d\varphi$, deduce that for *all* finite sequences s we have $\rho_s \leqslant \pi$, and that there is no finite sequence $(z_k) = s$ such that $\rho_s = \pi$.

b) Show that, for every $\varepsilon > 0$, there exists a finite sequence $(z_k) = s$ such that $\rho_s > \pi - \varepsilon$ (take the z_k to be the roots of a binomial equation of sufficiently high degree). *

2) Let (z_n) be an infinite sequence of complex numbers $z_n = x_n + iy_n$ such that $x_n \geqslant 0$ for all n. Show that if the series whose general terms are z_n and z_n^2 are convergent, then the series whose general term is z_n^2 is absolutely convergent. Give an example in which this result is false when the condition $x_n \geqslant 0$ (which can also be written $-\pi/2 \leqslant \mathrm{Am}(z_n) \leqslant \pi/2$) is replaced by

$$-(1-\varepsilon)\frac{\pi}{2} \leqslant \mathrm{Am}(z_n) \leqslant (1+\varepsilon)\frac{\pi}{2}$$

no matter how small the real number $\varepsilon > 0$ may be. (Here Am (z) denotes that radian measure of the amplitude of z_n which belongs to the interval $]-\pi, +\pi]$.)

3) Let $(z_\iota)_{\iota\in I}$ be a family of complex numbers such that $\prod_{\iota\in I}|z_\iota| = +\infty$, $\left(\text{resp. } \prod_{\iota\in I}|z_\iota| = 0\right)$. Show that the family (z_ι) is multipliable in the space $\tilde{\mathbf{C}}$ (§ 4, no. 3) and that its product is ∞ (resp. 0).

4) Show that the infinite product whose general factor is $1 + i/n$ is not convergent, but that the product of the absolute values of its factors is absolutely convergent.

5) Let (z_n) be an infinite sequence of non-zero complex numbers such that $\lim_{n\to\infty} z_n = 1$. Show that, if there exist permutations σ of $\mathbf{N}$ such that the infinite product whose general factor is $z_{\sigma(n)}$ is convergent, then the set of products

$$\mathop{\mathsf{P}}_{n=0}^{\infty} z_{\sigma(n)}$$

corresponding to all these permutations can be (i) a single point; or (ii) the whole group $\mathbf{C}^*$; or (iii) an open ray with origin 0; or (iv) a circle with centre 0; or (v) a "logarithmic spiral", the image of $\mathbf{R}$ under the mapping $t \to a^{t-t_0}(\cos t + i \sin t)$, where a is a real number > 0 and $\neq 1$. (Argue as in Exercise 2 of Chapter VII, § 3, using the fact that the multiplicative group $\mathbf{C}^*$ is isomorphic to the additive group $\mathbf{R} \times \mathbf{T}$.)

§ 4

1) Let f be a polynomial in n complex variables, with complex coefficients and not identically zero. Show that the complement in $\mathbf{C}^n$ of the set S of points $z = (z_i)$ such that $f(z_1, z_2, \ldots, z_n) = 0$ (the "algebraic variety" with equation $f = 0$) is connected. (If a, b are two distinct points of $\complement S$, consider the intersections of S and the complex line through a and b.)

¶ 2) Let K be a non-discrete Hausdorff topological division ring (or field).

a) Let E be a left vector space of dimension n over K. If $(a_i)_{1\leqslant i\leqslant n}$ is a basis of E, and if we transport to E the topology of K^n (the product of the topologies of the factors) by means of the bijective linear mapping

$$(x_i) \to \sum_{i=1}^{n} x_i a_i,$$

then the topology so defined on E is independent of the basis (a_i) chosen, is compatible with the additive group structure of E, and is such that the mapping $(t, x) \to tx$ of $K \times E$ into E is continuous. If F is a vector subspace of E, the topology induced on F by that of E is the same as the topology defined on F by starting from an arbitrary basis of F, as above; F is closed in E, and $\complement F$ is dense in E unless $F = E$.

b) Generalize Propositions 4, 5 and 6 of Chapter VI, § 1 to the division ring K. (To show that the mapping $X \to X^{-1}$ is continuous in the neighbourhood of every nonsingular square matrix of order n over K, when K is not commutative, use induction on n: considering a neighbourhood of the unit matrix I_n of order n, note that every matrix X sufficiently near to I_n can be put in the form

$$\begin{pmatrix} 1 & 0 & \dots & 0 \\ \lambda_2 & & & \\ \vdots & & I_{n-1} & \\ \lambda_n & & & \end{pmatrix} \cdot \begin{pmatrix} \mu_1 & \mu_2 & \dots & \mu_n \\ 0 & & & \\ \vdots & & Y & \\ 0 & & & \end{pmatrix},$$

where I_{n-1} is the unit matrix of order $n-1$ and Y is a nonsingular matrix of order $n-1$.)

c) If K is commutative and E is an algebra of rank n over K, then the topology of E is compatible with its ring structure. Furthermore, if E has an identity element, the group G of units of E is open and dense in E, and the topology induced on G by that of E is compatible with the group structure of G. (If e is the identity element of E, and if x is a unit of E, then each of the equations $xy = e$, $yx = e$ has *one* and *one only* solution y; for each of these equations consider the equivalent system of n linear equations which give the components of y with respect to a basis of E.)

d) If K is a non-complete field and E an algebra of rank n over K, and if the completion $\hat{K}$ of K is a field, then the completion $\hat{E}$ of the algebra E is that obtained by extending the ring of operators of E to $\hat{K}$. Hence construct examples of a topological division ring whose completion is not a division ring.

3) Using Proposition 10 of Chapter I, § 3, no. 6, prove that the real projective space $\mathbf{P}_n(\mathbf{R})$ is homeomorphic to the subspace of $\mathbf{P}_n(\mathbf{C})$ consisting of these points which have at least one system of real homogeneous coordinates. [$\mathbf{R}^*_{n+1}$ being considered as embedded in $\mathbf{C}^*_{n+1}$, let A be a closed set in $\mathbf{R}^*_{n+1}$, saturated with respect to $\Delta_n(\mathbf{R})$; let B be the set of all points ζx, where $x \in A$ and $\zeta \in \mathbf{U}$; show that $\bar{B}$ is saturated with respect to $\Delta_n(\mathbf{C})$, and that A is the trace of $\bar{B}$ on $\mathbf{R}^*_{n+1}$.]

4) In $\mathbf{P}_n(\mathbf{C})$ let H_n be the "quadric" defined by the equation

$$x_0^2 + x_1^2 + \dots + x_n^2 = 0.$$

Show that every point of H_n has an open neighbourhood homeomorphic to $\mathbf{C}^{n-1}$, that H_n is connected, and that the intersection of H_n with the complement of any complex projective hyperplane is connected (to prove this last assertion, reduce to the case $n = 2$).

Show that H_2 is homeomorphic to $\mathbf{S}_2$, and H_3 to $\mathbf{S}_2 \times \mathbf{S}_2$ (use the parametric representation of the quadric by means of its rectilinear generators).

5) The sphere $\mathbf{S}_5$ being considered as a subspace of $\mathbf{C}^3$, consider the mapping of $\mathbf{S}_5$ into $\mathbf{R}^7$ which sends each point $\boldsymbol{x} = (x_1, x_2, x_3)$ of $\mathbf{S}_5$ (x_1, x_2, x_3 complex) to the point $y = (y_i)_{1 \leqslant i \leqslant 7}$ of $\mathbf{R}^7$, where

$$y_1 = |x_1|^2 - |x_2|^2, \quad y_2 = \mathfrak{R}(x_1\bar{x}_2), \quad y_3 = \mathfrak{I}(x_1\bar{x}_2), \quad y_4 = \mathfrak{R}(x_1\bar{x}_3),$$
$$y_5 = \mathfrak{I}(x_1\bar{x}_3), \quad y_6 = \mathfrak{R}(x_2\bar{x}_3), \quad y_7 = \mathfrak{I}(x_2\bar{x}_3).$$

This function has the same value at two points $\boldsymbol{x}, \boldsymbol{x}'$ of $\mathbf{S}_5$ such that $\boldsymbol{x}' = \zeta\boldsymbol{x}$, where $|\zeta| = 1$. Show that, by passing to the quotient, it induces a homeomorphism of $\mathbf{P}_2(\mathbf{C})$ onto a subspace of $\mathbf{R}^7$.

¶ 6) Let K be a non-discrete Hausdorff topological division ring (or field).

a) Endow the left projective space $\mathbf{P}_n(\mathrm{K})$ with the quotient of the topology of K^*_{n+1} by the equivalence relation $\Delta_n(\mathrm{K})$. Extend Propositions 1, 4 and 5 of Chapter VI, § 3, to $\mathbf{P}_n(\mathrm{K})$ endowed with this topology. Show that $\mathbf{P}_n(\mathrm{K})$ is connected if K is connected, and otherwise is totally disconnected.

* *b*) If K is locally compact (but not discrete), show that $\mathbf{P}_n(\mathrm{K})$ is compact. (Let U be a compact neighbourhood of 0 in K; let $a \in \mathrm{K}$ be such that

$$\lim_{m \to \infty} a^m = 0;$$

let S be the subset of K^*_{n+1} consisting of points $\boldsymbol{x} = (x_i)$, all of whose coordinates x_i belong to U, and such that $x_k \in a . \complement \mathrm{U}$ for some index k; show that $\mathbf{P}_n(\mathrm{K})$ is the canonical image of S). *

c) Likewise, extend Definition 2 and Propositions 6 and 8 of Chapter VI, § 3 to the set $\mathbf{P}_{n,p}(\mathrm{K})$ of projective linear varieties of p dimensions in $\mathbf{P}_n(\mathrm{K})$; * also Proposition 7, assuming that K is locally compact [for Proposition 7 consider, for each sequence σ of indices, the subset S_σ of A_σ consisting of matrices all of whose rows belong to the set S defined in *b*); show that $\mathbf{P}_{n,p}(\mathrm{K})$ is the canonical image of the union of the sets S_σ]. * Generalize Proposition 9 of Chapter VI, § 3, when K is a *field*.

d) If K is *not* commutative, let $\mathbf{P}'_{n,p}(\mathrm{K})$ denote the space of p-dimensional projective linear varieties in the *right* projective space of n dimensions over K [the topology of $\mathbf{P}'_{n,p}(K)$ being defined in the same way as that of $\mathbf{P}_{n,p}(K)$]. Show that $\mathbf{P}_{n,p}(K)$ and $\mathbf{P}'_{n,n-p-1}(K)$ are homeomorphic [to every $(p+1)$-dimensional vector subspace V of the left vector space $E = K_s^{n+1}$ make correspond the orthogonal complement of V, which is a vector subspace of the dual E^* of E].

7) Extend Exercise 6 of Chapter VI, § 3 to spaces of projective linear varieties over **C** and **H**.

¶ 8) Extend Exercises 7, 8 and 9 of Chapter VI, § 3 to the spaces $W_{n,p}$ (no. 4).
In the vector space $\mathbf{H}^n$ over the division ring of quaternions, we again denote by $W_{n,p}$ the set of all sequences $(\boldsymbol{x}_k)_{1 \leqslant k \leqslant p}$ of p vectors $\boldsymbol{x}_k = (x_{kj})_{1 \leqslant j \leqslant n}$ such that

$$\sum_{j=1}^{n} x_{ij}\bar{x}_{ij} = 1 \qquad (1 \leqslant i \leqslant p)$$

and

$$\sum_{j=1}^{n} x_{ij}\bar{x}_{ij} = 0 \qquad (i \neq k)$$

($\bar{x}$ denoting the conjugate of the quaternion x). Extend Exercises 7, 8 and 9 of Chapter VI, § 3, to these spaces $W_{n,p}$.

HISTORICAL NOTE

(Numbers in brackets refer to the bibliography at the end of this note.)

We shall not repeat here the complete history of the development of the theories of complex numbers and quaternions, since these theories belong essentially to algebra; but we shall say something about the geometrical representation of complex numbers, which in many respects was a decisive step forward in the history of mathematics.

Without any doubt the first to have a clear conception of the one-to-one correspondence between complex numbers and points in the plane was C. F. Gauss (*), who moreover applied this idea to the theory of complex numbers and foresaw the uses to which it would be put by the analysts of the 19th century. During the 17th and 18th centuries, mathematicians had gradually come to the conviction that imaginary numbers, which allowed them to solve all quadratic equations, could also be used to solve algebraic equations of any degree. Many attempts to prove this theorem were published during the course of the 18th century; but, not mentioning those which were based only on a vicious circle, there was not one which was not open to serious objection. Gauss, after a detailed examination of these attempts and a closely reasoned criticism of their deficiencies, proposed in his inaugural dissertation (written in 1797, published in 1799) to give at last a rigorous proof. Taking up an idea thrown off in passing by d'Alembert [in his proof published in 1746 (**)],

(*) The first to have had the idea of such a correspondence was undoubtedly Wallis, in his *Treatise on Algebra* published in 1685; but his ideas on this topic were confused and had no influence on his contemporaries.

(**) This proof (in which in fact d'Alembert does not make any use of the remark which served as Gauss's starting point) is historically the first which does not reduce to blatantly begging the question. Gauss, who justly criticized its flaws, nevertheless recognized the value of d'Alembert's fundamental idea: "*the essence of the proof*

Gauss remarked that the points (a, b) of the plane which are such that $a + ib$ is a root of the polynomial $P(x + iy) = X(x, y) + iY(x, y)$ are the intersections of the curves $X = 0$ and $Y = 0$; by means of a qualitative study of these curves, he then shows that a continuous arc of one of them joins points of two distinct regions bounded by the other, and concludes from this that the curves intersect ([1], vol. 3, p. 3; see also [1 *a*]). In its clarity and originality this proof was a considerable advance on previous attempts, and was certainly one of the first examples of purely topological reasoning applied to a problem of algebra (*).

In his dissertation, Gauss does not explicitly define the correspondence between points of the plane and complex numbers; as to the latter, and the questions of "existence" to which they had given rise for two hundred years, he reserves his position and deliberately presents all his arguments in a form which involves only real quantities. But the march of ideas in his proof would be wholly unintelligible if it did not presuppose a fully conscious identification of points of the plane with complex numbers; and his research at the same period in number theory and elliptic functions, which also involve complex numbers, reinforces this supposition. The way in which the geometrical conception of imaginaries became familiar to him, and the results it could lead to in his hands, are clearly shown by the notes (published only recently) in which he applies complex numbers to the solution of problems of elementary geometry ([1], vol. 4, p. 396 and vol. 8, p. 307). Even more explicit is the letter to Bessel in 1811 ([1], vol. 8, p. 90-91) in which he sketches the essentials of the theory of integration of functions of a complex variable: "*Just as the whole domain of real quantities can be represented by means of an infinite line, so the* complete *domain of* all *quantities, both real and imaginary, can be realized by means of an infinite plane, in which each point, determined by its abscissa* a *and its ordinate* b, *represents at the same time the quantity* $a + ib$. *The continuous passage from one value of* x *to another consequently takes place along a curve, and can therefore be achieved in an infinity of ways...*".

But it was not until 1831 that Gauss (*à propos* the introduction of "Gaussian integers" $a + ib$, where a and b are integers) publicly expounded his ideas on this point so precisely ([1], vol. 2, *Theoria Residuo-*

---

seems to me not to be affected by all these objections" ([1], vol. 3, p. 11); a little later, he sketched a method which would make d'Alembert's argument rigorous, and this is effectively the line of argument used by Cauchy in one of his proofs of the same theorem (cf. § 2, Exercise 2).

(*) Gauss published altogether four proofs of the "Theorem of d'Alembert-Gauss"; of these, the last is a variant of the first and, like it, appeals to intuitive topological properties of the plane; but the second and third are based on completely different principles. The proof we have given in § 1 is essentially Gauss's second proof, which itself is the realization of an idea of Euler and de Foncenex.

rum Biquadraticorum, Commentatis Secunda, art. 38, p. 109, and *Anzeige*, p. 174 *et seq.*). In the intervening years, the idea of representing complex numbers geometrically had been independently rediscovered by two modest amateurs, both of them more or less self-taught, who made no other contribution to mathematics, and who were both without much contact with the scientific circles of their time. For this reason their work was in danger of passing completely unnoticed; this is precisely what happened to the first in point of date, a pamphlet written by a Dane, C. Wessel. Published in 1798, it was clearly conceived and written, but was not rescued from oblivion until a century later. The same mishap might have overtaken the second, by the Swiss J. Argand, who owed it only to chance that he saw his work, published in 1806, exhumed seven years later (*). This work provoked an active discussion in the *Annales de Gergonne*, and was the subject, in France and England, of several publications by obscure authors between 1820 and 1830. But the authority of a great name was needed to put an end to these controversies and to rally mathematicians to the new point of view, and it was not until the middle of the century that the geometrical representation of complex numbers at last became universally adopted, following the publications of Gauss (mentioned above) in Germany, the work of Hamilton and Cayley on hypercomplex systems in England, and finally the support of Cauchy (**) in France, only a few years before the genius of Riemann was to extend still further the role of geometry in the theory of analytic functions, and at the same stroke create the science of topology.

***

The measurement of angles by means of the arcs they cut off on a circle is as old as the notion of angle itself, and was already known to the Babylonians: their unit of angular measure was the degree, which we still use. The Babylonians used only measures of angles lying between 0 and 360°; this

---

(*) Unlike Gauss, Wessel and Argand were more concerned with *justifying* operations with complex numbers than with applying the geometrical representation they proposed to new investigations; Wessel gives no applications, and the only one given by Argand is a proof of the theorem of d'Alembert-Gauss, which is scarcely more than a variant of d'Alembert's proof, and is open to the same objections.

(**) In his first investigations on integrals of functions of complex variables (between 1814 and 1826), Cauchy considered complex numbers as "symbolic" expressions and did not identify them with the points of the plane; but this did not hinder him from constantly associating the number $x + iy$ with the point (x, y), and freely using the language of geometry in this context.

was adequate for their purposes, which were primarily to fix the positions of celestial objects at determinate points of their apparent orbits and to construct tables of these orbits for scientific or astrological purposes.

Among the Greek geometers of the classical era, the notion of angle (*Euclid's Elements*, I, Definitions 8 and 9) was even more restricted, because it applied only to angles smaller than two right angles; and since on the other hand their theory of proportion and measurement was based on the comparison of *arbitrarily large* multiples of the magnitudes being measured, angles could not be measurable quantities for them, although of course they had the concepts of equal angles, of one angle being greater or less than another, and of the sum of two angles, provided that the sum did not exceed two right angles. Just as with the addition of fractions, the measurement of angles must have been in their eyes an empirical procedure, devoid of scientific value. This attitude is well illustrated by the admirable essay of Archimedes on spirals ([3], vol. 2, pp. 1-121) in which, since he cannot define them by the proportionality of radius vector to angle, he gives a kinematic definition (Definition 1, p. 44; cf. the statement of Proposition 12, p. 46) from which he succeeds in extracting, as the rest of the work shows, everything that the general notion of angular measure would have given him had he been in possession of it. As to the Greek astronomers, they seem to have been content to follow their Babylonian predecessors on this point as on many others.

Here too, as in the evolution of the concept of real number (cf. the Historical Note on Chapter IV), the relaxation of the spirit of rigour during the decadence of Greek science brought a return to the "naïve" point of view which, in some respects, comes nearer to our own than does the rigid Euclidean conception. Thus an ill-advised interpolator inserted the following famous proposition in Euclid (*Euclid's Elements*, VI, 33): "Angles are proportional to the arcs which they cut out on a circle" (*), and an anonymous scholar who comments on the "proof" of this proposition does not hesitate to introduce, of course without justification, arcs equal to arbitrarily large multiples of a circumference, and the angles corresponding to these arcs (**). But even Vietà, in the 16th century, although he appeared to come close to our modern conception of angle

(*) That this really is an interpolation is shown without doubt by the absurdity of the proof, which is ineptly based on the classical paradigms of the method of Eudoxus; moreover, it is clear that this result has nothing to do with the end of Book VI. It is amusing to see Theon, in the 4th century A.D., congratulate himself for having grafted, onto this interpolation, another in which he purports to prove that "the areas of sectors of a circle are proportional to their angles at the centre" (*Euclid's Elements*, edited by Heiberg, vol. 5, p. XXIV). This was six centuries after Archimedes had determined the areas of sectors of spirals.

(**) *Euclid's Elements*, ed. Heiberg, vol. 5, p. 357.

when he discovered that the equation $\sin nx = \sin \alpha$ has several roots, obtained only those roots corresponding to angles smaller than two right angles ([4], p. 305). Only in the 17th century was this point of view definitively superseded; and, after the discovery by Newton of the power-series expansions for $\sin x$ and $\cos x$ had furnished expressions of these functions which were valid for all values of the variable, Euler finally formulated the precise conception of the notion of angular measure, in connection with logarithms of "imaginary" numbers ([5], (1), vol. 17, p. 220).

Of course, the classical definition of angular measure in terms of the length of a circular arc is not only intuitive but essentially correct; however it requires, to make it rigorous, the notion of the length of a curve, i.e., integral calculus. From the point of view of the structures which come into play, this is a very long-winded procedure, and it is possible, as we have seen in the text, to arrive at the same end by no other means than those of the theory of topological groups; in this manner the real exponential and the complex exponential appear as arising from the same source, the theorem which characterizes the "one-parameter groups" (Chapter V, § 3, Theorem 1).

BIBLIOGRAPHY

[1] C. F. Gauss, *Werke*, vol. 2 (2nd edition, Göttingen, 1876); vol. 3 (2nd edition, *ibid.*, 1876); vol. 4 (*ibid.*, 1873) and vol. 8 (*ibid.*, 1900).

[1 *a*] *Die vier Gauss'schen Beweise für die Zerlegung ganzer algebraischer Functionen in reelle Factoren ersten oder zweiten Grades* [Ostwald's Klassiker, no. 14, Leipzig (Teubner), 1904].

[2] *Euclidis Elementa*, 5 vols., edited by J. L. Heiberg, Leipzig (Teubner) 1883-1888.

[3] *Archimedis Opera Omnia*, 3 vols., vol. 2, pp. 1-121, edited by J. L. Heiberg, 2nd edition, Leipzig (Teubner) 1913-15.

[3 *a*] *Les Œuvres complètes d'Archimède*, translated by P. Ver Eecke, Paris-Brussels (Desclée de Brouwer), 1921.

[4] Francisci Vietae, *Opera Mathematica*, p. 305, Leyden (Bonaventure et Abraham Elzevir), 1646.

[5] L. Euler, *Opera omnia* (1), vol. 17, p. 220, Leipzig (Teubner), 1915.

CHAPTER IX

Use of real numbers in general topology

1. GENERATION OF A UNIFORMITY BY A FAMILY OF PSEUDOMETRICS; UNIFORMIZABLE SPACES

1. PSEUDOMETRICS

DEFINITION 1. *If* X *is a set, a pseudometric on* X *is any mapping* f *of* $X \times X$ *into the interval* $[0, +\infty]$ *of the extended real line* $\overline{\mathbf{R}}$ *which satisfies the following conditions*:

(EC_I) *For all* $x \in X, f(x, x) = 0$.
(EC_{II}) *For all* $x \in X$ *and all* $y \in X, f(x, y) = f(y, x)$ (symmetry).
(EC_{III}) *For all* x, y, z *in* X,

$$f(x, y) \leqslant f(x, z) + f(z, y)$$

(triangle inequality).

Examples. 1) On real number space $\mathbf{R}^n$, Euclidean distance (Chapter VI, § 2, no. 1) is a pseudometric.

2) If X is any set, the function f defined on $X \times X$ by the conditions $f(x, x) = 0$ for all $x \in X$, $f(x, y) = +\infty$ if $x \neq y$ is a pseudometric on X.

3) If X is any set and if g is any finite real-valued function defined on X, then the function f defined on $X \times X$ by $f(x, y) = |g(x) - g(y)|$ is a pseudometric on X.

* 4) Let X be the set of all continuous mappings of the interval $[0, 1]$ of $\mathbf{R}$ into $\mathbf{R}$. If for each pair of elements x, y of X we put

$$f(x, y) = \int_0^1 |x(t) - y(t)|\, dt,$$

then f is a pseudometric on X. *

Remarks. 1) Example 2 above shows that a pseudometric can take the value $+\infty$ for certain pairs of elements of X.

2) If f is a pseudometric on X, we can in general have $f(x, y) = 0$ without $x = y$, as Example 3 above shows (cf. § 2).

From the triangle inequality it follows that if $f(x, z)$ and $f(y, z)$ are *finite* then so is $f(x, y)$; moreover, in this case we have

$$f(x, z) \leqslant f(y, z) + f(x, y) \quad \text{and} \quad f(y, z) \leqslant f(x, z) + f(x, y),$$

so that

$$|f(x, z) - f(y, z)| \leqslant f(x, y). \tag{1}$$

If f is a pseudometric on X, then so is λf for any finite real number $\lambda > 0$. If $(f_\iota)_{\iota \in \mathrm{I}}$ is any family of pseudometrics on X, the sum $\sum_{\iota \in \mathrm{I}} f_\iota(x, y)$ is defined for all $(x, y) \in \mathrm{X} \times \mathrm{X}$; if $f(x, y)$ denotes the value of this sum, then f is a pseudometric on X. Again, the ***upper envelope*** g of the family (f_ι) (Chapter IV, § 5, no. 5) is a pseudometric on X, for the relations $f_\iota(x, y) \leqslant f_\iota(x, z) + f_\iota(z, y)$ imply

$$\sup_{\iota \in \mathrm{I}} f_\iota(x, y) \leqslant \sup_{\iota \in \mathrm{I}} (f_\iota(x, z) + f_\iota(z, y)) \leqslant \sup_{\iota \in \mathrm{I}} f_\iota(x, z) + \sup_{\iota \in \mathrm{I}} f_\iota(z, y)$$

[Chapter IV, § 5, no. 7, formula (17)].

2. DEFINITION OF A UNIFORMITY BY MEANS OF A FAMILY OF PSEUDOMETRICS

We have seen in Chapter VI, § 2, no. 3 that if, for each real number $a > 0$, we denote by U_a the set of all pairs $(\boldsymbol{x}, \boldsymbol{y})$ of points of $\mathbf{R}^n$ whose Euclidean distance apart is $\leqslant a$, then the U_a form a fundamental system of entourages of the uniformity of $\mathbf{R}^n$ as a runs through the set of real numbers > 0.

More generally, let f be a pseudometric on a set X; for each $a > 0$, let U_a denote $\overset{-1}{f}([0, a])$, and let us show that, as a runs through the set of all real numbers > 0, the U_a form a *fundamental system of entourages* of a uniformity on X. Axiom $(\mathrm{U}'_{\mathrm{I}})$ is satisfied by reason of $(\mathrm{EC}_{\mathrm{I}})$; if $a \leqslant b$, we have $\mathrm{U}_a \subset \mathrm{U}_b$ and therefore the U_a satisfy $(\mathrm{B}_{\mathrm{I}})$; by $(\mathrm{EC}_{\mathrm{II}})$, we have $\overset{-1}{\mathrm{U}}_a = \mathrm{U}_a$ and therefore $(\mathrm{U}'_{\mathrm{II}})$ is satisfied; finally, by $(\mathrm{EC}_{\mathrm{III}})$ we have $\overset{2}{\mathrm{U}}_a \subset \mathrm{U}_{2a}$, so that $(\mathrm{U}'_{\mathrm{III}})$ is satisfied. [Cf. Chapter II, § 1, no. 1, Definition 2]. Consequently we may make the following definition :

DEFINITION 2. *Given a pseudometric f on a set X, the uniformity defined by f is the uniformity on X which has as a fundamental system of entourages the family of sets $\overset{-1}{f}([0, a])$, where a runs through the set of all real numbers > 0.*

Two pseudometrics on X are said to be equivalent if they define the same uniformity.

Remarks. 1) If (a_n) is any sequence of numbers > 0 and tending to 0, the U_{a_n} form a fundamental system of entourages of the uniformity defined by f.

2) The definition of a uniformity by a pseudometric f consists in taking as a fundamental system of entourages the *inverse image* under f of the neighbourhood filter of 0 in the subspace $[0, +\infty]$ of $\overline{\mathbf{R}}$. Note that this procedure is quite analogous to that which allowed us to define the uniformities on a topological group (Chapter III, § 3, no. 1).

Let f and g be two pseudometrics on X. From Definition 2 it follows that the uniformity defined by f is *coarser* than the uniformity defined by g if and only if, for each $a > 0$ there exists $b > 0$ such that the relation $g(x, y) \leqslant b$ implies $f(x, y) \leqslant a$. A necessary and sufficient condition for f and g to be *equivalent* pseudometrics is that for each $a > 0$ there exists $b > 0$ such that $g(x, y) \leqslant b$ implies $f(x, y) \leqslant a$, and $f(x, y) \leqslant b$ implies $g(x, y) \leqslant a$.

In particular, if there exists a constant k such that $f \leqslant kg$, the uniformity defined by f is coarser than the uniformity defined by g.

Let φ be a mapping of the interval $[0, +\infty]$ into itself, satisfying the following conditions: 1) $\varphi(0) = 0$, and φ is continuous at 0; 2) φ is increasing in $[0, +\infty]$ and is strictly increasing in a neighbourhood of 0; 3) for all $u \geqslant 0$ and $v \geqslant 0$, we have $\varphi(u + v) \leqslant \varphi(u) + \varphi(v)$. Then if f is any pseudometric on a set X, the composition $g = \varphi \circ f$ is a pseudometric *equivalent* to f.

The reader may easily verify that we may, for example, take φ to be any one of the following functions:

$$\sqrt{u}, \qquad \log(1 + u), \qquad \frac{u}{1 + u}, \qquad \inf(u, 1).$$

The last two examples show that there always exist *bounded* pseudometrics equivalent to any given pseudometric (finite or not).

DEFINITION 3. *If $(f_\iota)_{\iota \in I}$ is a family of pseudometrics on a set X, then the least upper bound of the set of uniformities defined on X by the pseudometrics f_ι is called the uniformity defined by the family (f_ι).*

Two families of pseudometrics on X are said to be equivalent if they define the same uniformity on X.

From the definition of the least upper bound of a set of uniformities (Chapter II, § 2, no. 5), the filter of entourages of the uniformity $\mathcal{U}$ defined on X by a family of pseudometrics $(f_\iota)_{\iota \in I}$ is the filter *generated* (Chapter I, § 6, no. 2) by the family of sets $\overset{-1}{f_\iota}([0, a])$, where ι runs through I and a runs through the set of real numbers > 0. In other words, we obtain a fundamental system of entourages of $\mathfrak{U}$ by proceeding as follows: we take at random a finite number of indices $\iota_1, \iota_2, \ldots, \iota_n$ and, corresponding to each ι_k, a number $a_k > 0$; then we consider the set V of pairs $(x, y) \in X \times X$ such that $f_{\iota_k}(x, y) \leqslant a_k$ for $1 \leqslant k \leqslant n$; these sets V (for all possible choices of n, the ι_k and the a_k) form a fundamental system of entourages for $\mathfrak{U}$. Moreover, we may restrict ourselves to the case in which all the a_k are equal to the *same* number $a > 0$, since the entourage consisting of all pairs (x, y) such that

$$\sup_{1 \leqslant k \leqslant n} (f_{\iota_k}(x, y)) \leqslant \inf_{1 \leqslant k \leqslant n} a_k$$

is evidently contained in V.

For each finite subset H of I, let g_H denote the upper envelope of the family $(f_\iota)_{\iota \in H}$. As H runs through the set of all finite subsets of I and a runs through the set of real numbers > 0, the sets $\overset{-1}{g_H}([0, a])$ form a *fundamental system of entourages* of the uniformity $\mathfrak{U}$. Now the g_H are *pseudometrics* on X (no. 1), and the upper envelope of a finite number of functions of the family (g_H) belongs to this family, by definition; we express this property by saying that the family of pseudometrics (g_H) is *saturated.* The family of pseudometrics (g_H) is therefore *equivalent* to the family (f_ι), and is said to be the family of pseudometrics obtained by *saturating* (f_ι). From what has just been said it follows that we may always restrict ourselves to considering uniformities defined by *saturated* families of pseudometrics.

In the particular case where I is a *finite* set, this argument shows that the uniformity defined by the family of pseudometrics $(f_\iota)_{\iota \in I}$ is also defined by the *single* pseudometric $g = \sup_{\iota \in I} f_\iota$.

Let $\mathfrak{U}$, $\mathfrak{U}'$ be two uniformities on X, defined respectively by two *saturated* families $(f_\iota)_{\iota \in I}$, $(g_\varkappa)_{\varkappa \in K}$. Then $\mathfrak{U}$ is *coarser* than $\mathfrak{U}'$ if and only if, for each index $\iota \in I$ and each real number $a > 0$, there is an index $\varkappa \in K$ and a number $b > 0$ such that the relation $g_\varkappa(x, y) \leqslant b$ implies $f_\iota(x, y) \leqslant a$.

Example of a uniformity defined by a family of pseudometrics. Let $(f_\iota)_{\iota \in I}$ be an arbitrary family of (finite) *real-valued functions* defined on a set X. Let $\mathfrak{U}$ be the coarsest uniformity on X with respect to which the f_ι are uniformly continuous (Chapter II, § 2, no. 3). Then it follows from the definition of the entourages of $\mathfrak{U}$ (*loc. cit.*) that $\mathfrak{U}$ is the uniformity defined on X by the pseudometrics

$$g_\iota(x, y) = |f_\iota(x) - f_\iota(y)|.$$

3. PROPERTIES OF UNIFORMITIES DEFINED BY FAMILIES OF PSEUDOMETRICS

Let $\mathfrak{U}$ be a uniformity defined on a set X by a family of finite pseudometrics (f_ι). If we endow $X \times X$ with the uniformity which is the product of $\mathfrak{U}$ by itself, then each of the real-valued functions f_ι is *uniformly continuous* on $X \times X$; for by (1) we have

$$|f_\iota(x, y) - f_\iota(x', y')| \leqslant f_\iota(x, x') + f_\iota(y, y'),$$

and therefore the relations $f_\iota(x, x') \leqslant \varepsilon/2$, $f_\iota(y, y') \leqslant \varepsilon/2$ imply

$$|f_\iota(x, y) - f_\iota(x', y')| \leqslant \varepsilon.$$

For $\mathfrak{U}$ to be *Hausdorff* it is necessary and sufficient, from the definition of the entourages of $\mathfrak{U}$, that for each pair of *distinct* points x, y of X there is an index ι such that $f_\iota(x, y) \neq 0$.

In particular, if $\mathfrak{U}$ is defined by a *single* pseudometric f, then $\mathfrak{U}$ is Hausdorff if and only if the relation $f(x, y) = 0$ implies $x = y$ (cf. § 2). If $\mathfrak{U}$ is not Hausdorff, the intersection of all the entourages of $\mathfrak{U}$ is the subset of $X \times X$ consisting of pairs (x, y) such that $f_\iota(x, y) = 0$ for all ι; this subset is the graph of an equivalence relation R on X, and the Hausdorff uniformity associated with $\mathfrak{U}$ is defined on X/R (cf. Chapter II, § 3, no. 8). It is then easily verified that the functions f_ι are compatible (in x and in y) with the relation R (*Set Theory*, R, § 5, no. 7) and that the functions $\bar{f}_\iota$, obtained from f_ι by passing to the quotient (with respect to x and y), are pseudometrics on X/R which define the Hausdorff uniformity associated with $\mathfrak{U}$ (cf. § 2, no. 1).

If A is a non-empty subset of X, the restriction to $A \times A$ of a pseudometric on X is clearly a pseudometric on A. The uniformity *induced* by $\mathfrak{U}$ on A is clearly that defined by the family of restrictions to $A \times A$ of the pseudometrics f_ι.

Let us now look at the *completion* of the uniform space X when $\mathfrak{U}$ is Hausdorff.

PROPOSITION 1. *Let* X *be a Hausdorff uniform space whose uniformity* $\mathfrak{U}$ *is defined by a family of finite pseudometrics* (f_ι), *and let* $\hat{X}$ *be the completion of* X. *Then the functions* f_ι *can be extended by continuity to* $\hat{X} \times \hat{X}$; *the extended functions* $\bar{f}_\iota$ *are finite pseudometrics on* $\hat{X}\times\hat{X}$, *and the family* $(\bar{f}_\iota)$ *defines the uniformity of* $\hat{X}$.

First, the f_ι can be extended by continuity to $\hat{X}\times\hat{X}$, because they are uniformly continuous on $X\times X$; and the extended functions $\bar{f}_\iota$ are uniformly continuous on $\hat{X}\times\hat{X}$ (Chapter II, § 3, no. 6, Theorem 2); moreover, they are pseudometrics on $\hat{X}$ by virtue of the principle of extension of inequalities (Chapter IV, § 5, no. 2, Theorem 1). Let $\mathfrak{U}_1$ denote the uniformity on $\hat{X}$ obtained by completion, and let $\mathfrak{U}_2$ denote the uniformity defined by the family of pseudometrics $(\bar{f}_\iota)$. Then $\mathfrak{U}_2$ is *coarser* than $\mathfrak{U}_1$; for each $\bar{f}_\iota$ is uniformly continuous on $\hat{X}\times\hat{X}$ with respect to $\mathfrak{U}_1$, and hence for each $a > 0$ there exists an entourage V of $\mathfrak{U}_1$ such that, whenever $(x, y) \in V$, we have $|\bar{f}_\iota(x, y) - \bar{f}_\iota(x, x)| \leqslant a$, that is [since $\bar{f}_\iota(x, x) = 0$], $V \subset \overset{-1}{\bar{f}_\iota}([0, a])$; hence every entourage of $\mathfrak{U}_2$ is an entourage of $\mathfrak{U}_1$. On the other hand, $\mathfrak{U}_1$ and $\mathfrak{U}_2$ induce the *same* uniformity $\mathfrak{U}$ on X. As $\hat{X}$ is complete with respect to $\mathfrak{U}_1$, it follows that $\mathfrak{U}_1$ and $\mathfrak{U}_2$ are identical (Chapter II, § 3, no. 7, Proposition 14).

4. CONSTRUCTION OF A FAMILY OF PSEUDOMETRICS DEFINING A UNIFORMITY

The significance of defining a uniformity by means of a family of pseudometrics lies in the fact that *all uniformities can be so obtained.* Namely:

THEOREM 1. *Given a uniformity* $\mathfrak{U}$ *on a set* X, *there is a family of pseudometrics on* X *such that the uniformity defined by this family is identical with* $\mathfrak{U}$.

For each entourage V of the uniformity $\mathfrak{U}$, define inductively a sequence of symmetric entourages (U_n) such that $U_1 \subset V$ and $\overset{3}{U}_{n+1} \subset U_n$ for all $n \geqslant 1$. The sequence (U_n) is a fundamental system of entourages of a uniformity $\mathfrak{U}_V$ coarser than $\mathfrak{U}$; moreover, it is clear that $\mathfrak{U}$ is the *least upper bound* of all the uniformities U_V as V runs through the filter of entourages of $\mathfrak{U}$. Theorem 1 is therefore a consequence of the following proposition:

PROPOSITION 2. *If a uniformity* $\mathfrak{U}$ *on* X *has a countable fundamental system of entourages, then there is a pseudometric* f *on* X *such that* $\mathfrak{U}$ *is identical with the uniformity defined by* f.

Let (V_n) be a countable fundamental system of entourages of $\mathfrak{U}$. Define inductively a sequence (U_n) of symmetric entourages of $\mathfrak{U}$ such that $U_1 \subset V_1$ and

$$\overset{3}{U}_{n+1} \subset U_n \cap V_n \quad \text{for} \quad n \geqslant 1.$$

Clearly (U_n) is another fundamental system of entourages of $\mathfrak{U}$, and we have in particular $\overset{3}{U}_{n+1} \subset U_n$ for $n \geqslant 1$. We define a real-valued function g on $X \times X$ as follows: $g(x, y) = 0$ if $(x, y) \in U_n$ for all n; $g(x, y) = 2^{-k}$ if $(x, y) \in U_n$ for $1 \leqslant n \leqslant k$, but $(x, y) \notin U_{k+1}$; $g(x, y) = 1$ if $(x, y) \notin U_1$. The function g is symmetric and positive, and we have $g(x, x) = 0$ for all $x \in X$. Put

$$f(x, y) = \inf \sum_{i=0}^{p-1} g(z_i, z_{i+1}),$$

the greatest lower bound being taken over the set of all finite sequences $(z_i)_{0 \leqslant i \leqslant p}$ (p arbitrary) such that $z_0 = x$ and $z_p = y$. We shall show that f is a *pseudometric* which satisfies the inequalities

$$\text{(2)} \qquad \frac{1}{2} g(x, y) \leqslant f(x, y) \leqslant g(x, y).$$

It follows immediately from the definition that f is symmetric and positive and satisfies the triangle inequality. Also it is clear that $f(x, y) \leqslant g(x, y)$, hence $f(x, x) = 0$ for all $x \in X$, and therefore f is a pseudometric. To prove the left-hand half of the inequalities (2), let us show by induction on p that, for every finite sequence $(z_i)_{0 \leqslant i \leqslant p}$ of $p + 1$ points of X such that $z_0 = x$ and $z_p = y$, we have

$$\text{(3)} \qquad \sum_{i=0}^{p-1} g(z_i, z_{i+1}) \geqslant \frac{1}{2} g(x, y).$$

This is clear if $p = 1$. Put $a = \sum_{i=0}^{p-1} g(z_i, z_{i+1})$; the inequality (3) is true if $a \geqslant 1/2$, because $g(x, y) \leqslant 1$. Suppose then that $a < 1/2$, and let h be the largest of the indices q such that

$$\sum_{i<q} g(z_i, z_{i+1}) \leqslant \frac{a}{2};$$

we have then $\sum_{i<h} g(z_i, z_{i+1}) \leqslant a/2$ and $\sum_{i<h+1} g(z_i, z_{i+1}) > a/2$, whence

$$\sum_{i>h} g(z_i, z_{i+1}) \leqslant \frac{a}{2}.$$

By the inductive hypothesis we have $g(x, z_h) \leqslant a$ and $g(z_{h+1}, y) \leqslant a$; on the other hand it is clear that $g(z_h, z_{h+1}) \leqslant a$. Let k be the smallest integer > 0 such that $2^{-k} \leqslant a$; then $k \geqslant 2$, and $(x, z_h) \in U_k$, $(z_h, z_{h+1}) \in U_k$, $(z_{h+1}, y) \in U_k$ by the definition of g; hence $(x, y) \in \overset{3}{U}_k \subset U_{k-1}$, which implies that $g(x, y) \leqslant 2^{1-k} \leqslant 2a$.

Hence the inequalities (2) are proved; they show that, for each $a > 0$, the set $\overset{-1}{f}([0, a])$ contains U_k for each index k such that $2^{-k} < a$, and conversely that each U_k contains the set $\overset{-1}{f}([0, 2^{-k-1}])$; hence the sets $\overset{-1}{f}([0, a])$ form a fundamental system of entourages of the structure $\mathcal{U}$.

Q.E.D.

Remark. A uniformity $\mathcal{U}$ on X is defined by the family Φ of *all pseudometrics* on X which are *uniformly continuous* on $X \times X$. For clearly the uniformity defined by the family Φ is *coarser* than $\mathcal{U}$; conversely, Theorem 1 shows that there is a subfamily of Φ which defines the uniformity $\mathcal{U}$ and therefore the uniformity defined by Φ is *finer* than $\mathcal{U}$.

5. UNIFORMIZABLE SPACES

In Chapter II, § 4, no. 1, we posed the problem of characterizing uniformizable topological spaces. The solution is given by the following theorem:

THEOREM 2. *A topological space* X *is uniformizable if and only if it satisfies the following axiom:*

(O_{IV}) *Given any point* $x_0 \in X$ *and any neighbourhood* V *of* x_0, *there exists a continuous real-valued function on* X *which takes its values in* $[0,1]$, *is equal to* 0 *at* x_0, *and is equal to* 1 *on* $\complement V$.

The condition is *necessary*. For if there is a uniformity on X compatible with the topology of X, then by Theorem 1 this uniformity can be defined by a family (f_ι) of pseudometrics on X, and we may assume with no loss of generality that this family is *saturated* (no. 2). From the definition of the entourages of the uniformity defined by such a family of pseudometrics, there is a pseudometric f_α of the family (f_ι), and a number $a > 0$, such that $f_\alpha(x_0, x) \geqslant a$ for all $x \in \complement V$. It follows that the function $g(x) = \inf\left(1, \frac{1}{a} f_\alpha(x_0, x)\right)$ satisfies all the conditions laid down in (O_{IV}).

The condition is *sufficient*. For let Φ be the set of all *continuous* mappings of X into [0, 1]. Axiom (O_{IV}) shows that *the coarsest uniformity with respect to which all the functions belonging to* Φ *are uniformly continuous is compatible* with the topology of X (Chapter II, § 2, no. 3).

DEFINITION 4. *A topological space is said to be completely regular if it is uniformizable and Hausdorff.*

Equivalently, in view of Theorem 2, a space is completely regular if it satisfies axioms (H) [cf. Chapter I, § 8, no. 1, Proposition 1] and (O_{IV}).

Remark. Axiom (O_{IV}) implies (O_{III}) (cf. Chapter I, § 8, no. 4), for if V is a neighbourhood of x_0 and if f is a continuous real-valued function on X with values in [0, 1], such that $f(x_0) = 0$, $f(x) = 1$ for all $x \in \complement V$, then the set $\overset{-1}{f}([0, 1/2])$ is a *closed* neighbourhood of x_0 contained in V. In particular, every *completely regular* space is *regular* (which justifies the terminology). But there are examples of regular spaces which are not completely regular (*), so that (O_{III}) does not imply (O_{IV}).

Every compact space is completely regular (Chapter II, § 4, no. 1, Theorem 1) and therefore so is every subspace of a compact space. We can now complete this proposition by proving its *converse*:

PROPOSITION 3. *A topological space* X *is completely regular if and only if it is homeomorphic to a subspace of a compact space.*

Consider the coarsest uniformity on X with respect to which all continuous mappings of X into [0, 1] are uniformly continuous; we have already used this uniformity in the proof of Theorem 2, where we saw that it is compatible with the topology of X if X is uniformizable. Furthermore, this uniformity is a structure of a *precompact* space, by the compactness of the interval [0, 1] and Proposition 3 of Chapter II, § 4, no. 2. If X is Hausdorff, the completion of X with respect to this is therefore compact, and the proposition is proved.

We can therefore say that a completely regular space can be *embedded* in a compact space. It is often convenient to present this result in the following way:

In general, a *cube* is a topological space K^I, the product of a family of topological spaces each identical with a *compact interval* K of **R**, indexed by an arbitrary set I. If I is finite and has n elements, we recover the notion of an *n-dimensional closed cube*, which was defined in Chapter VI, § 1, no. 1. A cube is a *compact* space (Chapter I, § 9, no. 5, Theorem 3).

PROPOSITION 4. *If a topological space* X *is completely regular, it is homeomorphic to a subspace of a cube.*

Let $(f_\iota)_{\iota \in I}$ denote the family of all continuous mappings of X into $K = [0, 1]$, and let g denote the mapping $x \to (f_\iota(x))$ of X into

(*) See A. TYCHONOFF, *Math. Ann.*, **102**, (1930), p. 553.

K^I. If x, y are any two distinct points of X, it follows from axioms (H) and (O_{IV}) that there is an index ι such that $f_\iota(x) \neq f_\iota(y)$, and therefore g is a *one-to-one* mapping of X into K^I. Moreover, it is immediate that g is an isomorphism of the coarsest uniformity on X for which all the f_ι are uniformly continuous, onto the uniformity induced on $g(X)$ by the product uniformity of K^I; *a fortiori*, g is a homeomorphism of X onto $g(X)$.

6. SEMI-CONTINUOUS FUNCTIONS ON A UNIFORMIZABLE SPACE

In Chapter IV, § 6, no. 2, Corollary to Theorem 4, we showed that the upper envelope of a family of continuous real-valued functions on a topological space is a lower semi-continuous function. If the space is *uniformizable*, there is a converse to this proposition:

PROPOSITION 5. *In order that every lower semi-continuous real-valued function* f (finite or not) *on a topological space* X *should be the upper envelope of the continuous real-valued functions on* X (finite or not) *which are* $\leqslant f$, *it is necessary and sufficient that* X *be uniformizable.*

The condition is *necessary*. Let x_0 be any point of X and let V be any open neighbourhood of x_0; then the characteristic function φ_V of the set V is lower semi-continuous (Chapter IV, § 6, no. 2, Corollary to Proposition 1); by hypothesis, there is therefore a continuous real-valued function g on X such that $g \leqslant \varphi_V$ and $g(x_0) = a > 0$. The continuous function $\inf\left(1, \frac{1}{a}g^+\right)$ takes its values in $[0, 1]$, is equal to 0 in $\complement V$, and equal to 1 at x_0. Hence (Theorem 2) X is uniformizable.

The condition is *sufficient*. Consider first the case in which f takes its values in $[-1, +1]$. We have to show that, for each $x_0 \in X$ and each number $a < f(x_0)$, there is a continuous real-valued function g on X such that $g \leqslant f$ and $g(x_0) \geqslant a$. If $a \leqslant -1$, we may take g to be the constant -1. If $-1 < a < f(x_0)$, there is a neighbourhood V of x_0 such that $f(x) \geqslant a$ for all $x \in V$. Since X is uniformizable, there is a continuous real-valued function h on X, with values in $[0, 1]$, such that $h(x_0) = 0$ and $h(x) = 1$ for all $x \in \complement V$. We may then take $g(x) = a - (a + 1)h(x)$, and we have a continuous function satisfying the stated conditions. Note that this function takes its values in $[-1, +1]$.

The general case follows by transfer of structure; for there is a strictly increasing homeomorphism of $[-1, +1]$ onto $\overline{\mathbf{R}}$ (Chapter IV,

§ 4, no. 2, Proposition 2), and the definition of a semi-continuous function involves only the order structure and the topology of $\overline{\mathbf{R}}$.

Remark. In the above proof we see that the function g does not take the value $+1$. By transfer of structure it follows that every lower semi-continuous real-valued function f on the uniformizable space X is the upper envelope of the continuous real-valued functions $g \leqslant f$ on X *which do not take the value* $+\infty$.

2. METRIC SPACES AND METRIZABLE SPACES

1. METRICS AND METRIC SPACES

DEFINITION 1. *A metric on a set* X *is a finite pseudometric* d *on* X *such that the relation* $d(x, y) = 0$ *implies* $x = y$. *A metric space is a set* X *endowed with the structure defined by a given metric on* X.

A metric space X is always considered as carrying the uniformity and the topology defined by the given metric on X.

Examples. 1) The Euclidean distance $d(x, y)$ (Chapter VI, § 2, no. 1) is a metric on real n-dimensional number space $\mathbf{R}^n$; so are the functions

$$\sup_{1 \leqslant i \leqslant n} |x_i - y_i| \quad \text{and} \quad \sum_{i=1}^{n} |x_i - y_i|.$$

All these metrices are *equivalent* (§ 1, no. 2).

2) On any set X the pseudometric d, defined by the relations $d(x, x) = 0$ and $d(x, y) = 1$ if $x \neq y$, is a metric. The uniformity it defines on X is the *discrete* uniformity.

We have a definition equivalent to Definition 1 if we say that a metric is a *finite* pseudometric such that the uniformity defined by this pseudometric is *Hausdorff*; a finite pseudometric which is equivalent to a metric is therefore a metric.

Uniform spaces defined by a *single pseudometric* (which we may assume to be *finite*) can be reduced to metric spaces when the pseudometric is not a metric. Let f be such a pseudometric on a set X, and let $\mathfrak{U}$ be the uniformity defined by f; $\mathfrak{U}$ is not Hausdorff, and the intersection of the entourages of $\mathfrak{U}$ is the subset of $X \times X$ defined by the equivalence relation $f(x, y) = 0$. Let R denote this relation. If $x \equiv x' \pmod{R}$, then by the triangle inequality we have

$$f(x, y) \leqslant f(x, x') + f(x', y) = f(x', y)$$

and similarly $f(x', y) \leqslant f(x, y)$, so that $f(x, y) = f(x', y)$; in other words, f is a function *compatible* (in x and y) with the equivalence relation R (*Set Theory*, R, § 5, no. 7). Let $\bar{f}$ be the function induced by f on the quotient set; $\bar{f}$ is defined on $(X/R) \times (X/R)$, and if x and y are any two points of X and if $\dot{x}$ and $\dot{y}$ denote the equivalence classes (mod R) of x and y respectively, then we have $\bar{f}(\dot{x}, \dot{y}) = f(x, y)$. It follows immediately that $\bar{f}$ is a *metric* on X/R; it is called the metric *associated* with the pseudometric f; furthermore, the uniformity it defines on X/R is precisely the Hausdorff uniformity *associated* with $\mathfrak{U}$ by the definition of this uniformity (Chapter II, § 3, no. 8, Remark). Thus, by passing to a suitable quotient space, the uniform structure defined by a single pseudometric can be reduced to the structure of a metric space.

Proposition 1 of § 1, no. 3 determines the structure of the *completion* of a metric space:

PROPOSITION 1. *Let* X *be a metric space and let* d *be its metric. If* $\hat{X}$ *is the completion of* X (with respect to the uniformity defined by d), *the function* d *can be extended by continuity to* $\hat{X} \times \hat{X}$; *the extended function* $\bar{d}$ *is a metric on* $\hat{X}$, *and the uniformity of* $\hat{X}$ *coincides with that defined by the metric* $\bar{d}$.

Proposition 1 of § 1, no. 3 shows that $\bar{d}$ is a finite pseudometric on $\hat{X}$, and that the uniformity defined by $\bar{d}$ on $\hat{X}$ is the uniformity obtained by completion; since this latter uniformity is Hausdorff, $\bar{d}$ is a *metric*. Whenever we consider the completion of a metric space X as a metric space, it is always to be understood that the metric on $\hat{X}$ is that obtained by extending the metric on X by continuity.

2. STRUCTURE OF A METRIC SPACE

Let X and X′ be two metric spaces, d the metric on X, d' the metric on X′. In accordance with the general definitions (*Set Theory*, R, § 8, no. 5) a one-to-one mapping f of X onto X′ is an *isomorphism* of the metric space structure of X onto that of X′ if

$$d(x, y) = d'(f(x), f(y)) \tag{1}$$

for all $x \in X$ and all $y \in X$.

Note that if f is a mapping of X onto X′ which satisfies the identity (1), then f must be *bijective* and therefore an isomorphism of X onto X′; such an isomorphism is also called an *isometry* (or an *isometric* mapping) of X onto X′.

An isometry of X onto X′ is of course an isomorphism of the uniformity (resp. topology) of X onto the uniformity (resp. topology) of X′; the converses of these statements are false, as is shown by the existence of distinct equivalent metrics (§ 1, no. 2).

Let X be a metric space, d the metric on X. For each $a > 0$ let V_a denote the subset of $X \times X$ consisting of all pairs (x, y) such that $d(x, y) < a$, and let W_a denote the subset of $X \times X$ consisting of all pairs (x, y) such that $d(x, y) \leqslant a$. As a runs through the set of all real numbers > 0 (or merely a sequence of numbers > 0 which tends to 0), the sets V_a (resp. W_a) form a fundamental system of *open* (resp. *closed*) entourages of the uniformity of X, because of the continuity of d (§ 1, no. 3). We have $\overline{V}_a \subset W_a$, but these two sets are not necessarily the same.

By analogy with the case of the Euclidean distance on $\mathbf{R}^n$, the set $V_a(x)$ [resp. $W_a(x)$] is called the *open* (resp. *closed*) ball with *centre* x and *radius* a; it is an *open* (resp. *closed*) set in X. Again, the set of all $y \in X$ such that $d(x, y) = a$ is called the *sphere* with centre x and radius a; it is a *closed* set. From what has been said, the open (resp. closed) balls with centre x and radius a form a fundamental system of neighbourhoods of x as a runs through the set of all real numbers > 0, or a sequence of numbers > 0 which tends to 0.

The reader should beware of assuming that balls and spheres in an arbitrary metric space enjoy the same properties as the Euclidean balls and spheres studied in Chapter VI, § 2. Thus the closure of an open ball need not be the closed ball of the same centre and radius; the frontier of a closed ball need not be the sphere of the same centre and radius; an open (or closed) ball need not be connected; and a sphere can be empty (cf. Exercise 4).

Let A and B be any two non-empty subsets of a metric space X. The number

$$d(A, B) = \inf_{x \in A, y \in B} d(x, y)$$

is called the *distance between the sets* A *and* B. In particular we denote by $d(x, A)$ the distance between the set $\{x\}$ and the set A; this is called the *distance from the point x to the set* A. Thus

$$d(x, A) = \inf_{y \in A} d(x, y)$$

whence

$$d(A, B) = \inf_{x \in A} d(x, B)$$

(Chapter IV, § 5, no. 4, Proposition 9).

Remark. If $d(x, A) = a$, it can happen that there is *no* point of A whose distance from x is equal to a. However, this situation can never arise if A is compact, for then Weierstrass's theorem (Chapter IV, § 6, no. 1, Theorem 1) shows that there exists $y \in A$ such that $d(x, A) = d(x, y)$.

PROPOSITION 2. *The statements* $d(x, A) = 0$ *and* $x \in \overline{A}$ *are equivalent.*

For $d(x, A) = 0$ expresses the fact that the ball $V_a(x)$ meets A whatever the value of $a > 0$; and this is equivalent to $x \in \overline{A}$.

PROPOSITION 3. *The function* $d(x, A)$ *is uniformly continuous on* X.

Let x and y be any two points of X; then given any $\varepsilon > 0$ there exists $z \in A$ such that $d(y, z) \leqslant d(y, A) + \varepsilon$, and therefore

$$d(x, z) \leqslant d(x, y) + d(y, z) \leqslant d(x, y) + d(y, A) + \varepsilon$$

by the triangle inequality.

A fortiori $d(x, A) \leqslant d(x, y) + d(y, A) + \varepsilon$, and since ε is arbitrary it follows that $d(x, A) \leqslant d(x, y) + d(y, A)$. Similarly we have

$$d(y, A) \leqslant d(x, y) + d(x, A),$$

so that

$$|d(x, A) - d(y, A)| \leqslant d(x, y), \tag{2}$$

whence the result follows.

Remark. We can have $d(A, B) = 0$ for two subsets A, B of X such that $\overline{A} \cap \overline{B} = \emptyset$, provided that neither subset consists of a single point. For example, on the real line **R**, the set of integers > 0 and the set of points of the sequence $(n + 1/2n)_{n \geqslant 1}$ are two disjoint closed sets whose distance a part is zero.

However, if A is *compact* and B is *closed*, the relation $d(A, B) = 0$ implies $A \cap B \neq \emptyset$, for by virtue of the relation

$$d(A, B) = \inf_{x \in A} d(x, B)$$

it follows from Proposition 3 and Weierstrass's theorem that there exists $x_0 \in A$ such that $d(x_0, B) = d(A, B) = 0$ and hence (Proposition 2) $x_0 \in B$.

The *diameter* of a non-empty subset A of X is the number (finite or equal to $+\infty$)

$$\delta(A) = \sup_{x \in A, y \in A} d(x, y).$$

The notion of a "W_a-small set" (Chapter II, § 3, no. 1) is identical with that of a set of diameter $\leqslant a$. A non-empty set A consists of a single point if and only if $\delta(A) = 0$.

A subset A of X is *bounded* (with respect to the metric d) if its diameter is *finite*; equivalently, if for each point $x_0 \in X$, A is contained in a ball with centre x_0. Every subset of a bounded set is bounded, and the union of a finite family of bounded sets is a bounded set.

Note that a subset of X can be bounded with respect to a metric d but unbounded with respect to a metric equivalent to d (cf. § 1, no. 2).

3. OSCILLATION OF A FUNCTION

Related to the notion of diameter is that of the *oscillation* of a function f, defined on an arbitrary set X and taking its values in a *metric space* X'; if A is any non-empty subset of X, the diameter $\delta(f(A))$ is called the *oscillation of* f *in* A.

If moreover X is a subset of a *topological space* Y, and if $x \in \overline{X}$, the number

$$\omega(x; f) = \inf \delta(f(V \cap X))$$

(as V runs through the neighbourhood filter of x in Y) is called the *oscillation of* f *at* $x \in \overline{X}$.

PROPOSITION 4. *The oscillation* $\omega(x; f)$ *of an arbitrary function* f, *defined on a subset* X *of a topological space* Y *and taking its values in a metric space* X', *is upper semi-continuous on* $\overline{X}$.

Let a be any point of $\overline{X}$; then for each $k > \omega(a; f)$ there exists an open neighbourhood V of a such that $\delta(f(V \cap X)) \leqslant k$; for each $x \in V \cap \overline{X}$, V is a neighbourhood of x and therefore

$$\omega(x; f) \leqslant \delta(f(V \cap X)) \leqslant k,$$

which shows that ω is upper semi-continuous at the point a.

In order that $\omega(x; f) = 0$ at a point $x \in \overline{X}$ it is necessary and sufficient that for each $\varepsilon > 0$ there should exist a neighbourhood V of x such that $f(V \cap X)$ is contained in a ball of radius ε; if $x \in X$, this condition expresses the fact that f is *continuous* at the point x (with respect to X); if $x \in \overline{X} \cap \complement X$, the image under f of the trace on X of the neighbourhood filter of x in Y is a *Cauchy filter base* on X'; in particular:

PROPOSITION 5. *Let* f *be a function defined on a subset* X *of a topological space* Y, *taking its values in a* complete *metric space* X'. *Then* f *has a limit relative to* X *at a point* $x \in \overline{X}$ *if and only if the oscillation of* f *at* x *is zero.*

4. METRIZABLE UNIFORM SPACES

DEFINITION 2. *A metric on a set* X *is said to be compatible with a uniformity* $\mathfrak{U}$ *on* X *if the uniformity defined by the metric coincides with* $\mathfrak{U}$.

A uniformity on a set X *is said to be metrizable if there is a metric on* X *compatible with this uniformity. A uniform space is said to be metrizable if its uniformity is metrizable.*

Distinct metrics can be compatible with the same uniformity; they are then *equivalent* (§ 1, no. 2, Definition 2).

THEOREM 1. *A uniformity is metrizable if and only if it is Hausdorff and the filter of entourages of the uniformity has a countable base.*

The condition is *necessary*, for (with the notation of no. 2) the entourages $V_{1/n}$ $(n \geqslant 1)$ form a base of the filter of entourages of the uniformity of a metric space.

The condition is *sufficient*, for, if it is satisfied, the uniformity under consideration is defined by a single pseudometric, by Proposition 2 of § 1, no. 4; since the uniformity is Hausdorff, this pseudometric is a metric.

COROLLARY 1. *A Hausdorff uniformity defined by a countable family of pseudometrics is metrizable.*

For if (f_n) is a sequence of pseudometrics defining such a structure, the filter of entourages is generated by the countable family of sets $\overset{-1}{f}_n([0, 1/m])$, where m and n each run through the set of integers > 0.

COROLLARY 2. *Every countable product of metrizable uniform spaces is metrizable.*

For such a space is Hausdorff and its uniformity has a countable fundamental system of entourages (Chapter II, § 2, no. 6).

5. METRIZABLE TOPOLOGICAL SPACES

DEFINITION 3. *A metric on a set* X *is said to be compatible with a topology* $\mathfrak{T}$ *on* X *if the topology defined by this metric coincides with* $\mathfrak{T}$. *A topological space is said to be metrizable if there exists a metric on* X *compatible with the topology of* X.

Two metrics on a set X which are both compatible with the same topology $\mathfrak{T}$ can be *inequivalent*.

The subspace $\mathbf{R}_+^*$ of $\mathbf{R}$ provides an example of this. Both the uniformity induced by the additive uniformity of $\mathbf{R}$ and the uniformity induced by the multiplicative uniformity of $\mathbf{R}^*$ are metrizable and are compatible with the topology of $\mathbf{R}_+^*$; but they are not comparable.

We remark also that there can exist *non-metrizable* uniformities compatible with the topology of a *metrizable* topological space (Exercise 7).

We shall content ourselves here with *necessary* conditions for the metrizability of a topological space (for a necessary and sufficient condition, cf. § 4, Exercise 22). In the first place, a space cannot be metrizable unless it is *completely regular* (indeed we shall see, in § 4, no. 1, Proposition 2, that a metrizable space is necessarily "normal", which is a stronger condition). On the other hand, Theorem 1 shows that:

PROPOSITION 6. *Every point of a metrizable space has a countable fundamental system of neighbourhoods.*

More generally:

PROPOSITION 7. *In a metrizable space, every closed set is the intersection of a countable family of open sets, and every open set is the union of a countable family of closed sets.*

Let d be a metric compatible with the topology of a metrizable space X. If A is a closed subset of X, it is the intersection of the open sets $V_{1/n}(A)$ [the set of all $x \in X$ such that $d(x, A) < 1/n$; cf. Proposition 2]. The second part of the proposition follows by taking complements.

Remarks. 1) These necessary conditions are not sufficient (cf. Exercise 13).
2) There are spaces in which every point has a countable fundamental system of neighbourhoods but in which there exist closed sets which are not countable intersections of open sets (Exercise 15); such spaces are not metrizable.

Corollary 2 of Theorem 1, no. 4, shows that a *countable* product of metrizable topological spaces is metrizable. Also the *sum* X (Chapter I, § 2, no. 4) of an *arbitrary* family $(X_\iota)_{\iota \in I}$ of metrizable spaces is metrizable. For if d_ι is a metric compatible with the topology of X_ι for each $\iota \in I$, we may assume that d_ι is bounded and that the diameter of X_ι is $\leqslant 1$; we can then define a distance d compatible with the topology of X by putting $d(x, y) = d_\iota(x, y)$ if x and y both belong to the same X_ι, and $d(x, y) = 1$ otherwise.

6. USE OF COUNTABLE SEQUENCES

Proposition 6 is the origin of the part played by *countable sequences of points* in the theory of metrizable spaces; for many problems, they can be used to advantage in place of *filters*. This is because the *neighbourhood filters* of points of a metrizable space (and therefore also *convergent filters*) are *determined* by *convergent sequences* of points of the space: for since the neighbourhood filter of a point has a countable base, it is the intersection of the *elementary filters* finer than itself (Chapter I, § 6, no. 8, Proposition 11), i.e., of the elementary filters associated with sequences which converge to the point in question.

On the other hand, the notion of a convergent sequence is not adapted to the study of topological spaces in which there are points whose neighbourhood filter has no countable base. In particular, Hausdorff non-discrete topological spaces can be constructed in which, at every point x, the intersection of any countable family of neighbourhoods of x is again a neighbourhood of x (*); in such a space the only convergent sequences are those in which all the terms are equal from a certain index onwards.

As examples of the use of countable sequences we give the following propositions:

PROPOSITION 8. *In a metrizable space* X, *a point* x *lies in the closure of a non-empty subset* A *of* X *if and only if there is a sequence of points of* A *which converges to* x.

We know already, from Chapter I, § 7, no. 3, that the condition is *sufficient.* To see that it is *necessary*, consider a countable fundamental system (V_n) of neighbourhoods of x such that $V_{n+1} \subset V_n$ for each n. If x lies in the closure of A then each V_n meets A, and if x_n lies in $V_n \cap A$, the sequence (x_n) converges to x (**).

From Proposition 8 we deduce:

PROPOSITION 9. *A metric space* X *is complete if and only if every Cauchy sequence in* X *is convergent.*

Let $\hat{X}$ be the completion of X. If there is a point $x \in \hat{X}$ which does not belong to X, then there is a sequence (x_n) of points of X which converges to x, and this is a non-convergent Cauchy sequence in X.

PROPOSITION 10. *Let* X *be a metrizable space and let* f *be a mapping of* X *into a topological space* X'. *Then* f *is continuous at a point* $x \in X$ *if and only if, whenever* (x_n) *is a sequence of points of* X *which converges to* x, *the sequence* $(f(x_n))$ *converges to* $f(x)$ *in* X'.

The condition is necessary, from Chapter I, § 7, no. 4, Proposition 9, Corollary 1. To show that it is sufficient, consider the filter $\mathfrak{B}'$ of neighbourhoods of $f(a)$ in X'; the hypothesis implies that $\overset{-1}{f}(\mathfrak{B}')$ is coarser than every elementary filter associated with a sequence which converges to a, that is to say every elementary filter which converges to a; but

(*) See e.g., J. DIEUDONNÉ, *Notes de Tératopologie* (I), *Revue scientifique* (Revue rose), 1939, p. 39.

(**) This proposition can still be valid in certain spaces in which at least one point does not have any countable fundamental system of neighbourhoods; for example, the space obtained by compactifying an uncountable discrete space by adjoining a point at infinity (Chapter I, § 9, no. 8, Theorem 4).

the intersection of these filters is the neighbourhood filter of a (Chapter I, § 6, no. 8, Proposition 11). Hence the result.

Note that Propositions 8 and 10 are valid in any space X in which every point has a countable fundamental system of neighbourhoods.

7. SEMI-CONTINUOUS FUNCTIONS ON A METRIZABLE SPACE

PROPOSITION 11. *Let* X *be a metrizable space and let* f *be a lower semi-continuous function on* X *which takes its values in a closed interval* $[a, b]$ *of* $\overline{\mathbf{R}}$. *Then* f *is the upper envelope of an increasing sequence of continuous functions on* X *which take their values in* $[a, b]$.

By transfer of structure we may assume that $a = 0$ and $b = 1$.

(i) Suppose first that $f = \varphi_A$, where A is an open subset of X. Then the function g_n defined by $g_n(x) = n.\inf(d(x, X - A), 1/n)$ is continuous and $\geqslant 0$ on X; also $g_n(x) = f(x)$ when $x \in X - A$ and when

$$d(x, X - A) \geqslant 1/n.$$

It follows immediately that $f = \sup_n (g_n)$.

(ii) In the general case consider, for each integer $n \geqslant 1$, the finite decreasing sequence of open sets

$$A_k = \overset{-1}{f}\left(\left]\frac{k}{n}, +\infty\right[\right) \quad (0 \leqslant k \leqslant n-1);$$

the function $g_n = \frac{1}{n}\sum_{k=1}^{n-1} \varphi_{A_k}$ is lower semi-continuous, and we have $0 \leqslant f(x) - g_n(x) \leqslant 1/n$ for all n; hence f is the upper envelope of the sequence (g_n). On the other hand, g_n is a linear combination with positive coefficients of a finite number of characteristic functions of open sets and is therefore the upper envelope of a countable sequence $(h_{mn})_{m \geqslant 0}$ of continuous functions $\geqslant 0$, by (i); hence $f = \sup_{m,n} h_{mn}$. If we put $f_n = \sup_{p \leqslant n, q \leqslant n} h_{pq}$, we see that the sequence (f_n) is an increasing sequence of continuous functions $\geqslant 0$, with f as upper envelope, and which do not take the value 1, since $g_n \leqslant n - 1/n$.

8. METRIZABLE SPACES OF COUNTABLE TYPE

DEFINITION 4. *A metrizable space is said to be of countable type* (*or separable*) *if its topology has a countable base.*

Clearly every subspace of a metrizable space of countable type is again of countable type. The definition of the base of the topology of a product space (Chapter I, § 4, no. 1) and Corollary 2 to Theorem 1 of no. 4 show that the product of a *countable* family of metrizable spaces of countable type is a metrizable space of countable type. Again, the sum of a *countable* family of metrizable spaces of countable type is a metrizable space of countable type (no. 5).

PROPOSITION 12. *If* X *is a metrizable topological space, the following statements are equivalent:*

a) X *is of countable type.*

b) X *has a countable dense subset.*

c) X *is homeomorphic to a subspace of the cube* I^N, *where* I *is the interval* $[0, 1]$ *in* **R**.

From the preceding remarks it is clear that c) implies a); a) implies b), for if (U_n) is a countable base of the topology of X and a_n is a point of U_n, the a_n form a dense subset of X. Finally let us show that b) implies c). Let (a_n) be a dense sequence of points of X, and for each $x \in X$ let $\varphi(x)$ be the point $(d(x, a_n))_{n \in N}$ of I^N (d being a metric compatible with the topology of X, with respect to which the diameter of X is $\leqslant 1$). We shall show that φ is a homeomorphism of X onto a subspace of I^N. Indeed, φ is continuous, because each of the functions $x \to d(x, a_n)$ is continuous; and φ is injective, because each point of X is the limit of some subsequence of (a_n) (Proposition 8). Let B be a ball with centre x_0 and radius r in X, and let n be an integer such that

$$d(x_0, a_n) < \tfrac{1}{3} r.$$

The image under φ of the set W of points $x \in X$ such that

$$|d(x_0, a_n) - d(x, a_n)| < \tfrac{1}{3} r$$

is by definition a neighbourhood of $\varphi(x_0)$ in $\varphi(X)$. But for each $x \in W$ we have $d(x, a_n) < d(x_0, a_n) + \frac{1}{3} r < \frac{2}{3} r$, whence

$$d(x, x_0) \leqslant d(x_0, a_n) + d(x, a_n) < r,$$

which shows that W is a neighbourhood of x_0 contained in B, and therefore that φ is a homeomorphism of X onto $\varphi(X)$.

Note that for an arbitrary topological space X, property b) does not necessarily imply the existence of a countable base, even if X is compact and every point of X has a countable fundamental system of neighbourhoods (Exercise 13; cf. Chapter I, § 1, Exercise 7).

PROPOSITION 13. *Let* X *be a topological space which has a countable base* (U_n); *then for each open covering* $(V_\iota)_{\iota \in I}$ *of* X *there exists a countable subset* J *of* I *such that* $(V_\iota)_{\iota \in J}$ *is a covering of* X.

Let H be the subset of N consisting of indices n such that U_n is contained in at least one of the V_ι; the sequence $(U_n)_{n \in H}$ is a covering of X, because every point $x \in X$ belongs to some V_ι, and since V_ι is open, there is an index n such that $x \in U_n \subset V_\iota$. Hence there is a mapping ψ of H into I such that $U_n \subset V_{\psi(n)}$ for each $n \in H$; taking $J = \psi(H)$, which is countable, the proposition is proved.

9. COMPACT METRIC SPACES; COMPACT METRIZABLE SPACES

The criterion for precompactness of a uniform space (Chapter II, § 4, no. 2, Theorem 3) gives rise to the following proposition for metric spaces:

PROPOSITION 14. *A metric space X is precompact if and only if, for each $\varepsilon > 0$, there is a finite covering of X by sets of diameter $\leqslant \varepsilon$.*

If we adjoin the hypothesis that X is *complete* we have a criterion for *compactness* of metric spaces.

From Proposition 14 we get a *topological* criterion for compactness, applicable to metrizable spaces:

PROPOSITION 15. *A metrizable topological space X is compact if and only if every infinite sequence of points of X has a cluster point in X.*

Axiom (C) of Chapter I, § 9, no. 1 shows that the condition is *necessary*. To show *sufficiency*, let d be a metric compatible with the topology of X. We show first that the metric space X so defined is *complete*: every Cauchy sequence in X has a cluster point and is therefore convergent (Chapter II, § 3, no. 2, Proposition 5, Corollary 2); hence X is complete, by Proposition 9. Next we shall show that X is *precompact*; if this were not so, then by Proposition 14 there would exist a real number $\alpha > 0$ such that X could not be covered by any finite number of subsets of X of diameter $\leqslant \alpha$. We could then define by induction on n an infinite sequence (x_n) of points of X by the condition $d(x_p, x_n) > \frac{1}{2}\alpha$ for all $p < n$; and such a sequence can have no cluster point, since every ball of radius $< \frac{1}{2}\alpha$ contains at most one point of the sequence.

COROLLARY. *A subset A of a metrizable topological space X is relatively compact if and only if every infinite sequence of points of A has a cluster point in X.*

Let d be a metric compatible with the topology of X. We shall show that the space $\overline{A}$ is compact, by applying the criterion of Proposition 15. Let (x_n) be a sequence of points of $\overline{A}$; then for each index n there exists $y_n \in A$ such that $d(x_n, y_n) < 1/n$; the sequence (y_n) has, by hypothesis, a cluster point $a \in X$, and a is also a cluster point of the sequence (x_n), for if y_m lies in the ball with centre a and radius $1/n$, for some $m > n$, then x_m lies in the ball with centre a and radius $2/n$.

It should be remarked that Proposition 15 is not a consequence of the existence of a countable fundamental system of neighbourhoods at each point of X; there are examples of non-metrizable, non-compact spaces in which every point has a countable fundamental system of neighbourhoods and every infinite sequence of points has a cluster point (Exercise 15).

PROPOSITION 16. *A compact space* X *is metrizable if and only if its topology has a countable base.*

The condition is *necessary*. By Proposition 14, for each integer $n \geqslant 1$ there is a finite subset A_n of X such that the distance of A_n from every point of X is $\leqslant 1/n$; the countable set $A = \bigcup_n A_n$ is therefore dense in X, and the result follows from Proposition 12 of no. 8.

The condition is *sufficient*. Let (U_n) be a countable base of the topology of X. Every neighbourhood of a point of the diagonal Δ of $X \times X$ therefore contains a neighbourhood of the form $U_n \times U_n$; applying the Borel-Lebesgue axiom to the compact subset Δ of $X \times X$, it follows that every neighbourhood of Δ contains a finite union of sets of the form $U_n \times U_n$, which is a neighbourhood of Δ. Hence the neighbourhoods of Δ which are finite unions of sets of the form $U_n \times U_n$ form a fundamental system of entourages of the uniformity of X (Chapter II, § 4, no. 1, Theorem 1), and the result therefore follows from Theorem 1 of no. 4.

COROLLARY. *Let* X *be a locally compact space and let* X′ *be the compact space obtained by adjoining a point at infinity* ω *to* X (Chapter I, § 9, no. 8). *Then the following statements are equivalent:*

a) *The topology of* X *has a countable base.*

b) X′ *is metrizable.*

c) X *is metrizable and* σ*-compact.*

a)$\Longrightarrow$b): Let (U_n) be a countable base of the topology of X. Each neighbourhood of a point $x \in X$ contains a compact neighbourhood of x, which in turn contains a neighbourhood of x equal to some U_n. Hence the relatively compact U_n form a base of the topology of X, and we may therefore suppose that all the U_n are relatively compact. X is therefore a countable union of compact sets $\overline{U}_n$, i.e. it is σ-compact; this implies that in X′ the point ω has a countable fundamental system (V_n) of open neighbourhoods (Chapter I, § 9, no. 9, Proposition 15, Corollary 2). Hence each neighbourhood of a point $y \in X'$ contains either one of the U_n or one of the V_n, which is a neighbourhood of y, and so the U_n and the V_n form a countable base of the topology of X'. Hence X' is metrizable, by Proposition 16.

b) $\Longrightarrow$ c): If X' is metrizable, then ω has a countable fundamental system of neighbourhoods, and therefore X is σ-compact by Chapter I, § 9, no. 9, Proposition 15, Corollary 2.

c) $\Longrightarrow$ a): By hypothesis, there is an increasing sequence (V_n) of relatively compact open sets which cover X and are such that $\overline{V}_n \subset V_{n+1}$ (Chapter I, § 9, no. 9, Proposition 15). The subspace $\overline{V}_n$ is compact and metrizable and therefore has a countable base (Proposition 16), and therefore so does V_n. Let $(U_{mn})_{m \geqslant 1}$ be a base of the topology of V_n. For each $x \in X$ and each neighbourhood W of x, there exists n such that $x \in V_n$, hence there exists m such that $x \in U_{mn} \subset V_n \cap W$. Hence the sets U_{mn} $(m \geqslant 1, n \geqslant 1)$ form a base of the topology of X.

10. QUOTIENT SPACES OF METRIZABLE SPACES

If X is a metrizable space and R is an equivalence relation on X, the quotient space X/R is not necessarily metrizable (even if X is locally compact * and X/R is normal $_*$). However:

Proposition 17. *Every Hausdorff quotient space of a compact metrizable space is compact and metrizable.*

Equivalently, if f is a continuous mapping of a compact metrizable space X into a Hausdorff space X', then $f(X)$ *is a metrizable subspace of* X' (Chapter I, § 9, no. 4, Theorem 2, Corollary 4).

Let X be a compact metrizable space, and let R be an equivalence relation on X such that X/R is Hausdorff. Then X/R is compact (Chapter I, § 9, no. 4, Theorem 2), hence by Proposition 16 it is enough to show that the topology of X/R has a countable base. To do this, we use the facts that R is *closed* (Chapter I, § 10, no. 4, Proposition 8) and that the classes mod R are compact. Let φ be the canonical mapping of X onto X/R, and let (U_n) be a countable base of the topology of X. Let z be any point of X/R and let V be a neighbourhood of z in X/R; then $\overset{-1}{\varphi}(V)$ is a neighbourhood in X of the compact set $\overset{-1}{\varphi}(z)$. If x is any point of $\overset{-1}{\varphi}(z)$, there is a set U_n containing x and contained in $\overset{-1}{\varphi}(V)$, and therefore by the Borel-Lebesgue axiom there is a finite open covering $(U_{n_k})_{1 \leqslant k \leqslant r}$ of $\overset{-1}{\varphi}(z)$ such that, if W denotes $\bigcup_k U_{n_k}$, W is a neighbourhood of $\overset{-1}{\varphi}(z)$ contained in $\overset{-1}{\varphi}(V)$. Since R is closed, it follows that $\varphi(W)$ is a neighbourhood of z in X/R, contained in V (Chapter I, § 5, no. 4, Proposition 10). Let $\mathfrak{B}$ denote the set of interiors of sets of the form $\varphi(W)$, where W runs through the set $\mathfrak{F}$

of all finite unions of sets of the form U_n; we have then shown that $\mathfrak{B}$ is a base of the topology of X/R, and since $\mathfrak{F}$ is countable, so is $\mathfrak{B}$.

PROPOSITION 18. *Let* X *be a complete metric space, let* R *be an open equivalence relation on* X *such that* X/R *is Hausdorff, and let* $\varphi : X \to X/R$ *be the canonical mapping. Then if* K *is any compact subset of* X/R, *there is a compact subset* K′ *of* X *such that* $\varphi(K') = K$.

Let $\mathfrak{B}_1$ be the set of all open balls of radius 1/2 in X. As B runs through $\mathfrak{B}_1$, the sets $\varphi(B)$ form an open covering of K, and therefore there exists a finite number of points $x_1, \ldots, x_m$ of X such that the images under φ of the open balls with radius 1/2 and centre x_i $(1 \leqslant i \leqslant m)$ form an open covering of K. Let $H_1 = \{x_1, \ldots, x_m\}$ and suppose that we have defined a finite set H_i, for $1 < i \leqslant n$, such that:

(i) $H_i \subset H_{i+1}$ and each point of H_{i+1} is at a distance $< 1/2^i$ from H_i, for $1 \leqslant i \leqslant n-1$;

(ii) the images under φ of the open balls of radius $1/2^i$, with centres at the points of H_i, form an open covering of K, for $1 \leqslant i \leqslant n$.

Let $\mathfrak{B}_{n+1}$ be the set of all open balls of radius $1/2^{n+1}$ whose centre x is such that $d(x, H_n) < 1/2^n$ (d being the metric on X). The properties of H_n show that the sets $\varphi(B)$, for $B \in \mathfrak{B}_{n+1}$, form an open covering of K; hence there is a finite set $L_{n+1} \subset X$ such that the images under φ of the open balls of radius $1/2^{n+1}$ whose centre belongs to L_{n+1} form an open covering of K. Taking $H_{n+1} = H_n \cup L_{n+1}$, we see that we can define inductively an infinite sequence (H_n) of finite subsets of X with properties (i) and (ii) above. Let $H = \bigcup_n H_n$, and let us show that H is *precompact*. For each $p > 0$ and each point $z_{n+p} \in H_{n+p}$, there exists a sequence of points $z_{n+i} \in H_{n+i}$ $(0 \leqslant i \leqslant p-1)$ such that

$$d(z_{n+i}, z_{n+i+1}) < 1/2^{n+i} \quad \text{for} \quad 0 \leqslant i \leqslant p-1;$$

it follows that $d(z_n, z_{n+p}) \leqslant \sum_{i=0}^{p-1} 1/2^{n+i} \leqslant 1/2^{n-1}$, and consequently $d(y, H_n) \leqslant 1/2^{n-1}$ for all $y \in H$, which proves our assertion. Since X is complete, $\overline{H}$ is compact, hence $\varphi(\overline{H})$ is compact. Next, let us show that $K \subset \varphi(\overline{H})$. If $z \in K$, then by definition $d(\overline{H}_n, \overset{-1}{\varphi}(z)) \leqslant 1/2^n$ for all n, and therefore $d(\overline{H}, \overset{-1}{\varphi}(z)) = 0$; but $\overset{-1}{\varphi}(z)$ is closed and $\overline{H}$ is compact, so that this implies $\overline{H} \cap \overset{-1}{\varphi}(z) \neq \emptyset$ (no. 2, Remark following Proposition 3); hence the assertion. Thus if $K' = \overline{H} \cap \overset{-1}{\varphi}(K)$, then K′ is closed in $\overline{H}$ and therefore compact, and from what has been proved we have $\varphi(K') = K$. Q.E.D.

3. METRIZABLE GROUPS, VALUED FIELDS, NORMED SPACES AND ALGEBRAS

1. METRIZABLE TOPOLOGICAL GROUPS

PROPOSITION 1. *The left and right uniformities of a topological group* G *are metrizable if and only if* G *is Hausdorff and the identity element* e *of* G *has a countable fundamental system of neighbourhoods.*

The condition is clearly necessary. Conversely, if it is satisfied, let (V_n) be a fundamental system of neighbourhoods of e; if U_n denotes the set of pairs $(x, y) \in G \times G$ such that $x^{-1}y \in V_n$, then the U_n form a countable fundamental system of entourages of the left uniformity of G; since this uniformity is Hausdorff, it follows from § 2, no. 4, Theorem 1 that it is metrizable. Similarly for the right uniformity of G.

A topological group G is said to be *metrizable* if its topology is metrizable. Proposition 1 then shows that its two uniformities are metrizable.

This result can be sharpened with the help of the following notion:

DEFINITION 1. *A metric* d *on a group* G *(written multiplicatively) is said to be left-invariant* (resp. *right-invariant*) *if we have*

$$d(zx, zy) = d(x, y) \quad [\text{resp. } d(xz, yz) = d(x, y)]$$

for all x, y, z *in* G.

PROPOSITION 2. *The left* (resp. *right*) *uniformity of a metrizable group* G *can be defined by a left-invariant* (resp. *right-invariant*) *metric on* G.

Suppose that the fundamental system (V_n) of neighbourhoods of e consists of symmetric neighbourhoods such that $V_{n+1}^3 \subset V_n$ for each n. Then the corresponding entourages U_n of the left uniformity are symmetric entourages such that $\overset{3}{U}_{n+1} \subset U_n$. The method used in the proof of Proposition 2 of § 1, no. 4 allows us to construct, from the sequence of entourages (U_n), a metric d on G compatible with the left uniformity of G; and since for each $z \in G$ the mapping $(x, y) \to (zx, zy)$ leaves each of the U_n invariant, the definition of d shows that it is a left-invariant metric. This method also holds for the right uniformity.

Note that, if the two uniformities of G are distinct, the metric d is not right-invariant, and hence in general $d(x^{-1}, y^{-1}) \neq d(x, y)$.

In particular, if G is a metrizable *abelian* group, its uniformity is defined by an invariant metric d; if G is written additively, we have

$$d(x, y) = d(0, y - x) = d(0, x - y).$$

We shall often write $|x|$ (or $\|x\|$) for $d(0, x)$; we have then

$$d(x, y) = |x - y|.$$

The function $|x|$ satisfies the following three conditions:

a) $|-x| = |x|$ for all $x \in G$.

b) $|x + y| \leqslant |x| + |y|$ for all $x \in G$ and all $y \in G$.

c) $|x| = 0$ if and only if $x = 0$.

Conversely:

PROPOSITION 3. *Let* G *be an abelian group, written additively, and let* $x \to |x|$ *be a mapping* $G \to \mathbf{R}_+$ *which satisfies conditions a), b) and c) above. Then the function* $d(x, y) = |x - y|$ *is an invariant metric on* G; *the topology* $\mathfrak{T}$ *which it defines on* G *is compatible with the group structure of* G, *and the uniformity defined by* d *is the same as the uniformity of the topological group obtained by endowing* G *with the topology* $\mathfrak{T}$.

The function $d(x, y)$ is a metric on G, for the relation $d(x, y) = 0$ is equivalent to $x = y$ by *c*); we have $d(x,y) = d(y, x)$ by *a*); and

$$d(x, y) = |(x - z) + (z - y)| \leqslant |x - z| + |z - y| = d(x, z) + d(z, y)$$

by *b*). Moreover, d is invariant, since $(x + z) - (y + z) = x - y$. For each real number $\alpha > 0$, let V_α be the set of all $x \in G$ such that $|x| < \alpha$; then the V_α form a fundamental system $\mathfrak{S}$ of neighbourhoods of 0 for the topology $\mathfrak{T}$, and since d is invariant, $a + \mathfrak{S}$ is a fundamental system of neighbourhoods of a for the topology $\mathfrak{T}$, for each $a \in G$. By *a*), the V_α are symmetric, and by *b*), we have $V_\alpha + V_\alpha \subset V_{2\alpha}$; hence the topology $\mathfrak{T}$ is compatible with the group structure of G (Chapter III, § 1, no. 2). The last part of the proposition follows immediately.

Conditions *a*), *b*) and *c*) are equivalent to *c*) together with the condition

b') $$|x - y| \leqslant |x| + |y|.$$

For *a*) and *b*) clearly imply *b'*); conversely, taking $x = 0$ in *b'*) and using *c*), we see that $|-y| \leqslant |y|$; replacing y by $-y$ it follows that $|-y| = |y|$, which is *a*); replacing y by $-y$ in *b'*) we then get *b*).

PROPOSITION 4. *If* G *is a metrizable group, then every Hausdorff quotient group* G/H *of* G *is metrizable. If* G *is also complete, then* G/H *is complete* (*).

The first part of the proposition is a consequence of the fact that the identity element of G/H has a countable fundamental system of neighbourhoods in G/H; for if (V_n) is a fundamental system of neighbourhoods of e in G, then the canonical images $\dot{V}_n$ of the sets V_n in G/H form a fundamental system of neighbourhoods of the identity element of G/H (Chapter III, § 2, no. 6, Proposition 17).

To show that G/H is complete if G is complete, it is enough, by § 2, no. 6, Proposition 9, to show that every Cauchy sequence $(\dot{x}_n)$ (with respect to the left uniformity of G/H) is convergent. We may assume, by passing to a subsequence of $(\dot{x}_n)$ if necessary, that for each pair of indices p, q such that $p \geqslant n$ and $q \geqslant n$, we have $\dot{x}_p^{-1}\dot{x}_q \in \dot{V}_n$; this means that for each pair of points $y \in \dot{x}_p$, $z \in \dot{x}_q$, we have $y^{-1}z \in HV_n = V_nH$; and therefore, for each $y \in \dot{x}_p$, the intersection of $\dot{x}_q$ and the neighbourhood yV_n of y is not empty. Suppose then that the sequence (V_n) has been chosen so that $V_{n+1}^2 \subset V_n$, and define inductively a sequence (x_n) of points of G, such that $x_n \in \dot{x}_n$ and $x_{n+1} \in x_nV_n$; this is possible by what has been said. It follows then by induction that for each $p > 0$ we have $x_{n+p} \in x_nV_nV_{n+1}\ldots V_{n+p-1} \subset x_nV_{n-1}$. The sequence (x_n) is therefore a Cauchy sequence in G, so it converges to a point a; and it follows immediately that the canonical image $\dot{a}$ of a in G/H is the limit of the sequence $(\dot{x}_n)$.

COROLLARY 1. *Let* G *be a complete metrizable group, let* G_0 *be a dense subgroup of* G *and let* H_0 *be a closed normal subgroup of* G_0. *If* H *is the closure of* H_0 *in* G, *the quotient group* G_0/H_0 *has a completion isomorphic to* G/H.

H is a normal subgroup of G (Chapter III, § 2, no. 3, Proposition 8) and Proposition 4 shows that G/H is complete. Also if φ is the canonical mapping of G onto G/H, it is clear that $\varphi(G_0)$ is dense in G/H. The result therefore follows from Chapter III, § 2, no. 7, Proposition 21.

Let G, G′ be two Hausdorff abelian topological groups, and $\hat{G}$, $\hat{G}'$ their respective completions. We recall (Chapter III, § 3, no. 3, Proposition 5) that if u is a continuous homomorphism of G into G′, then u is uniformly continuous and extends uniquely to a continuous homomorphism of $\hat{G}$ into $\hat{G}'$, which we shall denote by $\hat{u}$ in the remainder

(*) There exist non-metrizable complete groups G containing a closed subgroup H such that G/H is not complete.

of this subsection. The diagram

$$\begin{array}{ccc} G & \xrightarrow{u} & G' \\ {\scriptstyle i}\downarrow & & \downarrow{\scriptstyle i'} \\ \hat{G} & \xrightarrow{\hat{u}} & \hat{G}' \end{array}$$

(in which i, i' are the canonical injections) is commutative. If v is a continuous homomorphism of G' into a Hausdorff abelian topological group G'', and if $w = v \circ u$, then it is follows immediately that $\hat{w} = \hat{v} \circ \hat{u}$.

Let H be a closed subgroup of G and let $E = G/H$. Let $j : H \to G$ and $p : G \to E$ be the canonical mappings. Let $\overline{H}$ be the closure of H in $\hat{G}$; $\overline{H}$ is a complete group and we identify it with the completion $\hat{H}$ of H. The continuous extension $\hat{j}$ of j to $\hat{H}$ is evidently the canonical injection of $\hat{H}$ in $\hat{G}$.

Suppose from now on that G *is metrizable.* Then the canonical mapping of $E = G/H$ into $\hat{G}/\hat{H}$ is a topological isomorphism of E onto a dense subgroup of the complete group $\hat{G}/\hat{H}$ (Corollary 1), and thus we may identify $\hat{G}/\hat{H}$ with $\hat{E}$ and the continuous extension $\hat{p}$ of p to $\hat{G}$ with the canonical mapping of $\hat{G}$ onto $\hat{G}/\hat{H}$.

COROLLARY 2. *Let* G, G' *be two metrizable abelian topological groups; let* $u : G \to G'$ *be a strict morphism with kernel* N *and image* P. *Then* $\hat{u} : \hat{G} \to \hat{G}'$ *is a strict morphism with kernel* $\hat{N}$ *and image* $\hat{P}$.

Let $u = j \circ v \circ p$ be the canonical factorization of u, where v is an isomorphism of the topological group G/N onto the topological group $u(G) = P$. We have $\hat{u} = \hat{j} \circ \hat{v} \circ \hat{p}$, and we have seen that $\hat{p}$ is the canonical mapping of $\hat{G}$ onto $\hat{G}/\hat{N}$, and that $\hat{j}$ is the canonical mapping of $\hat{P}$ into $\hat{G}'$. On the other hand, $\hat{v}$ is an isomorphism of $\hat{G}/\hat{N}$ onto $\hat{P}$ (Chapter III, § 3, no. 4, Proposition 5), whence the result.

COROLLARY 3. *Let* G, G', G'' *be three metrizable abelian topological groups and let* $u : G \to G'$ *and* $v : G' \to G''$ *be two strict morphisms such that the sequence* $G \xrightarrow{u} G' \xrightarrow{v} G''$ *is exact* [i.e., $u(G) = \overset{-1}{v}(0)$]. *Then the sequence* $\hat{G} \xrightarrow{\hat{u}} \hat{G}' \xrightarrow{\hat{v}} \hat{G}''$ *is exact.*

For if we put $N = u(G) = \overset{-1}{v}(0)$, it follows from Corollary 2 that $\hat{N}$ is both the image of $\hat{u}$ and the kernel of $\hat{v}$.

Remarks. 1) Let G be a *non-Hausdorff* topological group such that the Hausdorff group associated with G is metrizable; equivalently, such that the identity element of G has a countable fundamental system of neighbourhoods in G. The proof of Proposition 4 applies to this case without modification, H being an arbitrary normal subgroup of G, and shows that the Hausdorff group associated with G/H is metrizable, and that G/H is a complete group (in general not Hausdorff) whenever G is complete.

2) Let d be a left-invariant metric which defines the topology of a metrizable group G, and let H be a closed normal subgroup of G. If $\dot{x}$ and $\dot{y}$ are any two points of G/H, consider the distance $d(\dot{x}, \dot{y})$ of the two closed subsets $\dot{x}, \dot{y}$ in G (§ 2, no. 2); we shall see that this function is a *left-invariant metric* on G/H and defines the topology of this quotient group.

Notice first that if $x \in \dot{x}$ and $y \in \dot{y}$ we have $d(\dot{x}, \dot{y}) = d(x, \mathrm{H}y)$; for $d(x, \mathrm{H}y) = \inf_{h \in \mathrm{H}} d(x, hy)$, and therefore $d(h'x, \mathrm{H}y) = d(x, \mathrm{H}y)$ for all $h' \in \mathrm{H}$, since d is left-invariant; this proves the assertion (§ 2, no. 2). Hence for each $\dot{z} \in \mathrm{G/H}$ we have [§ 2, no. 2, formula (2)]

$$|d(\dot{x}, \dot{z}) - d(\dot{y}, \dot{z})| = |d(x, \dot{z}) - d(y, \dot{z})| \leqslant d(x, y);$$

and since this inequality is valid for all $x \in \dot{x}$ and all $y \in \dot{y}$, we have $|d(\dot{x}, \dot{z}) - d(\dot{y}, \dot{z})| \leqslant d(\dot{x}, \dot{y})$, which shows that $d(\dot{x}, \dot{y})$ is a metric on G/H. Moreover, for any $z \in \dot{z}$, we have

$$d(\dot{z}\dot{x}, \dot{z}\dot{y}) = \inf_{h \in \mathrm{H}} d(zx, hzy)$$

from above; but since $hzy = z(z^{-1}hz)y$ and since $z^{-1}hz$ runs through H as h runs through H (H being normal), the left-invariance of $d(x, y)$ shows that we have $d(zx, \mathrm{H}zy) = d(x, \mathrm{H}y) = d(\dot{x}, \dot{y})$. Finally, if V is a neighbourhood of e in G defined by $d(e, x) < \alpha$, the image V of V in G/H is the set defined by $d(\dot{e}, \dot{x}) < \alpha$; this completes the proof.

2. VALUED DIVISION RINGS

DEFINITION 2. *An absolute value on a division ring* K *is a mapping* $x \to |x|$ *of* K *into* $\mathbf{R}_+$ *which satisfies the following conditions:*

($\mathrm{VM_I}$) $|x| = 0$ *if and only if* $x = 0$.

($\mathrm{VM_{II}}$) $|xy| = |x| . |y|$ *for all* x, y *in* K.

($\mathrm{VM_{III}}$) $|x + y| \leqslant |x| + |y|$ *for all* x, y *in* K.

By ($\mathrm{VM_{II}}$) we have $|x| = |1| . |x|$, and since by ($\mathrm{VM_I}$) there is at least one x such that $|x| \neq 0$, we have $|1| = 1$; it follows that $1 = |-1|^2$, hence $|-1| = 1$ and consequently

$$|-x| = |-1|\,|x| = |x|;$$

therefore $|x - y| \leqslant |x| + |y|$ for all x, y in K. We can therefore say that $d(x, y) = |x - y|$ is an *invariant metric* on the additive group K, and that the mapping $x \to |x|$ is a *homomorphism* of the multiplicative group K^* of non-zero elements of K into the multiplicative group $\mathbf{R}_+^*$ of real numbers > 0.

The invariant metric $|x-y|$ defines a metric space topology on K, compatible with its additive group structure (no. 1, Proposition 3); but, moreover, this topology is compatible with the *division ring structure* of K. For the continuity of xy on $K \times K$ follows from the relation

$$xy - x_0y_0 = (x-x_0)(y-y_0) + (x-x_0)y_0 + x_0(y-y_0),$$

which gives

$$|xy - x_0y_0| \leqslant |x-x_0|.|y-y_0| + |x_0|.|y-y_0| + |y_0|.|x-x_0|.$$

Likewise, the continuity of x^{-1} at every point $x_0 \neq 0$ follows from the identity $x^{-1} - x_0^{-1} = x^{-1}(x_0 - x)x_0^{-1}$, which gives, by $(\mathrm{VM_{II}})$,

$$|x^{-1} - x_0^{-1}| = \frac{|x-x_0|}{|x_0|.|x|};$$

now if $\varepsilon > 0$ is such that $\varepsilon < |x_0|$, the relation $|x-x_0| \leqslant \varepsilon$ implies $|x| \geqslant |x_0| - \varepsilon$, whence $|x^{-1} - x_0^{-1}| \leqslant \varepsilon / |x_0|(|x_0| - \varepsilon)$; and this establishes the continuity of x^{-1} at the point x_0.

Definition 3. *A valued division ring is a division ring* K *endowed with the structure defined by a given absolute value on* K.

A valued division ring will always be considered as endowed with the topology defined by its absolute value, which makes it a *topological* division ring. If K_0 is a division subring of a valued division ring K, the restriction to K_0 of the absolute value on K is an absolute value on K_0, which defines on K_0 the topology induced by the topology of K.

Examples. 1) Let K be an arbitrary division ring. For each $x \in K$, put $|x| = 1$ if $x \neq 0$, and $|0| = 0$. The mapping $x \to |x|$ so defined is an absolute value on K, called the *improper* absolute value. The topology defined by an absolute value $|x|$ on a division ring K is *discrete* if and only if $|x|$ is the improper absolute value. This condition is clearly sufficient; conversely, if the topology of K is discrete, $|x|$ can take no value $\alpha > 0$ other than 1; for if we had $|x_0| = \alpha < 1$, the sequence (x_0^n) would consist of non-zero terms and would converge to 0; to deal with the case $\alpha > 1$, consider x_0^{-1} in place of x_0.

2) The absolute value of a real number (Chapter IV, § 1, no. 6) satisfies axioms $(\mathrm{VM_I})$, $(\mathrm{VM_{II}})$ and $(\mathrm{VM_{III}})$, and the topology it defines on the field $\mathbf{R}$ is the topology of the real line. On the field $\mathbf{C}$ of complex numbers (identified with $\mathbf{R}^2$) [resp. the division ring $\mathbf{H}$ of quaternions (identified with $\mathbf{R}^4$)] the Euclidean norm is again an absolute value and

defines the topology of the field **C** (resp. the division ring **H**) (Chapter VIII, § 1, nos. 2 and 4).

3) On a division ring K, a *real valuation* is a function v defined on K^* with values in **R** which satisfies the following conditions: *a*) if $x \in K^*$ and $y \in K^*$, then $v(xy) = v(x) + v(y)$; *b*) if in addition $x + y \neq 0$, then $v(x+y) \geqslant \inf(v(x), v(y))$. If a is any real number > 1, we can then define an *absolute value* on K by putting $|x| = a^{-v(x)}$ for $x \neq 0$, and $|0| = 0$. For the relation $v(xy) = v(x) + v(y)$ for $x \neq 0$ and $y \neq 0$ implies the relation $|xy| = |x|.|y|$ for these values of x and y, and this relation is trivially true if one of x, y is zero; likewise, from the relation $v(x+y) \geqslant \inf(v(x), v(y))$ for $x \neq 0, y \neq 0$ and $x + y \neq 0$ we deduce $|x+y| \leqslant \sup(|x|, |y|) \leqslant |x| + |y|$, and these inequalities are still satisfied if one of $x, y, x+y$ is zero. In particular, if $v_p(x)$ is the p-adic valuation on the field **Q** of rational numbers (the exponent of p in the decomposition of x into a product of prime factors), then the corresponding absolute value $|x|_p = p^{-v_p(x)}$ is called the *p-adic absolute value* on the field **Q** (cf. Chapter III, § 6, Exercise 23).

Remark. If x is a root of unity in a valued division ring, then $|x| = 1$, for $x^n = 1$ implies $|x|^n = 1$ and hence $|x| = 1$. In particular, the only absolute value on a *finite* field is the *improper* absolute value, since every element $\neq 0$ of such a field is a root of unity.

DEFINITION 4. *Two absolute values on a division ring* K *are said to be equivalent if they define the same topology on* K.

PROPOSITION 5. *Two absolute values* $|x|_1$, $|x|_2$ *on a division ring* K, *neither of which is the improper absolute value, are equivalent if and only if the relation* $|x|_1 < 1$ *implies* $|x|_2 < 1$. *There exists then a real number* $\rho > 0$ *such that* $|x|_2 = |x|_1^\rho$ *for all* $x \in K$.

The condition is necessary, for the set of all $x \in K$ such that $|x|_1 < 1$ is the same as the set of all x such that, with respect to the topology defined by the absolute value $|x|_1$, $\lim_{n\to\infty} x^n = 0$.

Suppose conversely that $|x|_1 < 1 \Longrightarrow |x|_2 < 1$. Then $|x|_1 > 1 \Longrightarrow |x|_2 > 1$, because $|x^{-1}|_1 < 1$ and therefore $|x^{-1}|_2 < 1$. Since by hypothesis the absolute value $|x|_1$ is not improper, there exists $x_0 \in K$ such that $|x_0|_1 > 1$. Let $a = |x_0|_1$, $b = |x_0|_2$ and let $\rho = \log b/\log a > 0$. Let $x \in K^*$ and put $|x|_1 = |x_0|_1^\gamma$. If m and n are integers such that $n > 0$ and $m/n > \gamma$, then $|x|_1 < |x_0|_1^{m/n}$, and therefore $|x^n x_0^{-m}|_1 < 1$; hence $|x^n x_0^{-m}|_2 < 1$, $|x|_2 < |x_0|_2^{m/n}$. Similarly, if $m/n < \gamma$, we see that $|x|_2 > |x_0|_2^{m/n}$; it follows therefore that $|x|_2 = |x_0|_2^\gamma$; in other words

$$\log |x|_2 = \gamma \log b = \gamma\rho \log a = \rho \log |x|_1,$$

i.e., $|x|_2 = |x|_1^\rho$. It is now clear that the neighbourhoods of zero for the topologies defined on K by $|x|_1$ and $|x|_2$ are identical.

Conversely, if $|x|$ is any absolute value on K, the function $|x|^\rho$ is an absolute value on K (equivalent to $|x|$) for all ρ such that $0 < \rho \leqslant 1$. We have only to verify the inequality $|x+y|^\rho \leqslant |x|^\rho + |y|^\rho$; and since $|x+y|^\rho \leqslant (|x| + |y|)^\rho$ it is enough to show that, if $a > 0$ and $b > 0$, we have $(a+b)^\rho \leqslant a^\rho + b^\rho$ for any ρ such that $0 < \rho \leqslant 1$. If we put $c = a/(a+b)$ and $d = b/(a+b)$, we have $c + d = 1$, and the inequality to be proved is $c^\rho + d^\rho \geqslant 1$; but this follows immediately from the relations $c^\rho \geqslant c$ and $d^\rho \geqslant d$, which are valid since $0 < c \leqslant 1$, $0 < d \leqslant 1$, $0 < \rho \leqslant 1$.

Hence the set of values of $r > 0$ such that $|x|^r$ is an absolute value is a finite or infinite interval of **R** with left-hand end-point 0; if it is finite, it is evidently closed; for if we have $|x+y|^r \leqslant |x|^r + |y|^r$ for any x, y in K and all r such that $0 < r < r_0$, then by continuity the inequality is still valid for $r = r_0$. If $|x|^r$ is an absolute value for *all* $r > 0$, then we have

$$|x+y| \leqslant (|x|^r + |y|^r)^{1/r}$$

for all x and y in K and all $r > 0$. Now, if a, b are two real numbers $\geqslant 0$, we have $\lim_{r\to\infty} (a^r + b^r)^{1/c} = \sup(a, b)$; for, supposing for example that $a \geqslant b$, we have $a \leqslant (a^r + b^r)^{1/r} \leqslant 2^{1/r} a$, and the result follows by letting $r \to +\infty$.

Thus, if $|x|^r$ is an absolute value for all $r > 0$, we have

$$|x+y| \leqslant \sup(|x|, |y|)$$

which can be expressed by saying that $v(x) = -\log|x|$ $(x \neq 0)$ is a *valuation* on K.

Remark. The proof of Proposition 5 shows that, if the topology defined by $|x|_2$ is *coarser* than that defined by $|x|_1$, and if $|x|_1$ is not *improper*, then $|x|_1$ and $|x|_2$ are *equivalent*, for the relation $|x|_2 < 1$ then implies $|x|_2 < 1$. Thus the topologies defined by two absolute values on K, neither of which is improper, cannot be *comparable* without being *identical*.

PROPOSITION 6. *The completion* $\hat{K}$ *of a division ring* K *endowed with an absolute value* $|x|$ *is a division ring, and the function* $|x|$ *can be extended by continuity to an absolute value on* $\hat{K}$, *which defines the topology of* $\hat{K}$.

Let $\mathfrak{F}$ be a Cauchy filter on K (with respect to the additive uniformity) which does not have 0 as a cluster point; to show that $\hat{K}$ is a division ring it is enough to establish that the image of $\mathfrak{F}$ under the mapping $x \to x^{-1}$ is a Cauchy filter base (Chapter III, § 6, no. 8, Proposition 7). Now, by hypothesis there exists a real number $\alpha > 0$ and a set $A \in \mathfrak{F}$ such that $|x| \geqslant \alpha$ for all $x \in A$; on the other hand, for each $\varepsilon > 0$ there exists a set $B \in \mathfrak{F}$ such that $B \subset A$ and $|x - y| \leqslant \varepsilon$ for all $x \in B$

and $y \in B$; hence

$$|x^{-1} - y^{-1}| = \frac{|x - y|}{|x| \cdot |y|} \leqslant \frac{\varepsilon}{\alpha^2},$$

and the first part of the proposition follows. The invariant metric $|x - y| = d(x, y)$ extends by continuity to a metric on $\hat{K}$ (§ 2, no. 1, Proposition 1) which defines the topology of $\hat{K}$ and is invariant by the principle of extension of identities; we continue to denote this invariant metric by $d(x, y)$. If we put $|x| = d(0, x)$ for $x \in \hat{K}$, it is clear that $|x|$ is the extension by continuity of the function $|x|$ on K and is therefore an absolute value on $\hat{K}$ by the principle of extension of identities.

3. NORMED SPACES OVER A VALUED DIVISION RING

DEFINITION 5. *If* E *is a (left) vector space over a non-discrete valued division ring* K, *a norm on* E *is a mapping* $\mathbf{x} \to p(\mathbf{x})$ *of* E *into* $\mathbf{R}_+$ *which satisfies the following axioms:*

(NO_I) $p(\mathbf{x}) = 0$ *if and only if* $\mathbf{x} = 0$;

(NO_{II}) $p(\mathbf{x} + \mathbf{y}) \leqslant p(\mathbf{x}) + p(\mathbf{y})$ *for all* $\mathbf{x}, \mathbf{y}$ *in* E;

(NO_{III}) $p(t\mathbf{x}) = |t| p(\mathbf{x})$ *for all* $t \in K$ *and all* $\mathbf{x} \in E$.

The normed spaces most frequently met with have either **R** or **C** as field of scalars (with the usual absolute value).

From (NO_{III}) it follows in particular that $p(-\mathbf{x}) = p(\mathbf{x})$; hence if we put $d(\mathbf{x}, \mathbf{y}) = p(\mathbf{x} - \mathbf{y})$, d is an *invariant metric* on the additive group E, and defines a metric space topology compatible with the additive group structure of E (no. 1, Proposition 3); moreover, the mapping

$$(t, \mathbf{x}) \to t\mathbf{x}$$

is *continuous* on $K \times E$; for we have

$$t\mathbf{x} - t_0\mathbf{x}_0 = (t - t_0)(\mathbf{x} - \mathbf{x}_0) + (t - t_0)\mathbf{x} + t_0(\mathbf{x} - \mathbf{x}_0)$$

and therefore

$$p(t\mathbf{x} - t_0\mathbf{x}_0) \leqslant |t - t_0| p(\mathbf{x} - \mathbf{x}_0) + |t - t_0| p(\mathbf{x}) + |t_0| p(\mathbf{x} - \mathbf{x}_0),$$

which shows that the left-hand side can be made as small as we please by taking $|t - t_0|$ and $p(\mathbf{x} - \mathbf{x}_0)$ sufficiently small.

DEFINITION 6. *If* K *is a non-discrete valued division ring, a vector space* E *over* K, *endowed with the structure defined by a given norm on* E, *is called a normed space over* K.

A normed space will always be considered as endowed with the topology and the uniformity defined by its norm.

Examples. 1) On a non-discrete valued division ring K, considered as a (left or right) vector space over itself, the absolute value $|x|$ is a norm.

2) The expression $\|x\| = \sqrt{\sum_{i=1}^{n} x_i^2}$, which we have called the *Euclidean norm* on the space $\mathbf{R}^n$ (Chapter VI, § 2) is evidently a norm in the sense of Definition 5. So are the functions $\sup_{1 \leqslant i \leqslant n} |x_i|$ and $\sum_{i=1}^{n} |x_i|$.

3) Let $\mathscr{B}(E; K)$ be the set of all functions f on a set E which take their values in a non-discrete valued division ring K and are such that the real-valued function $x \to |f(x)|$ is *bounded* on E. This set is clearly a vector subspace of the (left or right) vector space K^E of all mappings of E into K. If we put $p(f) = \sup_{x \in E} |f(x)|$, then p is a *norm* on the vector space $\mathscr{B}(E; K)$ (cf. Chapter X, § 1).

* 4) On the vector space $\mathscr{C}(I; \mathbf{R})$ of all finite continuous real-valued functions defined on the interval $I = [0, 1]$ of $\mathbf{R}$, the function

$$p(x) = \int_0^1 |x(t)|\, dt$$

is a norm. *

In a normed space E, the (closed) ball B with centre 0 and radius 1, that is to say the set of all $x \in E$ such that $p(x) \leqslant 1$, will be called the *unit ball* in E. Let us show that a fundamental system of neighbourhoods of 0 in E is formed by the transforms of the unit ball by the *homotheties* $x \to tx$, where t runs through the set of non-zero elements of K. The image of B under this homothety is the closed ball with centre 0 and radius $|t|$; hence it is enough to show that for each real number $r > 0$ there exists $t \in K$ such that $0 < |t| < r$. Now since the absolute value of K is not improper, there exists $t_0 \in K$ such that $0 < |t_0| < 1$; it is therefore enough to take $t = t_0^n$, where n is a sufficiently large integer, in order that $|t| = |t_0|^n < r$.

DEFINITION 7. *Two norms on a vector space* E (over a non-discrete valued division ring K) *are said to be equivalent if they define the same topology on* E.

PROPOSITION 7. *Two norms p, q on a vector space* E *are equivalent if and only if there exist two numbers $a > 0$, $b > 0$ such that*

$$a.p(x) \leqslant q(x) \leqslant b.p(x) \tag{1}$$

for all $x \in E$.

These inequalities are *sufficient*, for it follows from the relation $a.p(x) \leqslant q(x)$ that, for each $r > 0$, the closed ball with centre 0 and radius ar (relative to the norm q) is contained in the closed ball with centre 0 and radius r (relative to the norm p); hence the topology defined by q is *finer* than the topology defined by p. Similarly the inequality $q(x) \leqslant b.p(x)$ shows that the topology defined by p is finer than that defined by q, and hence p and q are equivalent.

Let us now show that the inequalities (1) are *necessary*. If the topology defined by q is finer than the topology defined by p, then the unit ball with respect to p contains a closed ball with centre 0 and radius $\alpha > 0$ with respect to q; i.e., the relation $q(x) \leqslant \alpha$ implies $p(x) \leqslant 1$. If $t_0 \in K$ is such that $0 < |t_0| < 1$, then for each $x \neq 0$ in E there is a unique rational integer k such that $\alpha|t_0| < q(t_0^k x) \leqslant \alpha$; therefore $p(t_0^k x) \leqslant 1$, so that

$$p(x) \leqslant \frac{1}{|t_0|^k} \leqslant \frac{1}{\alpha|t_0|} q(x);$$

putting $a = \alpha|t_0|$ we have therefore $a.p(x) \leqslant q(x)$ for all $x \neq 0$, and this inequality is also valid when $x = 0$. Similarly we show that if the topology defined by p is finer than that defined by q, there exists $b > 0$ such that $q(x) \leqslant b.p(x)$.

Example. In the space $\mathbf{R}^n$, the three norms

$$\sqrt{\sum_{i=1}^{n} x_i^2}, \qquad \sup_{1 \leqslant i \leqslant n} |x_i| \qquad \text{and} \qquad \sum_{i=1}^{n} |x_i|$$

are equivalent, because we have

$$\sup_{1 \leqslant i \leqslant n} |x_i| \leqslant \sqrt{\sum_{i=1}^{n} x_i^2} \leqslant \sum_{i=1}^{n} |x|_i \leqslant n . \sup_{1 \leqslant i \leqslant n} |x_i|. \tag{2}$$

PROPOSITION 8. *Let* E *be a normed space over a non-discrete valued division ring, let p be the norm on* E, *and let* $\hat{E}$ *be the additive topological group which is the completion of the additive group* E. *Then the function $(t, x) \rightarrow tx$ can be extended by continuity to* $\hat{K} \times \hat{E}$ *and defines on* $\hat{E}$ *a vector space structure over* $\hat{K}$; *the norm p can be extended by continuity to a norm $\bar{p}$ on* $\hat{E}$ *which defines the topology of* $\hat{E}$.

The extension of tx by continuity is a particular case of the theorem of extension of a continuous bilinear mapping of a product of two abelian groups into a third (Chapter III, § 6, no. 5, Theorem 1); we have $1 . x = x$ and $t(ux) = (tu)x$ for $t \in \hat{K}$, $u \in \hat{K}$ and $x \in \hat{E}$, by the principle of extension of identities; hence the external law $(t, x) \to tx$ indeed defines on $\hat{E}$ a structure of a vector space over $\hat{K}$. On the other hand, the invariant metric $\bar{d}(x, y) = \bar{p}(x - y)$ extends to an invariant metric $\bar{d}$ on $\hat{E}$ (§ 2, no. 1, Proposition 1) which defines the topology of $\hat{E}$; if we set $\bar{p}(x) = \bar{d}(0, x)$, then $\bar{p}$ is the extension of p by continuity, and satisfies axioms (NO_I) and (NO_{II}); by virtue of the continuity of tx on $\hat{K} \times \hat{E}$, $\bar{p}$ also satisfies (NO_{III}) (principle of extension of identities) and is therefore a *norm* on $\hat{E}$.

When we have to consider a definite normed space structure on a vector space E, we shall usually denote the norm of a vector x by $\|x\|$, unless this notation is likely to lead to confusion.

4. QUOTIENT SPACES AND PRODUCT SPACES OF NORMED SPACES

PROPOSITION 9. *Let* E *be a normed space over a non-discrete valued division ring* K, *and let* H *be a closed vector subspace of* E. *If, for each class* $\dot{x} \in E/H$, *we put* $\|\dot{x}\| = \inf_{x \in \dot{x}} \|x\|$, *the function* $\|\dot{x}\|$ *is a norm on the vector space* E/H, *and the topology defined by this norm is the quotient by* H *of the topology of* E.

By Remark 2 of no. 1, $d(\dot{x}, \dot{y}) = \|\dot{x} - \dot{y}\|$ is an invariant metric on E/H which defines the quotient by H of the topology of E. It remains only to show that $\|t\dot{x}\| = |t| . \|\dot{x}\|$, and this follows immediately from the definition of $\|\dot{x}\|$ [Chapter IV, § 5, no. 7, formula (23)].

The norm $\|\dot{x}\|$ may also be interpreted as follows: it is the *distance* (in E) *of every point* $x \in \dot{x}$ *from the subspace* H, for the points of $\dot{x}$ are the points $x - z$, where z runs through H.

PROPOSITION 10. Let $(E_i)_{1 \leqslant i \leqslant n}$ *be a finite family of normed spaces over a non-discrete valued division ring* K, *and let* $E = \prod_{i=1}^{n} E_i$ *be the product vector space. If for each* $x = (x_i) \in E$ *we put* $\|x\| = \sup_{1 \leqslant i \leqslant n} \|x_i\|$, *then the function* $\|x\|$ *is a norm on* E, *and the topology it defines on* E *is the product of the topologies of the* E_i.

For if $x = (x_i)$ and $y = (y_i)$, we have $x + y = (x_i + y_i)$ and therefore

$$\|x + y\| = \sup_i \|x_i + y_i\| \leqslant \sup_i (\|x_i\| + \|y_i\|)$$
$$\leqslant \sup_i \|x_i\| + \sup_i \|y_i\| = \|x\| + \|y\|.$$

On the other hand, it is clear that $\|tx\| = |t|.\|x\|$, and that if $\|x\| = 0$, then $\|x_i\| = 0$ and therefore $x_i = 0$ for $1 \leqslant i \leqslant n$, so that $x = 0$; hence $\|x\|$ is a norm on E. Also the relation $\|x\| < a$ is equivalent to the n relations $\|x_i\| < a$, and therefore the norm $\|x\|$ defines the product topology on E.

Similarly we can show that the functions $\sum_{i=1}^{n} \|x_i\|$ and $\sqrt{\sum_{i=1}^{n} \|x_i\|^2}$ are norms on E; the inequalities (2) show that all three norms are *equivalent*.

In particular, in the (left or right) vector space K^n, if we put

$$p_1(x) = \sup_i |x_i|, \qquad p_2(x) = \sum_{i=1}^{i=1} |x_i|, \qquad p_3(x) = \sqrt{\sum_{i=1}^{n} |x_i|^2}$$

for $x = (x_i)_{1 \leqslant i \leqslant n}$, then the three functions p_1, p_2, p_3 are equivalent norms which define on K^n the topology which is the product of the topologies of the factors K.

5. CONTINUOUS MULTILINEAR FUNCTIONS

THEOREM 1. *Let* $E_i (1 \leqslant i \leqslant n)$ *and* F *be normed spaces over a non-discrete valued division ring* K, *and let* f *be a multilinear mapping of* $\prod_{i=1}^{n} E_i$ *into* F. *Then* f *is continuous on* $\prod_{i=1}^{n} E_i$ *if and only if there exists a real number* $a > 0$ *such that, for all* $x_i \in E_i$ $(1 \leqslant i \leqslant n)$, *we have*

$$\|f(x_1, x_2, \ldots, x_n)\| \leqslant a.\|x_1\|.\|x_2\|\ldots\|x_n\|. \tag{3}$$

The condition is *necessary*. For if f is continuous at the point $(0, 0, \ldots, 0)$ there exists a number $b > 0$ such that the relations $\|x_i\| \leqslant b$ $(1 \leqslant i \leqslant n)$ imply $\|f(x_1, \ldots, x_n)\| \leqslant 1$. Let t_0 be an element of K such that $0 < |t_0| < 1$; then for *every* point $(x_i) \in \prod_{i=1}^{n} E_i$ such that none of the x_i is zero, there exist n rational integers k_i such that $b|t_0| < \|t_0^{k_i} x_i\| \leqslant b$; consequently we have

$$|t_0|^{k_1+k_2+\cdots+k_n}\|f(x_1, x_2, \ldots, x_n)\| \leqslant 1;$$

on the other hand we have $\dfrac{1}{|t_0|^{k_i}} \leqslant \dfrac{1}{b|t_0|}\|x_i\|$, and the relation (3) therefore follows, with $a = (1/b|t_0|)^n$. This relation is evidently still valid when one of the x_i is zero.

The condition is *sufficient*. We shall show that, if it is satisfied, f is continuous at every point $(\boldsymbol{a}_i)$ of $\prod_{i=1}^{n} \mathrm{E}_i$. We can write

$$f(\boldsymbol{x}_1, \ldots, \boldsymbol{x}_n) - f(\boldsymbol{a}_1, \ldots, \boldsymbol{a}_n) = \sum_{i=1}^{n} f(\boldsymbol{a}_1, \ldots, \boldsymbol{a}_{i-1}, \boldsymbol{x}_i - \boldsymbol{a}_i, \boldsymbol{x}_{i+1}, \ldots, \boldsymbol{x}_n).$$

Now, using (3), the conditions $\|\boldsymbol{x}_i - \boldsymbol{a}_i\| \leqslant r$ $(1 \leqslant i \leqslant n)$ imply that

$$\|f(\boldsymbol{a}_1, \ldots, \boldsymbol{a}_{i-1}, \boldsymbol{x}_i - \boldsymbol{a}_i, \boldsymbol{x}_{i+1}, \ldots, \boldsymbol{x}_n)\| \leqslant ar \prod_{k \neq i}^{n} (\|\boldsymbol{a}_k\| + r);$$

hence, if c is the maximum of the numbers $\|\boldsymbol{a}_i\|$ $(1 \leqslant i \leqslant n)$, we have

$$\|f(\boldsymbol{x}_1, \ldots, \boldsymbol{x}_n) - f(\boldsymbol{a}_1, \ldots, \boldsymbol{a}_n)\| \leqslant nar(c + r)^{n-1}.$$

Since the right-hand side of this inequality is a polynomial in r with zero constant term, it tends to 0 as r tends to 0; hence f is continuous.

Remark. Two of the propositions proved earlier are consequences of this theorem: the continuity of the bilinear function $t\boldsymbol{x}$, by virtue of the relation $\|t\boldsymbol{x}\| = |t| \cdot \|\boldsymbol{x}\|$; and Proposition 7, by applying Theorem 1 to the identity mapping of E, considered as a linear mapping of the space E, endowed with the norm p into the space E endowed with the norm q (or *vice versa*).

6. ABSOLUTELY SUMMABLE FAMILIES IN A NORMED SPACE

DEFINITION 8. *In a normed space* E, *a family* $(\boldsymbol{x}_\iota)$ *of points of* E *is said to be absolutely summable if the family* $(\|\boldsymbol{x}_\iota\|)$ *of norms of the* $\boldsymbol{x}_\iota$ *is summable in* **R**.

This concept appears to depend on the norm chosen on E; but by Proposition 7 of no. 3 and the comparison principle for summable families of real numbers, a family which is absolutely summable with respect to a norm p on E is absolutely summable with respect to any norm on E which is *equivalent* to p.

If $(\boldsymbol{x}_\iota)_{\iota \in \mathrm{I}}$ is a family of points of E which is summable and absolutely summable, we have

$$\left\| \sum_{\iota \in \mathrm{I}} \boldsymbol{x}_\iota \right\| \leqslant \sum_{\iota \in \mathrm{I}} \|\boldsymbol{x}_\iota\|. \tag{4}$$

Indeed, for each finite subset J of I we have $\left\| \sum_{\iota \in \mathrm{J}} \boldsymbol{x}_\iota \right\| \leqslant \sum_{\iota \in \mathrm{J}} \|\boldsymbol{x}_\iota\|$, and the inequality (4) follows by passing to the limit with respect to the directed set of finite subsets of I.

PROPOSITION 11. *In a complete normed space* E, *every absolutely summable family is summable.*

For if (x_ι) is an absolutely summable family in E, then for each $\varepsilon > 0$ there is a finite subset J of the index set I such that, for each finite subset H of I which does not meet J, we have $\sum_{\iota \in H} \|x_\iota\| \leqslant \varepsilon$; hence *a fortiori* $\left\| \sum_{\iota \in H} x_\iota \right\| \leqslant \varepsilon$, and this proves the proposition, since E is complete (Cauchy's criterion, Chapter III, § 5, no. 2, Theorem 1).

A *series* whose general term is x_n is said to be *absolutely convergent* in E if the series whose general term is $\|x_n\|$ is convergent in $\mathbf{R}$, or (equivalently) if the family (x_n) is absolutely summable; consequently (Chapter III, § 5, no. 7, Proposition 9):

COROLLARY. *In a complete normed space* E, *every absolutely convergent series is commutatively convergent.*

The converse of Proposition 11 is in general *false.*

Consider for example the space $\mathcal{B}(\mathbf{N}; \mathbf{R})$ of bounded sequences $x = (x_n)_{n \in \mathbf{N}}$ of real numbers, with the norm $\|x\| = \sup_n |x_n|$. Let x_m be the sequence $(x_{mn})_{n \in \mathbf{N}}$ such that $x_{mn} = 0$ if $m \neq n$ and $x_{mm} = 1/m$ for $m \geqslant 1$. It is immediately verified that the sequence $(x_m)_{m \in \mathbf{N}}$ is summable in $\mathcal{B}(\mathbf{N}; \mathbf{R})$ and that its sum is the element $y = (y_n)$ such that $y_0 = 0$ and $y_n = 1/n$ if $n \geqslant 1$; but since $\|x_m\| = 1/m$, the sequence of norms of the x_m is not summable in $\mathbf{R}$.

However, we have seen in Chapter VII, § 3, no. 1, that every summable family in $\mathbf{R}^n$ is absolutely summable.

7. NORMED ALGEBRAS OVER A VALUED FIELD

DEFINITION 9. *If* A *is an algebra over a non-discrete valued field* K, *a norm* $p(x)$ *on* A (A being considered as a vector space over K) *is said to be compatible with the algebra structure of* A *if the topology it defines is compatible with the ring structure of* A. *An algebra over* K, *endowed with the structure defined by a norm compatible with the algebra structure, is called a normed algebra.*

If A is a normed algebra over K, and if $p(x)$ is the norm on A, the bilinear mapping $(x, y) \to xy$ of $A \times A$ into A is continuous, by hypothesis; hence by Theorem 1 of no. 5 there exists a real number $a > 0$ such that $p(xy) \leqslant a.p(x)p(y)$. Replacing $p(x)$ by the equivalent norm $a.p(x)$, we may therefore always assume that the norm $\|x\|$ on a normed algebra A is such that

$$\|xy\| \leqslant \|x\|.\|y\|. \tag{5}$$

It follows from (5) by induction that, for each integer $n > 0$, we have

$$\|x^n\| \leqslant \|x\|^n. \tag{6}$$

Examples. 1) Let K be a valued division ring and let K′ be a subfield of the centre of K such that the trace on K′ of the absolute value $|x|$ of K is not the improper absolute value on K′. Then K, with $|x|$ as norm, is a normed algebra over K′.

2) Let K be a non-discrete valued field and let $\mathbf{M}_n(K)$ be the ring of square matrices of order n over K. Regarded as a vector space over K, $\mathbf{M}_n(K)$ is isomorphic to K^{n^2}. If for each $X = (x_{ij}) \in \mathbf{M}_n(K)$ we define $\|X\| = \sup_{i,j} |x_{ij}|$, then $\|X\|$ is a norm on $\mathbf{M}_n(K)$, and the topology defined by this norm is the product topology on K^{n^2} (Proposition 10); from this it follows (because of the continuity of polynomials in any number of variables over K) that this norm is compatible with the K-algebra structure of $\mathbf{M}_n(K)$.

3) The set $\mathscr{B}(X; K)$ of all functions f on a set X with values in a non-discrete valued field K, such that $x \to |f(x)|$ is bounded on X, is an algebra over K; the norm $\|f\| = \sup_{x \in X} |f(x)|$ is compatible with the ring structure of $\mathscr{B}(X; K)$, because we have $\|fg\| \leqslant \|f\| . \|g\|$ (cf. Chapter X, § 1).

Let $\mathfrak{a}$ be a closed two-sided ideal in a normed algebra A. If in the quotient algebra $A/\mathfrak{a}$ we put $\|\dot{x}\| = \inf_{x \in \dot{x}} \|x\|$, we get a norm on $A/\mathfrak{a}$ which defines the topology which is the quotient by $\mathfrak{a}$ of the topology of A (Proposition 9); since this quotient topology is compatible with the quotient ring structure of $A/\mathfrak{a}$ (Chapter III, § 6, no. 4) it follows that the quotient algebra $A/\mathfrak{a}$, with the norm $\|\dot{x}\|$, is a normed algebra.

Likewise, if $(A_i)_{1 \leqslant i \leqslant n}$ is a family of n normed algebras over a valued field K, and if in the product algebra $A = \prod_{i=1}^{n} A_i$ we put $\|x\| = \sup_i \|x_i\|$, where $x = (x_i)$, we have a norm on A which defines the product of the topologies of the A_i (Proposition 10); since this topology is compatible with the ring structure of A (Chapter III, § 6, no. 4), it follows that the product algebra A, with the norm $\|x\|$, is a normed algebra.

Let A be a normed algebra over a valued field K. The completion $\hat{A}$ of A (Chapter III, § 6, no. 5, Proposition 6) is also endowed with a $\hat{K}$-vector space structure (no. 3, Proposition 8), and it is clear from the principle of extension of identities that we have $t(xy) = (tx)y = x(ty)$ for all $t \in \hat{K}$ and all $x, y \in \hat{A}$. Hence $\hat{A}$ is an algebra over $\hat{K}$; on the other hand (no. 3, Proposition 8), the norm on A extends by continuity to a norm which defines the topology of $\hat{A}$, and therefore $\hat{A}$, endowed with this norm, is a *normed algebra* over the field $\hat{K}$.

If $(x_\lambda)_{\lambda \in L}$ and $(y_\mu)_{\mu \in M}$ are two absolutely summable families in a normed algebra A, the family $(x_\lambda y_\mu)_{(\lambda, \mu) \in L \times M}$ is absolutely summable, because $\|x_\lambda y_\mu\| \leqslant \|x_\lambda\| \cdot \|y_\mu\|$ (Chapter IV, § 7, no. 3, Proposition 1); if in addition A is complete, all three families are summable and we have

$$\sum_{(\lambda, \mu) \in L \times M} x_\lambda y_\mu = \left(\sum_{\lambda \in L} x_\lambda\right)\left(\sum_{\mu \in M} y_\mu\right)$$

by the associativity of the sum on the left-hand side (Chapter III, § 5, no. 3, formula (2)).

If the normed algebra A has an *identity element* $e \neq 0$, the mapping $t \rightarrow te$ is an isomorphism of the field structure of K onto that of the subfield Ke of A; this isomorphism is also an isomorphism of the *topological field* structure of K onto that of Ke (the topology of the latter being induced by that of A), for the restriction $\|te\|$ of the norm of A to Ke is a norm equivalent to the absolute value

$$|t| = \frac{1}{\|e\|}\|te\|.$$

If $\|e\| = 1$ we have $\|te\| = |t|$, and we can then *identify* the valued field K with the normed subfield Ke of A, and in particular we may denote the identity element of A by the symbol 1.

In what follows we shall be concerned only with normed algebras which have an identity element e, and in which the norm satisfies the inequality (5); putting $x = y = e$ in this inequality, it follows that $\|e\| \geqslant 1$.

Proposition 12. *If the series whose general term is* z^n *is convergent in* A, *then* $e - z$ *is a unit of* A *and we have*

$$(e - z)^{-1} = \sum_{n=0}^{\infty} z^n. \tag{7}$$

Conversely, if $\|z\| < 1$ *and if* $e - z$ *is a unit in* A, *then the series whose general term is* z^n *is convergent and formula* (7) *is valid.*

For each $p > 0$ we have

$$(e - z)\sum_{n=0}^{p} z^n = e - z^{p+1}. \tag{8}$$

If the series whose general term is z^n is convergent and if y is its sum, then z^n tends to 0 as $n \rightarrow +\infty$; hence by passing to the limit in (8) we have $(e - z)y = e$; similarly we prove that $y(e - z) = e$, and hence $y = (e - z)^{-1}$ (note that this part of the argument is valid in any topological ring which has an identity element).

Conversely, if $||z|| < 1$, then since $||z^{p+1}|| \leqslant ||z||^{p+1}$, it follows that z^{p+1} tends to 0 as $p \to +\infty$; multiplying both sides (8) on the left by $(e-z)^{-1}$ and letting p tend to infinity, we see that the series whose general term is z^n converges and has $(e-z)^{-1}$ as its sum.

COROLLARY. *Let* A *be a complete normed algebra. Then for each* $z \in A$ *such that* $||z|| < 1$, $e-z$ *is a unit in* A.

The series whose general term is z^n is absolutely convergent, since $||z^n|| \leqslant ||z||^n$ for $n > 0$, and is therefore convergent, since A is complete (no. 6, Proposition 11).

PROPOSITION 13. *Let* G *be the group of units of a complete normed algebra* A. *Then* G *is an open subset of* A; *the topology induced on* G *by the topology of* A *is compatible with the group structure of* G; *and* G, *endowed with this topology, is a complete group* (with respect to each of its two uniformities).

The corollary to Proposition 12 shows that G contains a neighbourhood V of e in A; hence, for each $x_0 \in G$, the elements of x_0V are units, and x_0V is a neighbourhood of x_0 in A, since $x \to x_0 x$ is a homeomorphism of A onto itself (x_0 being a unit of A). Hence G is open in A.

To show that the topology induced on G by the topology of A is compatible with the group structure of G, it is sufficient to show that the function x^{-1} is *continuous* on G. Let $x_0 \in G$, and for each $x \in G$, write x in the form $x = x_0(e+u)$, so that $u = x_0^{-1}(x-x_0)$; then $||u|| \leqslant ||x_0^{-1}||.||x-x_0||$, and thus if $||x-x_0|| \leqslant 1/||x_0^{-1}||$, we have $||u|| \leqslant 1$, $e+u = x_0^{-1}x$ is a unit, the series whose general term is $(-1)^n u^n$ is absolutely convergent, and

$$(9) \qquad x^{-1} = (e+u)^{-1}x_0^{-1} = x_0^{-1} + \left(\sum_{n=1}^{\infty} (-1)^n u^n\right) x_0^{-1},$$

from which it follows that

$$||x_0^{-1} - x_0^{-1}|| \leqslant \left\|\sum_{n=0}^{\infty} (-1)^n u^n\right\|.||u||.||x_0^{-1}||$$
$$\leqslant \left\|\sum_{n=0}^{\infty} (-1)^n u^n\right\|.||x_0^{-1}||^2.||x-x_0||.$$

As x tends to x_0, $||x-x_0||$ tends to 0, and since

$$\left\|\sum_{n=0}^{\infty} (-1)^n u^n\right\| \leqslant ||e|| + \frac{||u||}{1-||u||}$$

remains bounded, it follows that x^{-1} tends to x_0^{-1}.

Finally, to show that the left uniformity of G is complete, let us show that every Cauchy filter $\mathfrak{F}$ with respect to this uniformity is a Cauchy filter with respect to the *additive* uniformity of A and converges to a point of G. For each ε such that $0 < \varepsilon < 1$, there is a set $M \in \mathfrak{F}$ such that $\|x^{-1}y - e\| \leqslant \varepsilon$ for all x, y in M, i.e., such that

$$\|y - x\| \leqslant \varepsilon \, \|x\|.$$

Let a be a point of M; for each $x \in M$ we have $\|x - a\| \leqslant \varepsilon \, \|a\|$, and therefore $\|x\| \leqslant (1 + \varepsilon)\|a\|$. On the other hand, there exists a set $N \subset M$ belonging to $\mathfrak{F}$ and such that $\|x^{-1}y - e\| \leqslant \varepsilon/(1 + \varepsilon)\|a\|$ for all x and y in N; it follows that $\|y - x\| \leqslant \varepsilon\|x\|/(1 + \varepsilon)\,\|a\| \leqslant \varepsilon$, which shows that $\mathfrak{F}$ is a Cauchy filter with respect to the additive uniformity of A, and therefore converges to a point x_0, since A is complete. Since x_0 is the limit of $\mathfrak{F}$, we have $\|x^{-1}x_0 - e\| \leqslant \varepsilon$ for all $x \in M$ by the principle of extension of inequalities; since $\varepsilon > 1$, it follows that $x^{-1}x_0$ is a unit in A; hence x_0 is a unit, i.e., $x_0 \in G$.

PROPOSITION 14. *In a complete valued division ring, the multiplicative group of non-zero elements is a complete group.*

The proof is similar to that of Proposition 13; we have only to replace the norm of A by the absolute value of the division ring under consideration.

Note that we cannot apply Proposition 13 directly, because a (non-commutative) valued division ring is not necessarily an algebra over a *non-discrete* valued field (the restriction of the absolute value to the centre of the division ring might be improper).

Remark. Proposition 13 is not necessarily true for a non-complete normed algebra. For example, in the algebra $\mathcal{C}(I; \mathbf{R})$ of all finite continuous real-valued functions on $I = [0, 1]$ [the norm being $\|x\| = \sup_{t \in I} |x(t)|$], the subalgebra P of all *polynomials* in t (restricted to I) is *not* complete; if $x(t)$ is any non-constant polynomial, then $1 + \varepsilon x$ is arbitrarily close to the identity element 1 of P when ε is arbitrarily small, but $1 + \varepsilon x$ is not a unit *in* P. However, if A is a non-complete normed algebra, G its group of units, $\hat{A}$ the completion of A, then G is a subgroup of the group of units of $\hat{A}$, and therefore the topology induced on G by the topology of A is compatible with the group structure of G.

4. NORMAL SPACES

1. DEFINITION OF NORMAL SPACES

Axiom (O_{IV}) for uniformizable spaces (§ 1, no. 5) can be stated in the following form: *given any closed set* A *and any point* $x \in \complement A$, *there is a continuous mapping of* X *into* [0, 1] *which is equal to* 0 *at* x *and is equal to* 1 *at every point of* A; this property can again be expressed by saying that in a uniformizable space we can *separate a point and a closed set* (not containing the point) *by a continuous real-valued function.*

We shall now study spaces in which it is possible in the same way to *separate two disjoint closed sets by a continuous real-valued function*:

DEFINITION 1. *A topological space* X *is said to be normal if it is Hausdorff and satisfies the following axiom:*

(O_V) *If* A *and* B *are any two disjoint closed subsets of* X, *there exists a continuous mapping of* X *into* [0, 1] *which is equal to* 0 *at every point of* A *and to* 1 *at every point of* B.

Clearly every normal space is completely regular; but there are completely regular spaces which are not normal (see Exercises 9, 10, 13, 26 and § 5, Exercises 15 and 16).

The statement of Axiom (O_V), like that of (O_{IV}), involves the real line **R** as an auxiliary set. But there is a condition equivalent to (O_V) which involves no auxiliary set:

THEOREM 1 (Urysohn). *Axiom* (O_V) *is equivalent to the following*:

(O'_V) *If* A *and* B *are any two disjoint closed subsets of* X, *then there exist two disjoint open sets* U, V *such that* $A \subset U$ *and* $B \subset V$.

It follows immediately that (O_V) implies (O'_V), for if f is a continuous mapping of X into [0, 1] which is equal to 0 on A and to 1 on B, then the open sets $\overset{-1}{f}([0, 1/2[)$ and $\overset{-1}{f}(]1/2, 1])$ contain A and B respectively and do not intersect.

To prove the converse, notice first that (O'_V) is equivalent to the following axiom:

(O''_V) *Given any closed set* A *and any open neighbourhood* V *of* A, *there exists an open neighbourhood* W *of* A *such that* $\overline{W} \subset V$.

If there is a continuous mapping $f: X \to [-1, +1]$ which is equal to -1 on A and to $+1$ on B, and if we put $U(t) = \overset{-1}{f}([-1, t[)$ for each $t \in [0, 1]$, then we have defined a family of open sets in X, indexed by [0, 1], such that (i) $A \subset U(0)$, (ii) $B \subset \complement U(1)$ and (iii) for each pair of real numbers t, t' such that $0 \leqslant t < t' \leqslant 1$ we have

$$\overline{U}(t) \subset U(t'); \tag{1}$$

for $U(t)$ is contained in the closed set $\overset{-1}{f}([-1, t])$. Conversely, suppose that we have defined a family $(U(t))$ of open sets ($0 \leqslant t \leqslant 1$) with these three properties (i), (ii) and (iii). For each $x \in X$, put $g(x) = 1$

if $x \in \complement U(1)$, and if $x \in U(1)$ let $g(x)$ be the greatest lower bound of the values of t such that $x \in U(t)$. Clearly $0 \leqslant g(x) \leqslant 1$ for each $x \in X$, $g(x) = 0$ on A, $g(x) = 1$ on B; also g is *continuous* on X, for if we put $g(x) = a$, we have $|g(y) - g(x)| \leqslant \varepsilon$ for all $y \in U(a+\varepsilon) \cap \complement \overline{U}(a-\varepsilon)$, which is a neighbourhood of x by (1) [with the conventions that $U(a+\varepsilon) = X$ if $a + \varepsilon > 1$, and $U(a-\varepsilon) = \emptyset$ if $a - \varepsilon < 0$].

Hence Theorem 1 will be proved if we can define a family $(U(t))$ of open sets satisfying conditions (i), (ii) and (iii) above; to do this we use Axiom (O''_V). Take $U(1) = \complement B$; since $A \subset U(1)$ there exists an open set $U(0)$ such that $A \subset U(0)$ and $\overline{U}(0) \subset U(1)$ by (O''_V). Suppose then that for each *dyadic* number $k/2^n$ $(k = 0, 1, \ldots, 2^n)$ we have defined an open set $U(k/2^n)$, these sets being such that $\overline{U}(k/2^n) \subset U((k+1)/2)^n$ for $0 \leqslant k \leqslant 2^n - 1$. For each dyadic number

$$(2k+1)/2^{n+1} \quad (0 \leqslant k \leqslant 2^n - 1)$$

there exists by (O''_V) an open set $U((2k+1)/2^{n+1})$ such that

$$\overline{U}(k/2^n) \subset U((2k+1)/2^{n+1})$$

and

$$\overline{U}((2k+1)/2^{n+1}) \subset U((k+1)/2^n).$$

Hence for each dyadic number r such that $0 \leqslant r \leqslant 1$ we can define an open set $U(r)$, such that $A \subset U(0)$, $B \subset \complement U(1)$, and

$$\overline{U}(r) \subset U(r') \tag{2}$$

for each pair of dyadic numbers r, r' such that $0 \leqslant r < r' \leqslant 1$.

Now define, for each real number $t \in [0, 1]$,

$$U(t) = \bigcup_{r \leqslant t} U(r) \quad (r \text{ dyadic});$$

by (2), this definition agrees with the preceding one for t dyadic; also, if $0 \leqslant t < t' \leqslant 1$, then there exist two dyadic numbers r, r' such that $t \leqslant r < r' \leqslant t'$; by (2) we have $U(t) \subset U(r)$, hence

$$\overline{U}(t) \subset \overline{U}(r) \subset U(r') \subset U(t');$$

this proves (1) and therefore completes the proof.

Theorem 1 will enable us to show that two important categories of topological spaces are normal. In the first place:

PROPOSITION 1. *A compact space is normal.*

A compact space satisfies axiom (O'_V), by Proposition 2 of Chapter I, § 9, no. 2.

As to *locally compact* spaces, every point of such a space has a compact neighbourhood, which is a normal subspace; but there are examples of locally compact spaces which are *not normal* (cf. Exercises 9 and 26, and § 5, Exercise 15).

PROPOSITION 2. *A metrizable space is normal.*

Let X be a metrizable space and let d be a metric compatible with the topology of X. Let A, B be two disjoint closed subsets of X; since the functions $d(x, A)$ and $d(x, B)$ are continuous, the set U (resp. V) of points x such that $d(x, A) < d(x, B)$ [resp. $d(x, B) < d(x, A)$] is open; clearly $A \subset U$ and $B \subset V$ and $U \cap V = \emptyset$, hence Axiom (O'_V) is satisfied.

Remarks. 1) Proposition 2 gives another *necessary* condition for metrizability; but this condition, even in conjunction with all the necessary conditions given in § 2, does not give a set of sufficient conditions for metrizability (cf. Exercise 6 and § 5, Exercise 10).

2) There are examples of normal spaces which are neither metrizable nor locally compact (see § 5, Exercise 16).

By (O'_V), every *closed* subset of a normal space is a *normal subspace*; but this is not always the case for an *arbitrary* subset of a normal space.

For example, a completely regular space which is not normal is homeomorphic to a subspace of a compact space (§ 1, no. 5, Proposition 3), and the latter is normal.

Finally we record that the *product* of two normal spaces is not necessarily normal (see Exercise 9 and § 5, Exercise 16).

2. EXTENSION OF A CONTINUOUS REAL-VALUED FUNCTION

Let X and Y be two topological spaces and let $A \neq X$ be a *closed* subset of X. If f is a continuous mapping of A into Y, it is not always possible to *extend* f to a continuous mapping of the whole of X into Y. When $Y = \overline{\mathbf{R}}$, the possibility of such an extension is determined by the following theorem:

THEOREM 2 (Urysohn). *Axiom* (O_V) *is equivalent to the following property:*

(O'''_V) *Given any closed subset* A *of* X *and any continuous real-valued function* f *(finite or not) defined on* A, *there exists an extension* g *of* f *to the whole space* X, *which is a continuous mapping of* X *into* $\overline{\mathbf{R}}$.

It is easy to see that (O'''_V) implies (O_V); for if B and C are two disjoint closed subsets of X, then the function which is equal to 0 on B and equal to 1 on C is defined and continuous on the closed set $B \cup C$, hence by (O_V''') has a continuous extension f to X. If $g = \inf(f^+, 1)$, then g is continuous on X, takes its values in $[0, 1]$ and is equal to 0 on B and to 1 on C.

Let us show conversely that (O_V) implies (O_V'''). Since $\overline{\mathbf{R}}$ and the interval $[-1, +1]$ are homeomorphic, we need consider only the case where the continuous mapping $f: A \to \overline{\mathbf{R}}$ takes its values in $[-1, +1]$. We shall construct an extension g of f to X by forming a sequence (g_n) of continuous functions on X, such that the sequence $(g_n(x))$ converges for all $x \in X$ to a point of the interval $[-1, +1]$; this limit will, by definition, be the value of g at x, and it will follow from the choice of the g_n that the function g satisfies the required conditions.

The definition of the g_n rests on the following lemma:

Lemma 1. *Let u be a continuous mapping of* A *into $[-1, +1]$; then there is a continuous mapping v of* X *into $[-1/3, +1/3]$, such that $|u(x) - v(x)| \leqslant 2/3$ for all $x \in A$.*

Let H be the set of all $x \in A$ such that $-1 \leqslant u(x) \leqslant -1/3$, and let K be the set of all $x \in A$ such that $1/3 \leqslant u(x) \leqslant 1$; H and K are closed in A, and therefore in X, and do not intersect; hence by (O_V) there is a continuous mapping v of X into $[-1/3, +1/3]$ which is equal to $-1/3$ on H and to $+1/3$ on K. The mapping v satisfies the conditions of the lemma.

We now define the functions g_n by induction. Applying the lemma with $u = f$, we define g_0 to be a continuous mapping of X into $[-1/3, +1/3]$ such that $|f(x) - g_0(x)| \leqslant 2/3$ for all $x \in A$. Suppose now that a continuous mapping g_n of X into the interval

$$[-1 + (\tfrac{2}{3})^{n+1}, 1 - (\tfrac{2}{3})^{n+1}]$$

has been defined, such that $|f(x) - g_n(x)| \leqslant (\frac{2}{3})^{n+1}$ for all $x \in A$. Applying the lemma to the function $u(x) = (\frac{3}{2})^{n+1}(f(x) - g_n(x))$, we see that there exists a continuous mapping h_{n+1} of X into the interval $\left[-\frac{2^{n+1}}{3^{n+2}}, \frac{2^{n+1}}{3^{n+2}}\right]$ such that

$$|f(x) - g_n(x) - h_{n+1}(x)| \leqslant (2/3)^{n+2}$$

for all $x \in A$; the induction is completed by taking $g_{n+1} = g_n + h_{n+1}$, since this function satisfies the inequality $|g_{n+1}(x)| \leqslant 1 - (2/3)^{n+2}$ for all $x \in X$, by virtue of the definition of h_{n+1}.

From this definition it follows that, for $m \geqslant p$ and $n \geqslant p$, we have

$$|g_m(x) - g_n(x)| \leqslant \frac{2^{p+1}}{3^{p+2}} \sum_{k=0}^{\infty} (2/3)^k = (2/3)^{p+1}$$

at each point $x \in X$; hence the sequence $(g_n(x))$ is a Cauchy sequence for each $x \in X$, and therefore converges to a point $g(x)$ of the interval $[-1, +1]$; and since $f(x) - g_n(x)$ tends to 0 for all $x \in A$ as $n \to \infty$, g is an extension of f to X. It remains therefore only to show that g is *continuous* on X.

Now let x be any point of X; then, given any $\varepsilon > 0$, there exists an integer n_0 such that $|g_m(y) - g_n(y)| \leqslant \varepsilon$ for all $y \in X$ and all $m \geqslant n_0$ and all $n \geqslant n_0$; hence, letting m tend to $+\infty$, we have

$$|g(y) - g_n(y)| \leqslant \varepsilon.$$

Let V be a neighbourhood of x such that $|g_n(y) - g_n(x)| \leqslant \varepsilon$ for all $y \in V$; then, for each $y \in V$ we shall have

$$|g(y) - g(x)| \leqslant |g(y) - g_n(y)| + |g_n(y) - g_n(x)| + |g(x) - g_n(x)| \leqslant 3\varepsilon,$$

which shows that g is continuous at x, and completes the proof of Theorem 2. (The last part of the proof uses, in a particular case, the idea of *uniform convergence*, which we shall study in general in Chapter X, no. 1.)

COROLLARY. *If f is a* finite *continuous real-valued function defined on* A, *then there exists a* finite *continuous real-valued function g defined on* X, *which extends f.*

First consider the case in which $f(x) \geqslant 0$ for all $x \in A$; then there is a continuous extension g_1 of f to X, taking its values in $[0, +\infty]$. If we put $B = \overset{-1}{g_1}(+\infty)$, then B is closed and by hypothesis does not meet A; the function h which is equal to f on A and to 0 on B is therefore a continuous function on the closed set $A \cup B$. Let g_2 be a continuous extension of h to X, again taking its values in $[0, +\infty]$; the function $g = \inf(g_1, g_2)$ is then a continuous extension of f to X, whose values are $\geqslant 0$ and *finite* at every point of X.

To pass to the general case, it is enough to remark that, if f is finite and continuous on A, then so are f^+ and f^-; extending f^+ and f^- to X by finite continuous functions g_1 and g_2 respectively, we see that the function $g_1 - g_2$ is finite and continuous on X and extends f.

Remark. If X is a normal space and if A is a closed subset of X, there exists also a continuous extension to X of every continuous mapping f of A into a *cube* K^I (§ 1, no. 5); for we have then $f = (f_\iota)_{\iota \in I}$, f_ι being a continuous mapping of A into the compact interval K of $\mathbf{R}$; since there exists a continuous mapping $g_\iota : X \to K$ which extends f_ι, the mapping $g = (g_\iota)$ is a continuous extension of f to X.

3. LOCALLY FINITE OPEN COVERINGS OF A CLOSED SET IN A NORMAL SPACE

THEOREM 3. *Let* $(A_\iota)_{\iota \in I}$ *be a locally finite open covering of a closed set* Y *in a normal space* X. *Then there is an open covering* $(B_\iota)_{\iota \in I}$ *of* Y *such that* $\overline{B}_\iota \subset A_\iota$ *for each* $\iota \in I$.

Well-order the index set I (*Set Theory*, Chapter III, § 2, no. 3, Theorem 1). We shall define a family $(B_\iota)_{\iota \in I}$ of open sets in X, by transfinite induction, such that (i) $\overline{B}_\iota \subset A_\iota$ for each $\iota \in I$; (ii) for each $\iota \in I$, the family formed by the B_λ such that $\lambda \leqslant \iota$ and by the A_λ such that $\lambda > \iota$ is an open covering of Y. Suppose that we have defined the B_ι for $\iota < \gamma$ so that (i) and (ii) are satisfied *for all* $\iota < \gamma$, and let us show that we can define B_γ in such a way that (i) and (ii) are also satisfied for $\iota = \gamma$. Let us first show that the B_ι for which $\iota < \gamma$ and the A_ι for which $\iota \geqslant \gamma$ form a covering of Y. By hypothesis, for each $x \in Y$ there is only a finite number of indices $\lambda \in I$ such that $x \in A_\lambda$, say $\lambda_1 < \lambda_2 < \cdots < \lambda_n$; let λ_h be the greatest of the λ_i such that $\lambda_i < \gamma$; if $h < n$ we have $x \in A_{\lambda_n}$ and $\lambda_n \geqslant \gamma$, and if $h = n$ the inductive hypothesis shows that x belongs to some B_λ such that $\lambda \leqslant \lambda_n < \gamma$, and our assertion follows. Now put $C = (\complement Y) \cup \left(\bigcup_{\iota<\gamma} B_\iota\right) \cup \left(\bigcup_{\iota>\gamma} A_\iota\right)$; C is open, and from what has been said we have $\complement A_\gamma \subset C$; by virtue of Axiom (O''_V) for normal spaces, there is therefore an open set V such that $\complement A_\gamma \subset \overline{V} \subset V \subset C$. If we put $B_\gamma = \complement \overline{V}$, we have $\overline{B}_\gamma \subset \complement V \subset A_\gamma$ and $B_\gamma \cup C = X$, so that the B_ι such that $\iota \leqslant \gamma$ and the A_ι such that $\iota > \gamma$ cover Y.

Remark. Note that we have used only the fact that the covering (A_ι) is *point-finite*, i.e., that every point of X belongs to only a finite number of sets A_ι.

DEFINITION 2. *Let* X *be a topological space and let* f *be a real-valued function defined on* X. *The support of* f, *denoted by* Supp (f), *is the smallest closed set* S *in* X *such that* $f(x) = 0$ *for all* $x \notin S$.

In other words, Supp (f) is the closure in X of the set of all $x \in X$ such that $f(x) \neq 0$; or again, it is the set of all $x \in X$ such that every neighbourhood of x contains a point y for which $f(y) \neq 0$.

Let $(f_\iota)_{\iota \in I}$ be a family of finite real-valued functions on X whose supports form a *locally finite* family; then the sum $\sum_{\iota \in I} f_\iota(x)$ is defined for each $x \in X$ (since it contains only a finite number of non-zero terms). The finite real-valued function $x \to \sum_{\iota \in I} f_\iota(x)$ is called the *sum* of the family (f_ι), and is denoted by $\sum_{\iota \in I} f_\iota$. If each of the f_ι is *continuous*, then so is $f = \sum_{\iota \in I} f_\iota$; for if x is any point of X, there is a neighbourhood V of x which meets only a finite number of supports of the f_ι, and hence there is a finite subset H of I such that $f(y) = \sum_{\iota \in H} f_\iota(y)$ for all $y \in V$.

Definition 3. *Given a family* $(A_\iota)_{\iota \in I}$ *of subsets of a topological space* X, *a family* $(f_\iota)_{\iota \in I}$ *of real-valued functions defined on* X *is said to be subordinate to the family* $(A_\iota)_{\iota \in I}$ *if we have* $\mathrm{Supp}(f_\iota) \subset A_\iota$ *for each index* $\iota \in I$.

A continuous partition of unity on X *is any family* $(f_\iota)_{\iota \in I}$ *of real-valued continuous functions* $\geqslant 0$ *on* X *whose supports form a locally finite family and which are such that* $\sum_{\iota \in I} f_\iota(x) = 1$ *for all* $x \in X$.

Proposition 3. *Given any locally finite open covering* $(A_\iota)_{\iota \in I}$ *of a normal space* X, *there exists a continuous partition of unity* $(f_\iota)_{\iota \in I}$ *on* X, *subordinate to the covering* $(A_\iota)_{\iota \in I}$.

By Theorem 3 there exists an open covering $(B_\iota)_{\iota \in I}$ of X such that $\overline{B}_\iota \subset A_\iota$ for each $\iota \in I$, and it is clear that the covering (B_ι) is locally finite. By axiom (O''_V), for each $\iota \in I$ there exists an open set C_ι such that $\overline{B}_\iota \subset C_\iota \subset \overline{C}_\iota \subset A_\iota$. By axiom (O_V), for each $\iota \in I$ there exists a continuous mapping g_ι of X into $[0,1]$, such that $g_\iota(x) = 1$ on $\overline{B}_\iota$ and such that the support of $\overline{g}_\iota$ is contained in $\overline{C}_\iota$, and therefore contained in A_ι. Since (B_ι) is a covering of X, we have $\sum_{\iota \in I} g_\iota(x) > 0$ for each $x \in X$; if we put

$$f_\iota(x) = \frac{g_\iota(x)}{\sum_{\iota \in I} g_\iota(x)}$$

for all $\iota \in I$ and all $x \in X$, then the f_ι form a continuous partition of unity subordinate to the covering (A_ι).

Corollary. *Given any locally finite open covering* (A_ι) *of a closed set* F *in a normal space* X, *there exists a family* (f_ι) *of continuous real-valued functions* $\geqslant 0$ *on* X, *subordinate to the covering* $(A_\iota)_{\iota \in I}$ *and such that* $\sum_{\iota \in I} f_\iota(x) = 1$ *for all* $x \in F$ *and* $\sum_{\iota \in I} f_\iota(x) \leqslant 1$ *for all* $x \in X$.

The family of sets consisting of $\complement F$ and the A_ι is a locally finite open covering of X. There is therefore a continuous partition of unity subordinate to this covering, consisting of a family $(f_\iota)_{\iota \in I}$ such that $\mathrm{Supp}(f_\iota) \subset A_\iota$ for each $\iota \in I$, and a function g whose support is contained in the complement of F. The family (f_ι) clearly satisfies the required conditions.

4. PARACOMPACT SPACES

We recall (Chapter I, § 9, no. 10) that a topological space X is said to be *paracompact* if it is Hausdorff and if every open covering of X has a locally finite open refinement.

PROPOSITION 4. *Every paracompact space is normal.*

This is a consequence of the following lemma:

LEMMA 2. *Let* A, B *be two disjoint closed subsets of a paracompact space* X. *If for each* $x \in A$ *there is an open neighbourhood* V_x *of* x *and an open neighbourhood* W_x *of* B *which do not intersect, then there exists an open neighbourhood* T *of* A *and an open neighbourhood* U *of* B *which do not intersect.*

Assuming the truth of this lemma for the moment, we can apply it to the case where B consists of a single point, because X is Hausdorff, and it shows then that X is *regular*. We can then apply Lemma 2 again to any two disjoint closed subsets of X, and this shows that Axiom (O'_V) is satisfied.

To prove the lemma, consider the open covering of X consisting of $\complement A$ and the V_x, where $x \in A$; let $(T_\iota)_{\iota \in I}$ be a locally finite open refinement of this covering. Then, by definition, if $A \cap T_\iota \neq \emptyset$ there exists $x_\iota \in A$ such that $T_\iota \subset V_{x_\iota}$. Let T be the open set which is the union of the T_ι which meet A, and let us show that there is an open neighbourhood U of B which does not meet T. For each $y \in B$ there is an open neighbourhood S_y of y which meets only a finite number of sets T_ι; let J be the finite subset of I consisting of those indices ι such that T_ι meets both S_y and A; if we put $U_y = S_y \cap \bigcap_{\iota \in J} W_{x_\iota}$, then U_y is an open neighbourhood of y which meets none of the T_ι which meet A, and hence $U_y \cap T = \emptyset$. Let $U = \bigcup_{y \in B} U_y$; then U' is an open neighbourhood of B which does not meet T, and the lemma is proved.

There exist normal spaces which are not paracompact (Exercise 19).

COROLLARY 1. *Given any open covering* $(A_\iota)_{\iota \in I}$ *of a paracompact space* X, *there exists a continuous partition of unity* $(f_\iota)_{\iota \in I}$ *on* x, *subordinate to the covering* (A_ι).

Let $(U_\lambda)_{\lambda \in L}$ be a locally finite open covering of X which refines the covering $(A_\iota)_{\iota \in I}$; then there is a mapping $\varphi : L \to I$ such that $U_\lambda \subset A_{\varphi(\lambda)}$ for each $\lambda \in L$. By Propositions 3 and 4, there exists a continuous partition of unity $(g_\lambda)_{\lambda \in L}$ subordinate to (U_λ). For each $\iota \in I$, put

$$f_\iota = \sum_{\varphi(\lambda) = \iota} g_\lambda;$$

this sum is defined and continuous since the supports of the g_λ form a locally finite covering; moreover, the union B_ι of the supports of the g_λ such that $\varphi(\lambda) = \iota$ is closed, by Proposition 4 of Chapter I, § 1, no. 6, and is contained in A_ι. Since we have $f_\iota(x) = 0$ whenever $x \in \complement B_\iota$, the support of f_ι is contained in B_ι, and therefore in A_ι. On the other hand, the family B_ι is locally finite, because for each $x \in X$ there is a neighbourhood V of x and a finite subset H of L such that $V \cap U_\lambda = \varnothing$ for all $\lambda \notin H$, and it follows therefore that $V \cap B_\iota = \varnothing$ for all $\iota \notin \varphi(H)$. Finally, for each $x \in X$ we have

$$1 = \sum_{\lambda \in L} g_\lambda(x) = \sum_{\iota \in I} \Big(\sum_{\varphi(\lambda)=\iota} g_\lambda(x) \Big) = \sum_{\iota \in I} f_\iota(x),$$

and the proof is complete.

COROLLARY 2. *If* F *is a closed subset of a paracompact space* X, *then every neighbourhood of* F *in* X *contains a closed (and therefore paracompact) neighbourhood of* F.

By Proposition 16 of Chapter I, § 9, no. 10, any closed subspace of X is paracompact; the corollary therefore follows from Proposition 4 and Axiom (O_V'').

5. PARACOMPACTNESS OF METRIZABLE SPACES

The following theorem sharpens the result of Proposition 2 of no. 1:

THEOREM 4. *Every metrizable space is paracompact.*

The theorem is a consequence of the following four lemmas.

LEMMA 3. *Let* $\mathfrak{R} = (U_\alpha)_{\alpha \in A}$ *be an open covering of a metrizable space* X. *Then there is a sequence* $(\mathfrak{S}_n)$ *of locally finite families of open subsets of* X, *such that* $\mathfrak{S} = \bigcup_n \mathfrak{S}_n$ *is an open covering of* X *which refines* $\mathfrak{R}$.

Let d be a metric on X compatible with its topology. For each $\alpha \in A$ and each integer n, let $F_{n\alpha}$ denote the set of all $x \in U_\alpha$ such that $d(x, X - U_\alpha) \geqslant 2^{-n}$. Since $X - U_\alpha$ is closed, we have $U_\alpha = \bigcup_n F_{n\alpha}$. Well-order the set A; for each $\alpha \in A$ and each integer n, let $G_{n\alpha}$ be the set of all $x \in F_{n\alpha}$ such that $x \notin F_{n+1,\beta}$ for all $\beta < \alpha$, and let $V_{n\alpha}$ be the set of all $y \in X$ such that $d(y, G_{n\alpha}) — 2^{-n-3}$. $V_{n\alpha}$ is clearly an open set; on the other hand, $V_{n\alpha} \subset U_\alpha$, because for each $y \in V_{n\alpha}$ there exists $x \in G_{n\alpha}$ such that $d(x, y) \leqslant 2^{-n-1}$, and since $x \in F_{n\alpha}$, we have

$$d(y, X - U_\alpha) \geqslant d(x, X - U_\alpha) - d(x, y) \geqslant 2^{-n-1},$$

so that $y \in U_\alpha$. Furthermore, for each $x \in X$ let α be the smallest index in A such that $x \in U_\alpha$; then there is an integer n such that $x \in F_{n\alpha}$, and the definition of α implies that $x \in G_{n\alpha}$, so that $x \in V_{n\alpha}$. This shows that if we put $\mathfrak{S}_n = (V_{n\alpha})_{\alpha \in A}$, then $\mathfrak{S} = \bigcup_n \mathfrak{S}_n$ is an open covering of X which refines $\mathfrak{R}$; thus it remains to be shown that each of the families $\mathfrak{S}_n$ is *locally finite*. To this end we shall first show that $d(G_{n\alpha}, G_{n\beta}) \geqslant 2^{-n-1}$ if $\alpha \neq \beta$. Suppose that $\beta < \alpha$; then if $x \in G_{n\alpha}$ and $y \in F_{n\beta}$ we have $x \notin F_{n+1, \beta}$ by definition, hence $d(x, X - U_\beta) < 2^{-n-1}$ and $d(y, X - U_\beta) \geqslant 2^{-n}$, and therefore $d(x, y) \geqslant 2^{-n-1}$; since $G_{n\beta} \subset F_{n\beta}$, the assertion follows. From this it follows immediately, using the definition of $V_{n\alpha}$ and $V_{n\beta}$, that $d(V_{n\alpha}, V_{n\beta}) \geqslant 2^{-n-2}$. From this last inequality we deduce that, for each $z \in X$, the open ball with centre z and radius 2^{-n-3} meets at most one set of the family $\mathfrak{S}_n$; hence $\mathfrak{S}_n$ is a locally finite family, and the proof is complete.

Lemma 4. *Let* $(\mathfrak{S}_n)$ *be a sequence of locally finite families of open sets in a topological space* X, *such that*

$$\mathfrak{S} = \bigcup_n \mathfrak{S}_n$$

is a covering of X. *Then there exists a locally finite (but not necessarily open) covering* $\mathfrak{B}$ *of* X *which refines* $\mathfrak{S}$.

Let E_n be the open set in X which is the union of all the sets of $\mathfrak{S}_n$; let U_n denote $\bigcup_{k=1}^{n} E_k$ and let A_n denote $U_n - U_{n-1}$ (with $U_0 = \emptyset$). Consider the set $\mathfrak{B}$ of subsets $V \cap A_n$, where $V \in \mathfrak{S}_n$ and n is any integer; we shall show that $\mathfrak{B}$ satisfies the conditions of the lemma. For each $x \in X$ there is an integer n such that $x \in A_n$, since the A_n form a partition of X; thus $x \in E_n$ and there exists $V \in \mathfrak{S}_n$ such that $x \in V$; so $x \in V \cap A_n$, and we have proved that $\mathfrak{B}$ is a covering of X. Clearly, this covering refines $\mathfrak{S}$. On the other hand, for each $x \in X$ there exists an integer n such that $x \in U_n$; since U_n is open and the $\mathfrak{S}_m$ are locally finite families, there exists a neighbourhood W_m of x, for each m, contained in U_n, which meets only a finite number of sets of $\mathfrak{S}_m$; hence the neighbourhood $W = \bigcap_{m=1}^{n} W_m$ of x meets only a finite number of sets of $\mathfrak{B}$, because $W \cap A_p = \emptyset$ for $p > n$. Hence $\mathfrak{B}$ is locally finite, and Lemma 4 is proved.

Lemma 5. *Let* X *be a regular space such that, for each open covering* $\mathfrak{R}$ *of* X, *there exists a (not necessarily open) locally finite covering* $\mathfrak{A}$ *of* X *which refines* $\mathfrak{R}$. *Then for each open covering* $\mathfrak{R}$ *of* X *there exists a locally finite closed covering* $\mathfrak{F}$ *of* X *which refines* $\mathfrak{R}$.

Let $\mathfrak{R}$ be any open covering of X. For each $x \in X$ there is an open set $U \in \mathfrak{R}$ which contains x, and therefore (since X is regular) an open neighbourhood V_x of x such that $\overline{V}_x \subset U$. The family $\mathfrak{V}$ formed by the V_x is an open covering of X, hence by hypothesis there is a locally finite covering $\mathfrak{B}$ which is finer than $\mathfrak{V}$. Let $\mathfrak{F}$ be the family of closures of the sets of $\mathfrak{B}$. Since the covering $\mathfrak{V}'$ formed by the $\overline{V}_x$ is finer than $\mathfrak{R}$, and since $\mathfrak{F}$ is finer than $\mathfrak{V}'$, it follows that $\mathfrak{F}$ is a closed covering of X which refines $\mathfrak{R}$. But also $\mathfrak{F}$ is locally finite, because if an open set does not meet a set $B \in \mathfrak{B}$, then it does not meet its closure $\overline{B}$ either.

Lemma 6. *Let* X *be a Hausdorff space such that, given any open covering* $\mathfrak{R}$ *of* X, *there exists a locally finite closed covering* $\mathfrak{F}$ *of* X *which refines* $\mathfrak{R}$. *Then* X *is paracompact.*

Let $\mathfrak{R}$ be any open covering of X. We have to show that there is a locally finite open covering of X which refines $\mathfrak{R}$. Let $\mathfrak{A}$ be a locally finite covering (closed or not) of X which refines $\mathfrak{R}$; for each $x \in X$, let W_x be an open neighbourhood of x which meets only a finite number of sets of $\mathfrak{A}$. The family $\mathfrak{W}$ of sets W_x is an open covering of X. Let $\mathfrak{F}$ be a locally finite *closed* covering of X which refines $\mathfrak{W}$. For each $A \in \mathfrak{A}$, let U_A be a set of $\mathfrak{R}$ which contains A, and let C_A be the union of the sets $F \in \mathfrak{F}$ such that $A \cap F = \emptyset$. Since $\mathfrak{F}$ is locally finite, C_A is closed in X (Chapter I, § 1, no. 6, Proposition 4) and therefore $A' = U_A \cap (X - C_A)$ is open. Since we have $A \cap C_A = \emptyset$ and $A \subset U_A$, it follows that $A \subset A'$, and the family $\mathfrak{A}'$ of sets A', as A runs through $\mathfrak{A}$, is an open covering of X; moreover, since $A' \subset U_A \in \mathfrak{R}$, $\mathfrak{A}'$ refines $\mathfrak{R}$. It remains to show that $\mathfrak{A}'$ is *locally finite*. For each $x \in X$ there is a neighbourhood T of x which meets only a finite number of sets of $\mathfrak{F}$, say $F_1, \dots, F_n$. Since each F_i is contained in a set of the form W_{y_i}, by definition F_i meets only a finite number of sets of $\mathfrak{A}$; let A_{ij} $(1 \leqslant j \leqslant s_i)$ be these sets. If A is a set of $\mathfrak{A}$ other than one of the A_{ij} $(1 \leqslant i \leqslant n,\ 1 \leqslant j \leqslant s_i)$, it follows from the definitions that A' meets none of the F_i, and therefore does not meet $T \subset \bigcup_{i=1}^{n} F_i$. This completes the proof of Lemma 6 and of Theorem 4.

5. BAIRE SPACES

1. NOWHERE DENSE SETS

Definition 1. *A subset* A *of a topological space* X *is said to be nowhere dense if its closure has no interior points.*

Equivalently, A is nowhere dense in X if and only if the *exterior* of A is *dense* in X.

A *closed* set A is nowhere dense if and only if it has no interior points; that is, if and only if it *coincides with its frontier*. An arbitrary subset A is nowhere dense if and only if the closure of A is nowhere dense. Every subset of a nowhere dense set is nowhere dense.

Examples. 1) The empty subset of X is nowhere dense. In a Hausdorff space X, a set consisting of a single point is nowhere dense if and only if the point is not isolated in X. A dense subset is never nowhere dense (unless $X = \emptyset$).

2) The frontier of an *open* or of a *closed* set is always nowhere dense.

3) In the space $\mathbf{R}^n$, every linear subspace of dimension $p < n$ is a nowhere dense set (Chapter VI, § 1, no. 4, Proposition 2).

Remark. The frontier of an *arbitrary* subset need not be nowhere dense: for example, if A and $\complement A$ are both dense sets, then the frontier of A is the whole space.

PROPOSITION 1. *The union of a finite number of nowhere dense sets is nowhere dense.*

It is enough to show that the union of two nowhere dense sets A, B is nowhere dense, and without loss of generality we may assume that A and B are *closed.* The proposition is then equivalent to saying that the intersection of two dense open sets $\complement A$, $\complement B$ is dense. Now if U is a non-empty open set, then $U \cap \complement A$ is open and non-empty, hence

$$(U \cap \complement A) \cap \complement B = U \cap (\complement A \cap \complement B)$$

is open and non-empty.

Let Y be a subspace of a topological space X. A subspace A of Y is said to be *nowhere dense relative to* Y if A is nowhere dense when considered as a subset of the topological space Y.

PROPOSITION 2. *Let* Y *be a subspace of a topological space* X, *and let* A *be a subset of* Y. *If* A *is nowhere dense relative to* Y, *then* A *is nowhere dense relative to* X. *Conversely, if* Y *is open in* X *and* A *is nowhere dense relative to* X, *then* A *is nowhere dense relative to* Y.

Suppose that A is nowhere dense relative to Y. If the closure $\overline{A}$ of A in X contains a non-empty open set U, then $U \cap A$ is not empty (by the definition of closure); hence $U \cap Y$ is a non-empty open set relative to Y, and is contained in the closure $\overline{A} \cap Y$ of A with respect to Y, which is contrary to the hypothesis.

Now suppose that Y is open in X and that $A \subset Y$ is nowhere dense relative to X. If U is open in Y and is not empty, then U is open in X and therefore contains a non-empty set V which is open in X (and *a fortiori* in Y) and does not meet A; hence A is nowhere dense relative to Y.

The second part of Proposition 2 is clearly not valid if Y is not open in X; consider, e.g., the situation where $Y \neq \emptyset$ is nowhere dense in X, and $A = Y$.

2. MEAGRE SETS

Definition 2. *A subset A of a topological space X is said to be meagre if it is the union of a countable family of nowhere dense sets.*

Equivalently, A is meagre if it is contained in a countable union of closed sets each of which has no interior points.

A meagre set can perfectly well be *dense* in X; even the whole space X can itself be a meagre set.

An example of the latter possibility is provided by any *countable* Hausdorff space with no isolated points, e.g., the rational line $\mathbf{Q}$. But a topological space X which is a meagre set in X need not be countable (see Exercise 9).

Every subset of a meagre set in a space X is meagre, and the union of a *countable* family of meagre sets is meagre.

Let Y be a subspace of X. A subset A of Y is said to be *meagre relative to* Y if A is meagre when considered as a subset of the topological space Y. It follows from Proposition 2 of no. 1 that if A is a subset of Y which is meagre relative to Y, then A is meagre relative to X; and that if also Y is *open* in X, every subset A of Y which is meagre relative to X is meagre relative to Y.

3. BAIRE SPACES

Definition 3. *A topological space X is said to be a Baire space if it satisfies one or the other of the following two equivalent conditions:*

(EB) *Every countable intersection of dense open sets in X is dense in X.*

(EB′) *Every countable union of closed sets with no interior points in X has no interior point in X.*

Axiom (EB) can be stated in two other equivalent forms:

(EB″) *Every non-empty open set in X is non-meagre.*

Indeed a set is meagre if and only if it is contained in a countable union of closed sets with no interior points.

(EB‴) *The complement of a meagre set in* X *is dense in* X.

This signifies that a meagre set cannot contain a non-empty open set, and is therefore equivalent to (EB″).

PROPOSITION 3. *Every non-empty open subspace* Y *of a Baire space* X *is a Baire space.*

This follows from (EB″), for every open (resp. meagre) set in Y is open (resp. meagre) in X.

It follows from Proposition 3 that every point of a Baire space has a fundamental system of neighbourhoods, each of which is a Baire space. Conversely:

PROPOSITION 4. *If every point of a topological space* X *has a neighbourhood which is a Baire space, then* X *is a Baire space.*

Let A be a non-empty open set in X, let $x \subset A$ and let V be an open neighbourhood of x which is a Baire space. If A were meagre in X, then $V \cap A$ would be meagre in V and open in V, which is contrary to hypothesis.

PROPOSITION 5. *In a Baire space* X, *the complement of a meagre set is a Baire space.*

Let A be a meagre set in X; then its complement $Y = \complement A$ in X is dense in X. Let B be a meagre set relative to Y; B is also meagre relative to X, hence $A \cup B$ is meagre relative to X. Hence the complement of $A \cup B$ in X, which is also the complement of B in Y, is dense in X and *a fortiori* dense in Y. Hence Y is a Baire space.

THEOREM 1 (Baire). (i) *Every locally compact space* X *is a Baire space.* (ii) *Every topological space* X *on which there exists a metric, compatible with the topology of* X *and which defines on* X *the structure of a complete metric space, is a Baire space.*

We shall show that axiom (EB) is satisfied in each case. Let (A_n) be a sequence of dense open sets in X, and let G be any non-empty open set. We can then define inductively a sequence (G_n) of non-empty open sets such that $G_1 = G$ and $\overline{G}_{n+1} \subset G_n \cap A_n$: for since by hypothesis G_n is not empty, $G_n \cap A_n$ is a non-empty open set; and since X is *regular* in both cases envisaged, there is a non-empty open set G_{n+1} such that $\overline{G}_{n+1} \subset G_n \cap A_n$. Hence the set $G \cap \bigcap_{n=1}^{\infty} A_n$ contains the intersection of the sets G_n, which is equal to the intersection of the sets $\overline{G}_n$; hence it is enough to show that $\bigcap_{n=1}^{\infty} \overline{G}_n \neq \emptyset$. Now if X is locally compact, we may assume that $\overline{G}_2$ is compact; in the compact space $\overline{G}_2$, the $\overline{G}_n$ $(n \geqslant 2)$ form a decreasing sequence of non-empty closed sets, and their intersection is therefore not empty by axiom (C″) (cf. Chapter I, § 9, no. 1, Definition 1]. If X is a complete metric space (with respect to a metric compatible with its topology) we may suppose that $\overline{G}_n$ has been chosen so that its diameter (with respect to this metric) tends to 0 as n tends to $+\infty$; the $\overline{G}_n$ therefore form a Cauchy filter base which converges to a point x, and x necessarily belongs to the intersection of the sets $\overline{G}_n$. Q.E.D.

Remark. There are Baire spaces which belong to neither of these two categories, in particular Baire spaces which are neither metrizable nor locally compact (Exercise 16); there are also metrizable Baire spaces which possess no *complete* metric space structure compatible with their topology (Exercise 14).

4. SEMI-CONTINUOUS FUNCTIONS ON A BAIRE SPACE

THEOREM 2. *Let* X *be a Baire space and let* (f_α) *be a family of lower semi-continuous real-valued functions on* X *such that, at every point* x *of* X, *the upper envelope* $\sup_\alpha f_\alpha(x)$ *is finite. Then every non-empty open set in* X *contains a non-empty open set on which the family* (f_α) *is uniformly bounded above.*

This theorem may also be stated in the form that the set of points in the neighbourhood of which the family (f_α) is uniformly bounded above is a *dense open set.*

Let $f = \sup_\alpha f_\alpha$ be the upper envelope of the family (f_α). The function f is lower semi-continuous (Chapter IV, § 6, no. 2, Theorem 4) and finite at every point of X. It is therefore enough to carry through the proof in the case where the family (f_α) consists of a single function f. Let A_n be the set of points $x \in X$ such that $f(x) \leqslant n$; A_n is closed (Chapter IV, § 6, no. 2, Proposition 1), and the assumptions imply that X is the union of the sets A_n; hence at least one of the A_n has an interior point, and therefore there is a non-empty open set on which f is bounded above (by an integer n). If we apply this result to any non-empty open subset of X (this subspace is also a Baire space by Proposition 3 of no. 3), we have the theorem.

In applications of this theorem it is generally the case that the f_α are *continuous* on X.

Remark. The theorem may be false if we do not suppose that X is a Baire space. For example if, for each rational number p/q (p, q being coprime integers, $q > 0$) we put $f(p/q) = q$, we have a lower semi-continuous function f on the rational line **Q** which is finite at each point (cf. Chapter IV, § 6, no. 2); but there is no non-empty open set in **Q** on which f is bounded above.

6. POLISH SPACES, SOUSLIN SPACES, BOREL SETS

1. POLISH SPACES

DEFINITION 1. *A topological space* X *is said to be Polish if it is metrizable of countable type* (§ 2, no. 8) *and if there is a metric, compatible with the topology of* X, *with respect to which* X *is complete.*

PROPOSITION 1. a) *Every closed subspace of a Polish space is Polish.*

b) *The product of a countable family of Polish spaces is Polish.*

c) *The sum of a countable family of Polish spaces is Polish.*

Every subspace of a metrizable space of countable type is metrizable of countable type, and every closed subspace of a complete space is complete (Chapter II, § 3, no. 4, Proposition 8). Every countable product of metrizable spaces of countable type is again metrizable of countable type (§ 2, no. 8), and every countable product of complete metric spaces is a complete metric space with respect to a metric compatible with its topology (Chapter II, § 3, no. 5, Proposition 10 and Chapter IX, § 2, no. 4, Theorem 1, Corollary 2). Finally, let (X_n) be a sequence of non-empty Polish spaces, and consider the product space $Y = \mathbf{N} \times \prod_n X_n$, where **N** carries the discrete topology; Y is a Polish space by what has already been proved. On the other hand, let a_n be a point of X_n for each n, and let f_n be the mapping of X_n into Y such that for each $x \in X_n$ we have

$$f_n(x) = (n, (y_p)),$$

where $y_p = a_p$ if $p \neq n$ and $y_n = x$. If X is the topological sum of the X_n, it is clear that the mapping f of X into Y which agrees with f_n on X_n for each n is a homeomorphism of X onto $f(X)$; also, for each n, $f_n(X_n)$ is closed in Y, and the family $(f_n(X_n))$ is locally finite because **N** is discrete; therefore $f(X) = \bigcup_n f_n(X_n)$ is closed in Y (Chapter I, § 1, no. 5, Proposition 4), and thus $f(X)$ is a Polish space, by a).

Proposition 2. *Every open subspace of a Polish space is Polish.*

Let X be a Polish space, let d be a metric on X compatible with its topology, and let $U \neq X$ be an open subset of X. Let V be the subset of $\mathbf{R} \times X$ consisting of all points (t, x) such that

$$t.d(x, X - U) = 1;$$

V is closed by Proposition 3 of § 2, no. 2 and is therefore Polish (Proposition 1). Since the restriction to V of the projection $pr_2 : \mathbf{R} \times X \to X$ is a homeomorphism of V onto U (§ 2, no. 2, Proposition 3), U is a Polish subspace of X.

Corollary. *Every locally compact, σ-compact, metrizable space* X *is Polish.*

Let X′ be the compact space obtained by adjoining a point at infinity to X; X′ is metrizable and of countable type (§ 2, no. 9, Corollary to Proposition 16), and X′ is complete with respect to its unique uniformity (Chapter II, § 4, no. 1, Theorem 1). Hence X′ is a Polish space; since X is open in X′, it follows that X is Polish.

Proposition 3. *Let* X *be a Hausdorff topological space. Then the intersection of a sequence* (A_n) *of Polish subspaces of* X *is a Polish subspace.*

Let f be the diagonal mapping of X into $X^{\mathbf{N}}$ [*Set Theory*, Chapter II, § 5, no. 3; recall that $f(x) = (y_n)$ where $y_n = x$ for all $n \in \mathbf{N}$]. We shall use the following lemma:

Lemma 1. *Let* (A_n) *be a sequence of subsets of a Hausdorff topological space* X. *Then the restriction of the diagonal mapping* $f : X \to X^{\mathbf{N}}$ *to the subspace* $\bigcap_n A_n$ *of* X *is a homeomorphism of* $\bigcap_n A_n$ *onto a closed subspace of* $\prod_n A_n$.

This image is the intersection of $\prod_n A_n$ and the diagonal $\Delta = f(X)$, which is closed in $X^{\mathbf{N}}$ because X is Hausdorff (Chapter I, § 8, no. 1); and f is a homeomorphism of X onto Δ.

With the hypotheses of Proposition 3, $\prod_n A_n$ is a Polish space (Proposition 1), hence $\bigcap_n A_n$ is Polish by Lemma 1 and Proposition 1.

Corollary. *The set of irrational numbers, endowed with the topology induced by that of the real line* **R**, *is a Polish space.*

It is the intersection of a countable family of open sets in **R**, namely the complements of sets consisting of a single rational number.

THEOREM 1. *A subspace* Y *of a Polish space* X *is Polish if and only if* Y *is the intersection of a countable family of open sets in* X.

The sufficiency of the condition follows immediately from Propositions 2 and 3. To show necessity, let d be a metric compatible with the topology of Y and with respect to which Y is complete. Let $\overline{Y}$ be the closure of Y in X. For each integer $n > 0$, let Y_n be the set of all $x \in \overline{Y}$ which have an open neighbourhood U such that the diameter of $U \cap Y$ (with respect to the metric d) is $\leqslant 1/n$. Clearly Y_n is open in $\overline{Y}$ and contains Y. Let $x \in \bigcap_n Y_n$; then $x \in \overline{Y}$, and the trace on Y of the neighbourhood filter of x in X is a Cauchy filter (with respect to d); hence this filter converges to a point of Y, and thus $x \in Y$. Hence $Y = \bigcap_n Y_n$. For each n, let H_n be an open subset of X such that $H_n \cap \overline{Y} = Y_n$, and let (U_m) be a sequence of open subsets of X such that $\overline{Y} = \bigcap_m U_m$ (§ 2, no. 5, Proposition 7); then Y is the intersection of the countable family of open sets $(H_n \cap U_m)$.

COROLLARY 1. *A space* X *is Polish if and only if it is homeomorphic to a countable intersection of open sets in the cube* $I^{\mathbf{N}}$, *where* I *is the interval* $[0, 1]$ *of* **R**.

The condition is clearly sufficient, and it is necessary because every metrizable space of countable type is homeomorphic to a subspace of $I^{\mathbf{N}}$ (§ 2, no. 8, Proposition 12).

COROLLARY 2. *Let* X *and* Y *be two Polish spaces and let* $f: X \to Y$ *be a continuous mapping. If* Z *is a Polish subspace of* Y, *then* $\overset{-1}{f}(Z)$ *is a Polish subspace of* X.

For $Z = \bigcap_n Z_n$, where the Z_n are open subsets of Y; hence

$$\overset{-1}{f}(Z) = \bigcap_n \overset{-1}{f}(Z_n),$$

and the sets $\overset{-1}{f}(Z_n)$ are open in X.

2. SOUSLIN SPACES

DEFINITION 2. *A topological space* X *is said to be a Souslin space if it is metrizable and if there exists a Polish space* P *and a continuous mapping of* P *onto* X. *A subset* A *of a topological space* X *is called a Souslin set if the subspace* A *is a Souslin space.*

Clearly every Polish space is Souslin, and the image of a Souslin space X under a continuous mapping of X into a metrizable space Y is a Souslin space.

PROPOSITION 4. *Every Souslin space* X *is of countable type.*

Let P be a Polish space and f a continuous mapping of P onto X. Then the image under f of a countable dense subset of P is a countable dense subset of X.

PROPOSITION 5. *Every closed* (resp. *open*) *subspace of a Souslin space* X *is Souslin.*

For if f is a continuous mapping of a Polish space P onto X, then the inverse image under f of a closed (resp. open) subset of X is a closed (resp. open) subset of P, hence is a Polish subspace of P (no. 1, Propositions 1 and 2).

PROPOSITION 6. *Let* X *be a Souslin space, let* Y *be a Hausdorff space and let* $f: X \to Y$ *be a continuous mapping. Then the inverse image under* f *of a Souslin subspace* A *of* Y *is a Souslin subspace of* X.

Let P, Q be Polish spaces, let g be a continuous mapping of P onto X and let h be a continuous mapping of Q onto A. Let R be the set of all $(x, y) \in P \times Q$ such that $f(g(x)) = h(y)$; R is closed in $P \times Q$ and is therefore a Polish subspace of $P \times Q$ (no. 1, Proposition 1). Let φ be the restriction to R of the projection pr_1. Then the subspace $\overset{-1}{f}(A)$ of X is the image of R under the continuous mapping $g \circ \varphi$ and is therefore a Souslin space.

PROPOSITION 7. *The product and the sum of a countable family of Souslin spaces are Souslin spaces.*

For each integer n, let X_n be a metrizable space, P_n a Polish space, and f_n a continuous mapping of P_n onto X_n. The product (resp. sum) of the spaces P_n is Polish (no. 1, Proposition 1), and the image of this space under the mapping which is the product of the f_n (resp. the mapping which agrees with f_n on P_n for all n) is the product (resp. sum) of the spaces X_n; the latter is metrizable, and is therefore a Souslin space.

PROPOSITION 8. *Let* X *be a metrizable space and let* (A_n) *be a sequence of Souslin subspaces of* X. *Then the union and the intersection of the* A_n *are Souslin spaces.*

These subspaces are certainly metrizable. The existence of the canonical map of the sum of the A_n onto the subspace $\bigcup_n A_n$ of X shows that the latter is a Souslin space; and $\bigcap_n A_n$ is Souslin by virtue of Propositions 5 and 7 and Lemma 1 of no. 1.

In general, even in a Polish space, the complement of a Souslin subspace is not necessarily Souslin (cf. Exercise 6); see, however, no. 6, Theorem 2, Corollary.

PROPOSITION 9. *Let* X *be a metrizable space, and let* A *be a relatively compact Souslin subspace of* X. *Then there exists a compact metrizable space* K, *a decreasing sequence* (B_n) *of subsets of* K, *each of which is a countable union of compact sets, and a continuous mapping* $f : K \to X$, *such that* $A = f\left(\bigcap_n B_n\right)$.

Replacing X by $\overline{A}$ if necessary, we may suppose that X is compact and that A is dense in X. Since A is a Souslin space, there is a Polish space P and a continuous mapping $g : P \to X$ such that $g(P) = A$. By no. 1, Theorem 1, Corollary 1, we may assume that P is the intersection of a decreasing sequence (U_n) of open subsets of the cube $I^{\mathbf{N}}$. Let K be the space $I^{\mathbf{N}} \times X$, which is compact and metrizable (§ 2, no. 4, Theorem 1, Corollary 2). Let $G \subset P \times X$ be the graph of g, let $\overline{G}$ be the closure of G in K, and let f denote the projection of $K = I^{\mathbf{N}} \times X$ onto X; then clearly we have $f(G) = A$. Since g is continuous, G is closed in $P \times X$ (Chapter I, § 8, no. 1, Proposition 2, Corollary 2) and $G = \overline{G} \cap (P \times X)$; hence $G = \bigcap_n B_n$, where

$$B_n = \overline{G} \cap (U_n \times X).$$

Since each U_n is a countable union of closed sets in $I^{\mathbf{N}}$ (§ 2, no. 5, Proposition 7), each B_n is a countable union of compact sets and the proof is complete.

3. BOREL SETS

DEFINITION 3. *Let* X *be a set and let* $\mathfrak{T}$ *be a set of subsets of* X. $\mathfrak{T}$ *is said to be a* σ*-algebra on* X *if the following conditions are satisfied:*

a) *The complement of every set of* $\mathfrak{T}$ *belongs to* $\mathfrak{T}$.

b) *Every countable intersection of sets of* $\mathfrak{T}$ *belongs to* $\mathfrak{T}$.

If $\mathfrak{T}$ is a σ-algebra on X, every *countable union* of sets of $\mathfrak{T}$ belongs to $\mathfrak{T}$ (for the complement of this union is an intersection of sets of $\mathfrak{T}$).

The set $\mathfrak{P}(X)$ of all subsets of X is clearly a σ-algebra. Every intersection of σ-algebras on X is a σ-algebra on X. For any subset of $\mathfrak{P}(X)$, there is therefore a *smallest* σ-algebra containing $\mathfrak{F}$; it is called the σ-algebra *generated* by $\mathfrak{F}$.

Definition 4. *In a topological space* X, *the elements of the* σ*-algebra generated by the set of all closed subsets of* X *are called Borel sets in* X.

Proposition 10. *Let* f *be a continuous mapping of a topological space* X *into a topological space* Y. *Then the inverse image under* f *of every Borel set in* Y *is a Borel set in* X.

Let $\mathfrak{T}$ be the set of all subsets A of Y such that $\overset{-1}{f}(A)$ is a Borel set in X. It follows immediately that $\mathfrak{T}$ is a σ-algebra which contains all the closed subsets of Y; hence $\mathfrak{T}$ contains all Borel sets in Y.

Proposition 11. *In a Souslin space* X, *every Borel set is a Souslin set.*

Let $\mathfrak{T}$ be the set of all subsets A of X such that both A and $\complement A$ are Souslin sets. By Proposition 8 of no. 2, $\mathfrak{T}$ is a σ-algebra. Every closed subset F of X belongs to $\mathfrak{T}$, for both F and $\complement F$ are Souslin sets (no. 2, Proposition 5); hence $\mathfrak{T}$ contains all Borel sets of X (cf. no. 6, Theorem 2, Corollary).

Corollary. *Let* f *be a continuous mapping of a Souslin space* X *into a metrizable space* Y. *If* B *is any Borel set in* X, *then* $f(B)$ *is a Souslin set in* Y.

For B is a Souslin space, hence so is $f(B)$ by the remark following Definition 2 (no. 2).

Remarks. 1) Even when X and Y are Polish spaces it is not in general true that the image of a Borel set in X under a continuous mapping of X into Y is a Borel set in Y (cf. Exercise 6; and no. 7, Theorem 3 Corollary).

2) Let X be a topological space and let Y be a Borel subset of X. Then the Borel sets of the space Y are exactly the Borel sets of X which are contained in Y. For (i) the Borel sets in X which are contained in Y form a σ-algebra on Y which contains the closed sets in Y and hence contains all Borel sets of Y; (ii) the subsets A of X such that $A \cap Y$ is a Borel set of Y form a σ-algebra on X which contains all the closed sets of X and therefore contains all Borel sets of X.

4. ZERO-DIMENSIONAL SPACES AND LUSIN SPACES

Definition 5. *A topological space is said to be zero-dimensional if it is Hausdorff and if every point has a fundamental system of neighbourhoods which are both open and closed.*

Every zero-dimensional space X is *totally disconnected;* for the component of a point x is contained in all the sets containing x which are both open and closed (Chapter I, § 11, no. 5), and the intersection of these sets is just $\{x\}$ if X is zero-dimensional.

Conversely, a *locally compact* totally disconnected space is zero-dimensional (Chapter II, § 4, no. 4, Proposition 6); but there exist totally disconnected metrizable spaces which are not zero-dimensional [Exercise 3 *b*)].

Every subspace of a zero-dimensional space is zero-dimensional, and topological sums and products of zero-dimensional spaces are zero-dimensional.

DEFINITION 6. *A topological space* X *is a Lusin space if it is metrizable and if there exists a zero-dimensional Polish space* P *and a continuous* bijective *mapping of* P *onto* X.

Every Lusin space is clearly a Souslin space.

PROPOSITION 12. *A metrizable space is a Lusin space if and only if it is the image of a Polish space under a continuous* bijective *mapping.*

The condition is clearly necessary; let us show that it is also sufficient. If f is a continuous bijection of a Lusin space X onto a metrizable space Y, it follows from Definition 6 that Y is a Lusin space. Hence we need only show that *a Polish space is a Lusin space.*

Notice first that, if X is a Lusin space, every closed (resp. open) subspace A of X is a Lusin space (cf. no. 7, Theorem 3); for if f is a continuous bijection of a zero-dimensional Polish space P onto X, then $\overset{-1}{f}(A)$ is closed (resp. open) in P and is therefore a zero-dimensional Polish subspace of P (no. 1, Propositions 1 and 2).

Every *countable product* of Lusin spaces is a Lusin space; this follows from Proposition 1 of no. 1 and the fact that every product of zero-dimensional spaces is zero-dimensional. Every *countable intersection* of Lusin subspaces of a Hausdorff topological space is a Lusin subspace; this follows from the preceding remarks and from Lemma 1 of no. 1. Furthermore:

LEMMA 2. *If a metrizable space* X *is such that there exists a countable partition* (A_n) *of* X *formed of Lusin subspaces, then* X *is a Lusin space.*

For each n, let P_n be a zero-dimensional Polish space and let f_n be a continuous bijection of P_n onto A_n. If P is the topological sum of the P_n, then P is Polish (no. 1, Proposition 1) and zero-dimensional, and the mapping $f: P \to X$ which agrees with f_n on P_n for each n is a continuous bijection of P onto X; hence the result.

Let us now show that the interval $I = [0, 1]$ of $\mathbf{R}$ is a Lusin space. Let J be the subspace of I consisting of all irrational numbers; then J is Polish (no. 1, Corollary to Proposition 3); also J is zero-dimensional, for if x is any point of J, the traces on J of intervals of $\mathbf{R}$ of the form $]r, s[$, where r and s are rational and $r < x < s$, form a fundamental system of open and closed neighbourhoods of x in J (because the traces on J of $]r, s[$ and $[r, s]$ are the same). Hence J is a Lusin subspace of I. Now J and the subspaces of I which consist each of a single rational point form a countable partition of I, and therefore I is a Lusin space by Lemma 2.

Let P be an arbitrary Polish space. By Corollary 1 to Theorem 1 (no. 1), P is homeomorphic to a subspace of the cube $I^{\mathbf{N}}$ which is a countable intersection of open sets in $I^{\mathbf{N}}$. Since I is a Lusin space, the remarks at the beginning of the proof show that P is a Lusin space, and the proof of Proposition 12 is therefore complete.

5. SIEVES

DEFINITION 7. *A sieve is a sequence* $C = (C_n, p_n)_{n \geqslant 0}$ *such that, for each* n, C_n *is a countable set and* p_n *is a surjection of* C_{n+1} *onto* C_n.

For each pair of integers m, n such that $0 \leqslant m \leqslant n$, let p_{mn} denote the identity mapping of C_m onto itself if $m = n$, and the surjection $p_m \circ p_{m+1} \circ \cdots \circ p_{n-1}$ of C_n onto C_m if $m < n$. Clearly $p_{mq} = p_{mn} \circ p_{nq}$ whenever $m \leqslant n \leqslant q$, and we may therefore consider the *inverse limit* L(C) of the family (C_n) with respect to the family of mappings (p_{mn}) (*Set Theory*, Chapter III, § 7). If each C_n is endowed with the discrete topology, then L(C) is an inverse limit of topological spaces (Chapter I, § 4, no. 4); as such, L(C) is called the topological space *associated* with the sieve C. L(C) is a closed subspace of the topological product $\prod_n C_n$, and it follows immediately that L(C) is a *zero-dimensional Polish space* (no. 4).

A *sifting* of a metric space X consists of a sieve $C = (C_n, p_n)$ and for each integer $n \geqslant 0$ a mapping φ_n of C_n into the set of *non-empty closed subsets of* X *of diameter* $\leqslant 2^{-n}$, such that:

a) X is the union of the sets $\varphi_0(c)$ as c runs through C_0;

b) for each n and each $c \in C_n$, $\varphi_n(c)$ is the union of the sets $\varphi_{n+1}(c')$, where c' runs through $\overset{-1}{p}_n(c)$.

A sifting is said to be *strict* if in addition, for each n, the sets $\varphi_n(c)$, as c runs through C_n, are *mutually disjoint*.

LEMMA 3. *Every metric space* X *of countable type possesses a sifting. If also* X *is zero-dimensional, then* X *has a strict sifting.*

Note first that if Y is a metric space of countable type and if ε is a real number > 0, then there is a countable covering of Y by sets of diameter $\leqslant \varepsilon$ (§ 2, no. 8, Proposition 13). If, moreover, Y is zero-dimensional, there is such a covering (V_n) formed of sets which are both open and closed; if W_n is the intersection of V_n and $\bigcap_{k<n} (Y - V_k)$, we see that the W_n are closed, of diameter $\leqslant \varepsilon$, pairwise disjoint and cover X. In any case, the closures of the non-empty sets of the covering form a countable covering of Y whose elements are non-empty closed sets of diameter $\leqslant \varepsilon$.

Let X be a metric space of countable type. Let C_0 be the set of indices of a countable covering of X formed of non-empty closed sets of diameter $\leqslant 1$, which are pairwise disjoint if X is zero-dimensional; φ_0 shall be the mapping which associates with each index $c \in C_0$ the corresponding set of the covering. Suppose that we have already defined the C_i and the φ_i and the surjective mappings $p_i : C_{i+1} \to C_i$ for $i \leqslant n$ in such a way that condition b) is satisfied for these indices. If $c \in C_n$, the space $\varphi_n(c)$ has a countable covering by non-empty closed sets of diameter $\leqslant 1/2^{n+1}$, which are mutually disjoint if X [and therefore $\varphi_n(c)$] is zero-dimensional; if $I(c)$ denotes the index set of this covering, we take c_{n+1} to be the sum of the sets $I(c)$ as c runs through C_n; for each $c' \in C_{n+1}$, let $p_n(c')$ denote the element $c \in C_n$ such that $c' \in I(c)$, and let $\varphi_{n+1}(c')$ denote the set with index c' in the covering of $\varphi_n(c)$ under consideration. Clearly we thus define by induction a sifting of X, and this sifting is strict if X is zero-dimensional; hence Lemma 3.

Now suppose that X is a *complete* metric space of countable type, and consider a sifting of X by a sieve C and mappings φ_n. If $\gamma = (c_n)$ is a point of the space L(C) associated with C, the sequence $(\varphi_n(c_n))$ is a decreasing sequence of closed sets in X whose diameters tend to 0; the intersection of this sequence of sets consists therefore of a single point, which we denote by $f(\gamma)$. Thus we have defined a mapping $f : L(C) \to X$. If two points γ, γ' of L(C) have the same i-th coordinates for $i \leqslant n$, it is clear that the distance between $f(\gamma)$ and $f(\gamma')$ is $\leqslant 1/2^n$, and therefore f is *continuous*, by virtue of the definition of the topology of L(C). For each $x \in X$ it follows from the definition of a sifting that we can define by induction on n a sequence $\gamma = (c_n)$ such that $x \in \varphi_n(c_n)$ for each $n \geqslant 0$, and $c_n = p_n(c_{n+1})$; hence $x = f(\gamma)$ and so f is *surjective*. Furthermore, if the sifting is strict, the sequence $\gamma = (c_n)$ such that $x = f(\gamma)$ is unique, and so in this case f is *bijective*. f is said to be the mapping *induced* by the sifting considered.

PROPOSITION 13. *If* X *is any Lusin* (resp. *Souslin*) *space, there exists a sieve* C *and a continuous bijection* (resp. *surjection*) *of* L(C) *onto* X.

Referring to the definition of a Lusin space (no. 4, Definition 6) [resp. a Souslin space (no. 2, Definition 2)], we reduce to the case where

X is Polish and zero-dimensional (resp. Polish), and the preceding argument then proves the result.

6. SEPARATION OF SOUSLIN SETS

THEOREM 2. *Let* X *be a metrizable space. If we are given a sequence* (X_n) *of mutually disjoint Souslin subspaces of* X, *then there exists a sequence* (B_n) *of mutually disjoint Borel sets in* X *such that* $X_n \subset B_n$ *for each* n.

The proof rests on two lemmas:

LEMMA 4. *Let* (A_n), (A'_m) *be two sequences of subsets of a topological space* X. *Suppose that for each pair* (A_n, A'_m) *there is a Borel set* B_{nm} *of* X *such that* $B_{nm} \supset A_n$ *and* $B_{nm} \cap A'_m = \emptyset$. *Then there is a Borel set* B *of* X *which contains* $\bigcup_n A_n$ *and does not meet* $\bigcup_m A'_m$.

For the Borel set $B = \bigcup_n \left(\bigcap_m B_{nm} \right)$ satisfies these conditions.

LEMMA 5. *Let* X *be a Hausdorff space and let* A, A′ *be two disjoint Souslin subspaces of* X. *Then there is a Borel set* B *of* X *such that* $B \supset A$ *and* $B \cap A' = \emptyset$.

By Proposition 13 of no. 5 there exist two sieves C, C′ and continuous mappings f of L(C) onto A, f' of L(C′) onto A′, constructed by the method described in no. 5. For each $n \geqslant 0$ and each $c \in C_n$, let $q_n(c)$ denote the subspace of L(C) formed of all sequences $(c_k)_{k \geqslant 0}$ such that $c_n = c$; $q_n(c)$ is a closed subspace of L(C). For each $\gamma = (c_n) \in L(C)$ the sequence of closed sets $q_n(c_n)$ is decreasing and forms a filter base with γ as limit. Moreover, for each $c \in C_n$, the sets $q_{n+1}(d)$, where d runs through the set $\overset{-1}{p}_n(c)$ in C_{n+1}, form a partition of $q_n(c)$. Make the analogous definitions of notation for the sieve c'.

We shall argue by contradiction and assume that every Borel set which contains A meets A′. In the first place, it follows from Lemma 4 and the definition of a sifting that there exist $c_0 \in C_0$ and $c'_0 \in C'_0$ such that every Borel set containing $f(q_0(c_0))$ meets $f'(q'_0(c'_0))$. We can then define, by induction on n,

$$\gamma = (c_n) \in L(C) \qquad \text{and} \qquad \gamma' = (c'_n) \in L(C')$$

as follows: suppose that c_i and c'_i have already been defined for $i < n$, in such a way that for each index $i < n$ every Borel set containing $f(q_i(c_i))$ meets $f'(q'_i(c'_i))$; applying Lemma 4 and the definition of a sifting to the sets $f(q_{n-1}(c_{n-1}))$ and $f'(q'_{n-1}(c'_{n-1}))$, we see that there exist $c_n \in C_n$ and $c'_n \in C'_n$ such that $p_{n-1}(c_n) = c_{n-1}$ and $p_{n-1}(c'_n) = c'_{n-1}$

and such that every Borel set containing $f(q_n(c_n))$ meets $f'(q'_n(c'_n))$. Now the sequence $f(q_n(c_n))$ converges to a point $a = f(\gamma) \in A$, and the sequence $f'(q'_n(c'_n))$ converges to a point $a' = f'(\gamma') \in A'$. Since $A \cap A' = \emptyset$ and X is Hausdorff, there is a closed neighbourhood V of a which does not contain a', and thus for n sufficiently large, V contains $f(q_n(c_n))$ and does not meet $f'(q'_n(c'_n))$. This is a contradiction, since V is a Borel set.

To prove Theorem 2, let Y_n denote the union of the sets X_i such that $i \neq n$; then Y_n is a Souslin subspace of X (no. 2, Proposition 8). For each index n there exists a Borel set B'_n which contains X_n and does not meet Y_n, by Lemma 5. Let B_n be the intersection of B'_n and $\bigcap_{i<n} (X - B'_i)$. Then the B_n are Borel sets, are mutually disjoint, and are such that $B_n \supset X_n$ for each n.

Corollary. *If a countable partition of a metrizable space is formed of Souslin sets, then these sets are Borel sets. In particular, every Souslin set in a metrizable space, whose complement is a Souslin set, is a Borel set.*

7. LUSIN SPACES AND BOREL SETS

Theorem 3. *Let* X *be a Lusin space. Then a subspace of* X *is a Lusin space if and only if it is a Borel set.*

This theorem is a consequence of the following two lemmas:

Lemma 6. *In a Lusin space* X, *every Borel set is a Lusin subspace of* X.

Let $\mathfrak{T}$ be the set of all subsets A of X such that both A and $\complement A$ are Lusin subspaces of X. Since every closed set and every open set in X is a Lusin subspace of X (no. 4), $\mathfrak{T}$ contains all closed subsets of X, and the lemma will therefore be proved if we can show that $\mathfrak{T}$ is a σ-algebra on X. For this it is enough to show that if (A_n) is a sequence of sets of $\mathfrak{T}$, then $\bigcap_n A_n$ and $\bigcup_n A_n$ are Lusin subspaces of X. Now we have seen in no. 4 that every countable intersection of Lusin subspaces is a Lusin subspace. On the other hand, if B_n is the intersection of A_n and $\bigcap_{i<n} \complement A_i$, it follows from the hypothesis and the preceding remark that B_n is a Lusin subspace; and since $\bigcup_n A_n = \bigcup_n B_n$, the subspace $\bigcup_n A_n$ is a Lusin subspace by Lemma 2 of no. 4.

Lemma 7. *Every Lusin subspace* A *of a metrizable space* X *is a Borel set in* X.

By Proposition 13 of no. 5 there exist a sieve C and a *continuous bijection* f of L(C) onto A. With the notation of Lemma 5 of no. 6, for each integer n and each $c \in C_n$, let $g_n(c)$ denote the subspace $f(q_n(c))$ of X; it is a Lusin subspace and therefore a Souslin subspace of X. As c runs through C_n, the sets $g_n(c)$ are pairwise disjoint, because f is bijective; hence, by Theorem 2 of no. 6, there is a family $c \to g'_n(c)$ $(c \in C_n)$ of Borel sets in X, pairwise disjoint and such that $g'_n(c) \supset g_n(c)$ for all $c \in C_n$. Replacing $g'_n(c)$ by its intersection with the closure $\overline{g_n(c)}$ of $g_n(c)$ in X if necessary, we may suppose that $g'_n(c) \subset \overline{g_n(c)}$. Let c_{n-1}, c_{n-2}, ..., c_0 denote the images of c in C_{n-1}, C_{n-2}, ..., C_0, under the surjections

$$p_{n-1,n} = p_{n-1},\ p_{n-2,n} = p_{n-2} \circ p_{n-1},\ \ldots,\ p_{0n} = p_0 \circ p_1 \circ \cdots \circ p_{n-1}$$

respectively; and let $h_n(c)$ denote the intersection of the sets

$$g'_n(c),\ g'_{n-1}\ (c_{n-1}),\ \ldots,\ g'_0(c_0).$$

Since $q_i(c_i) \supset q_n(c)$ for $0 \leqslant i \leqslant n-1$, $h_n(c)$ contains $g_n(c)$; it is clear also that $h_n(c)$ is a Borel set and is contained in $\overline{g_n(c)}$, and that as c runs through C_n, the $h_n(c)$ are mutually disjoint sets; finally, by construction, for each $c' \in C_{n+1}$ we have $h_{n+1}(c') \subset h_n(p_n(c'))$. Let then B_n be the union of the sets $h_n(c)$ as c runs through C_n; B_n is a Borel set, and $B_{n+1} \subset B_n$; also B_n contains the union of the sets $g_n(c)$ $(c \in C_n)$, which is A. Let B be the intersection of the decreasing sequence of sets B_n; B is a Borel set and contains A. We shall show that B = A, and this will complete the proof.

Let x be a point of B. Then, for each integer n, there exists a unique $c \in C_n$ such that $x \in h_n(c)$; let us denote this c by $c_n(x)$. The sequence $(c_n(x))_{n \geqslant 0}$ belongs to L(C). The decreasing sequence $(g_n(c_n(x)))$ converges by definition to a point $a \in A$; the sequence of closures of these sets also converges to a in X, hence *a fortiori* so does the sequence $(h_n(c_n(x)))$. Now x belongs to all the sets $h_n(c_n(x))$, therefore $x = a \in A$. Lemma 7 is thus proved, and with it Theorem 3.

COROLLARY. *If f is a continuous injective mapping of a Lusin space (or, in particular, a Polish space) X into a metrizable space Y, then f(X) is a Borel set in Y.*

8. BOREL SECTIONS

THEOREM 4. *Let X be a Polish space and let R be an equivalence relation on X, such that the equivalence classes mod R are closed in X and the saturation (with respect to R) of each closed set in X is a Borel set. Then there is a Borel set in X which meets each equivalence class in exactly one point.*

Consider a metric on X compatible with its topology, and with respect to which X is complete. By Lemma 3 of no. 5, there exists a sifting of X, defined by a sieve $C = (C_n, p_n)$ and a sequence of mappings (φ_n). For each $c \in C_n$, let $g_n(c)$ be the saturation of the closed set $\varphi_n(c)$ with respect to R; by hypothesis, $g_n(c)$ is a Borel set in X.

Since each set C_n is countable, we can linearly order each C_n in such a way that the set of elements smaller than a given element is finite. For each $c \in C_n$ we define a set $h_n(c)$ by induction on n, as follows. In the first place, for $c \in C_0$, $h_0(c)$ is the intersection of $\varphi_0(c)$ and the sets $X - g_0(c')$, where $c' \in C_0$ and $c' < c$. For $c \in C_{n+1}$, $h_{n+1}(c)$ is the intersection of $\varphi_{n+1}(c)$, $h_n(P_n(c))$ and the sets $X - g_{n+1}(c')$ for $c' \in C_{n+1}$, $p_n(c') = p_n(c)$ and $c' < c$. The $h_n(c)$ are clearly Borel sets.

We shall prove the following assertion: for each integer $n \geqslant 0$ and each equivalence class H mod R, there is a unique element $c \in C_n$ such that $h_n(c)$ meets H, and we have

$$h(c_n) \cap H = \varphi_n(c) \cap H,$$

which is therefore a *closed* set. For $n = 0$, consider the smallest element $c \in C_0$ such that $\varphi_0(c)$ meets H; then $\varphi_0(c) \cap H$ does not meet any set $g_0(c')$ for which $c' \in C_0$ and $c' < c$; hence it is contained in $h_0(c) \cap H$ and consequently is equal to this set; moreover, we have $H \subset g_0(c)$ and therefore $h_0(c') \cap H$ is empty for $c' \in C_0$ and $c' > c$; thus the assertion is proved for $n = 0$. We continue by induction on n: if there exists $c \in C_{n+1}$ such that $h_{n+1}(c)$ meets H, then it follows from the relation $h_{n+1}(c) \subset h_n(p_n(c))$ and the inductive hypothesis that $p_n(c)$ is the unique element $d \in C_n$ such that $h_n(d)$ meets H. Observe that $h_n(d)$, which is contained in $\varphi_n(d)$, is contained in the union of the sets $\varphi_n(c)$ for which $c \in \overset{-1}{p_n}(d)$, by the definition of a sifting; there is therefore a smallest element $c \in \overset{-1}{p_n}(d)$ such that $\varphi_{n+1}(c)$ meets H. We have therefore

$$\varphi_{n+1}(c) \cap H \subset \varphi_n(d) \cap H = h_n(d) \cap H$$

by the inductive hypothesis. Hence

$$\varphi_{n+1}(c) \cap H \subset \varphi_{n+1}(c) \cap h_n(d),$$

and since by definition $\varphi_{n+1}(c) \cap H$ meets none of the sets $g_{n+1}(c')$ for which $c' \in \overset{-1}{p_n}(d)$ and $c' < c$, it follows from the definition of $h_{n+1}(c)$ that $\varphi_{n+1}(c) \cap H = h_{n+1}(c) \cap H$. Moreover, we have $H \subset g_{n+1}(c)$ and therefore, if $c' \in \overset{-1}{p_n}(d)$ is such that $c' > c$, $h_{n+1}(c') \cap H$ is empty. Hence the assertion is proved for all n.

For each integer n, let S_n be the union of the sets $h_n(c)$, where c runs through C_n. The set S_n is a Borel set, and we have $S_{n+1} \subset S_n$.

Let S be the intersection of the sets S_n, which is also a Borel set in X; we shall show that S meets each equivalence class H mod R in exactly one point. For each n, let $c_n(H)$ be the unique element $c \in C_n$ such that $h_n(c)$ meets H; then $S_n \cap H = \varphi_n(c_n(H)) \cap H$, and $S \cap H$ is the intersection of the sets $\varphi_n(c_n(H)) \cap H$. Since the sequence $(c_n(H))$ belongs to L(C), the decreasing sequence of closed sets $\varphi_n(c_n(H))$, whose diameter tends to 0, converges to a point $x \in X$, since X is complete. The intersection of the closed sets $\varphi_n(c_n(H)) \cap H$ therefore consists of the point x alone, and the proof of Theorem 4 is complete.

Remark. In particular a *closed* equivalence relation R satisfies the hypotheses of Theorem 4. When X is a compact metrizable space, Theorem 4 therefore applies to any *Hausdorff* equivalence relation R, since a Hausdorff equivalence relation on a compact space is closed (Chapter I, § 10, no. 4, Proposition 8).

9. CAPACITABILITY OF SOUSLIN SETS

DEFINITION 8. *Let* X *be a Hausdorff topological space. A capacity on* X *is a mapping* f *of the set* $\mathfrak{P}(X)$ *of all subsets of* X *into the extended real line* $\overline{\mathbf{R}}$, *satisfying the following conditions:*

(CA_I) *If* $A \subset B$, *then* $f(A) \leqslant f(B)$.

(CA_{II}) *If* (A_n) *is any increasing sequence of subsets of* X, *then*

$$f\Big(\bigcup_n A_n\Big) = \sup_n f(A_n).$$

(CA_{III}) *If* (K_n) *is any decreasing sequence of* compact *subsets of* X, *then*

$$f\Big(\bigcap_n K_n\Big) = \inf_n f(K_n).$$

Examples. * Let μ be a positive measure on a locally compact space X; then the corresponding *outer measure* μ^* is a capacity on X.
It can be shown that in Euclidean space $\mathbf{R}^n$ ($n \geqslant 3$), the "Newtonian outer capacity" is a capacity in the sense of Definition 8. *

DEFINITION 9. *Let* f *be a capacity on* X; *a subset* A *of* X *is said to be capacitable (with respect to* f*) if* $f(A) = \sup_k f(K)$, *where* K *runs through the set of compact subsets of* A.

* For example, if f is an outer measure μ^*, every open set is capacitable; the capacitable sets A such that $\mu^*(A) < +\infty$ are precisely the μ-integrable sets. *

PROPOSITION 14. *Let* K *be a compact space and let* f *be a capacity on* K. *If* A *is the intersection of a decreasing sequence* (A_n) *of subsets of* K, *each of which is a countable union of closed sets, then* A *is capacitable.*

It is enough to show that, for each $a < f(A)$, there is a closed set $C \subset A$ such that $f(C) \geqslant a$. Let us first show that there exists a sequence $(B_n)_{n \geqslant 1}$ of closed sets such that $B_n \subset A_n$ and such that, if we define a sequence (C_n) inductively by the conditions $C_0 = A$, $C_n = C_{n-1} \cap B_n$ for $n \geqslant 1$, then $f(C_n) > a$ for each $n \geqslant 0$. Suppose the B_i have been defined for $i < n$; by hypothesis we have $C_{n-1} \subset A \subset A_n$ and $f(C_{n-1}) > a$; since A_n is the union of an increasing sequence of closed sets D_j, it follows from (CA_{II}) that

$$f(C_{n-1}) = \sup_j f(C_{n-1} \cap D_j).$$

Hence there is an index j such that $f(C_{n-1} \cap D_j) > a$, and we may take $B_n = D_j$.

Now let $C = \bigcap_n C_n$. Since $A = \bigcap_n A_n$ and $B_n \subset A_n$, we have

$$C = \bigcap_n B_n;$$

the set C is therefore compact and contained in A.

Let $B'_n = B_1 \cap B_2 \cap \cdots \cap B_n$; (B'_n) is a decreasing sequence of compact subsets of K; as $C_n \subset C_i \subset B_i$ for $i < n$ we also have $C_n \subset B'_n$. By (CA_{III}), $f(C) = \inf_n f(B'_n)$, and since $C \subset C_n \subset B'_n$ we also have $f(C) = \inf_n f(C_n) \geqslant a$. This completes the proof.

THEOREM 5. *Let* X *be a metrizable space and let* Y *be a relatively compact Souslin subspace of* X. *Then* Y *is capacitable with respect to every capacity* f *on* X.

We have seen (no. 2, Proposition 9) that there exists a compact space K, a decreasing sequence (A_n) of subsets of K, each of which is a countable union of compact sets, and a continuous mapping $\varphi : K \to X$ such that $Y = \varphi\left(\bigcap_n A_n\right)$. By Proposition 14, $\bigcap_n A_n$ is capacitable with respect to every capacity on K. Theorem 5 is therefore a consequence of the following proposition:

PROPOSITION 15. *Let* φ *be a continuous mapping of a Hausdorff space* K *into a Hausdorff space* X, *and let* f *be a capacity on* X. *If for each subset* A *of* K *we put* $g(A) = f(\varphi(A))$, *then* g *is a capacity on* K; *moreover, if* A *is capacitable with respect to* g, *then* $\varphi(A)$ *is capacitable with respect to* f.

It is clear that g satisfies axioms (CA_I) and (CA_{II}). On the other hand, let (C_n) be a decreasing sequence of compact subsets of K, and let $C = \bigcap_n C_n$; the sets $\varphi(C)$ are compact, and their intersection is $\varphi(C)$; for this intersection certainly contains $\varphi(C)$, and if $x \in \varphi(C_n)$ for all n, then the sets $\overset{-1}{\varphi}(x) \cap C_n$ form a decreasing sequence of non-empty compact subsets of K, and therefore their intersection is not empty. We have therefore $f(\varphi(C)) = \inf_n f(\varphi(C_n))$, that is $g(C) = \inf_n g(C_n)$; thus g satisfies axiom (CA_{III}) and is therefore a capacity on K.

Now let A be a subset of K which is capacitable with respect to g; if $a < f(\varphi(A)) = g(A)$, then there is a compact set $C \subset A$ such that $g(C) \geqslant a$; thus $\varphi(C)$ is a compact set contained in $\varphi(A)$, and $f(\varphi(C)) \geqslant a$. This shows that $\varphi(A)$ is capacitable with respect to f, and completes the proof of Proposition 15 and hence of Theorem 5.

Remark. * If μ is a positive measure on a locally compact metrizable space X, then every Souslin subset A of X is μ-measurable. For if K is any compact subset of X, $K \cap A$ is a relatively compact Souslin set, hence is capacitable with respect to μ^* and consequently μ-integrable. Note that the complement in X of a Souslin set, although not in general a Souslin set, is μ-measurable *.

APPENDIX

INFINITE PRODUCTS IN NORMED ALGEBRAS

1. MULTIPLIABLE SEQUENCES IN A NORMED ALGEBRA

Let A be a normed algebra over a non-discrete valued field K (Chapter IX, § 3, no. 7, Definition 9); we shall denote by $\|x\|$ the norm of an element $x \in A$, and we shall assume that this norm satisfies the inequality $\|xy\| \leqslant \|x\| . \|y\|$; also we shall assume that A has an identity element e.

Let $(x_n)_{n \in \mathbf{N}}$ be an infinite sequence of points of A. Every *finite* subset J of $\mathbf{N}$, linearly ordered by the ordering of $\mathbf{N}$, defines a *sequence* $(x_n)_{n \in J}$ of points of A, and we define the *product*

$$p_J = \prod_{n \in J} x_n$$

of this sequence; this product is called the *finite partial product* of the sequence $(x_n)_{n \in \mathbf{N}}$, corresponding to the finite subset J of $\mathbf{N}$ (recall that if $J = \varnothing$ we put $\prod_{n \in \varnothing} x_n = e$).

DEFINITION 1. *The sequence* $(x_n)_{n \in \mathbf{N}}$ *is said to be multipliable in the normed algebra* A *if the mapping* $J \to p_J$ *has a limit with respect to the filter of sections of the set* $\mathfrak{F}(\mathbf{N})$ *of finite subsets of* $\mathbf{N}$, *ordered by the relation* $\subset$; *this limit is called the product of the sequence* $(x_n)_{n \in \mathbf{N}}$, *and is denoted by* $\prod_{n \in \mathbf{N}} x_n$ (or simply $\prod_n x_n$); *the* x_n *are called the factors of this product.*

Definition 1 is equivalent to the following: *the sequence* (x_n) *is multipliable and its product is* p *if for each* $\varepsilon > 0$ *there exists a finite subset* J_0 *of* $\mathbf{N}$ *suc that, for every finite subset* $J \supset J_0$ *of* $\mathbf{N}$, *we have* $\|p_J - p\| \leqslant \varepsilon$.

Remarks. 1) When A is a commutative algebra, Definition 1 is identical with that given in Chapter III, § 5, no. 1, Remark 3; but when A is not commutative, the *order* structure of the index set $\mathbf{N}$ is essentially involved in Definition 1. If σ is an arbitrary permutation of $\mathbf{N}$, we cannot in general assert that the sequence $(x_{\sigma(n)})$ is multipliable if the sequence

(x_n) is multipliable; and if both sequences are multipliable, their products will in general be different.

2) Definition 1 can be immediately generalized to the case of a family $(x_n)_{n \in I}$ whose index set I is a subset of $\mathbf{Z}$ (linearly ordered by the order induced by that of $\mathbf{Z}$). We leave it to the reader to extend to this case the results below (cf. Exercises 1 and 2).

2. MULTIPLIABILITY CRITERIA

From now on we shall assume that the normed algebra A is *complete*.

THEOREM 1. *Let* $(x_n)_{n \in \mathbf{N}}$ *be a sequence of points in a complete normed algebra* A.

a) *If* (x_n) *is multipliable and if its product is a unit of* A, *then for each* $\varepsilon > 0$ *there exists a finite subset* J_0 *of* $\mathbf{N}$ *such that, for every finite subset* L *of* $\mathbf{N}$ *which does not meet* J_0, *we have* $\|e - p_L\| \leqslant \varepsilon$.

b) *Conversely, if the sequence* (x_n) *satisfies this condition, it is multipliable. Moreover, if each* x_n *is a unit, then* $\prod_{n \in \mathbf{N}} x_n$ *is a unit.*

a) Let p be the product of the multipliable sequence (x_n), and suppose that p is a unit in A; then (Chapter IX, § 3, no. 7, Proposition 13) there exist $\alpha > 0$ and $a > 0$ such that, for all $y \in A$ for which we have

$$\|y - p\| \leqslant \alpha,$$

y is a unit and $\|y^{-1}\| \leqslant a$. By hypothesis, for every ε such that $0 < \varepsilon < \alpha$, there is a finite subset H_0 of $\mathbf{N}$ such that, for every finite subset H of $\mathbf{N}$ containing H_0, we have $\|p_H - p\| \leqslant \varepsilon$. Let $J_0 = [0, m]$ be an interval of $\mathbf{N}$ which contains H_0; for each finite subset L of $\mathbf{N}$ which does not meet J_0, the integers belonging to L are all greater than those belonging to H_0; hence, if $H = H_0 \cup L$, we have $p_H = p_{H_0} p_L$. Now, since $\|p_{H_0} - p\| \leqslant \varepsilon \leqslant \alpha$, p_{H_0} is a unit, and

$$\|e - p_{H_0}^{-1} p\| \leqslant \varepsilon \|p_{H_0}^{-1} p\| \leqslant a\varepsilon;$$

since $\|p_{H_0} p_L - p\| \leqslant \varepsilon$, we deduce

$$\|p_L - p_{H_0}^{-1} p\| \leqslant \varepsilon \|p_{H_0}^{-1}\| \leqslant a\varepsilon,$$

and finally $\|e - p_L\| \leqslant 2a\varepsilon$.

b) Suppose that, for each $\varepsilon > 0$, there exists a finite subset J_0 of $\mathbf{N}$ such that, for every finite subset L of $\mathbf{N}$ which does not meet J_0, we have $\|e - p_L\| \leqslant \varepsilon$. Let $H_0 = [0, p]$ be an interval of $\mathbf{N}$ which contains J_0; then every finite subset H of $\mathbf{N}$ which contains H_0 can be written in the form $H_0 \cup L$, where the integers in L are all greater than those in H_0; hence we have $p_H = p_{H_0} p_L$, and since L does not meet J_0, $\|p_H - p_{H_0}\| \leqslant \varepsilon \|p_{H_0}\|$, and consequently $\|p_H\| = (1 + \varepsilon)\|p_{H_0}\|$. If $p_{H_0} = 0$, the sequence (x_n) is evidently multipliable and its product is 0; excluding this trivial case, there is an interval $H_1 = [0, q]$ containing H_0 and such that, for every finite subset L of N which does not meet H_1, we have $\|e - p_L\| \leqslant \varepsilon (\|p_{H_0}\|)^{-1}$. As above, it follows that, for each finite subset $H \supset H_1$,

$$\|p_H - p_{H_1}\| \leqslant (\|p_{H_0}\|)^{-1}\|p_{H_1}\|\varepsilon \leqslant \varepsilon(1 + \varepsilon).$$

Cauchy's criterion therefore shows that $J \to p_J$ has a limit in A with respect to the directed set $\mathfrak{F}(N)$.

If all the x_n are units, then so are all the finite partial products p_J; hence for each finite subset H containing H_0 we have

$$\|e - p_{H_0}^{-1} p_H\| \leqslant \varepsilon,$$

and this shows that, in the multiplicative group G of units of A, the image under the mapping $J \to p_J$ of the section filter of $\mathfrak{F}(N)$ is a Cauchy filter base with respect to the left uniformity of G; but since G is *complete* (Chapter IX, § 3, no. 7, Proposition 13), the limit of the mapping $J \to p_J$ belongs to G.

Remark. If (x_n) is multipliable and its product is not a unit, the condition of Theorem 1 is not necessarily satisfied: for example, if all the x_n are equal to the same element x, where $\|x\| < 1$, the sequence (x_n) is multipliable and its product is 0, and for each non-empty finite subset H of N, we have $\|p_H\| \leqslant \|x\| < 1 \leqslant \|e\|$.

COROLLARY 1. *If (x_n) is a multipliable sequence whose product is a unit of* A, *then* $\lim_{n\to\infty} x_n = e$.

COROLLARY 2. *If (x_n) is a multipliable sequence whose product is a unit of* A, *then every subsequence* $(x_{n_k})_{k\in N}$ of (x_n) [(n_k) being a strictly increasing sequence of integers] *is multipliable.*

This follows immediately from the criterion of Theorem 1.

THEOREM 2. *Let* A *be a complete normed algebra. If (u_n) is an absolutely convergent series of elements of* A, *then the sequence $(e + u_n)$ is multipliable in* A; *and if all the elements $e + u_n$ are units in* A, *then so is* $\prod_{n\in N} (e + u_n)$.

Let us apply the criterion of Theorem 1. For every finite subset L of N, we have $p_L - e = \prod_{n\in L} (e + u_n) - e = \sum_M \left(\prod_{n\in M} u_n\right)$, where M runs through the set of all non-empty subsets of L (linearly ordered by the induced ordering). Since $\left\|\prod_{n\in M} u_n\right\| \leqslant \prod_{n\in M} \|u_n\|$, we may write

$$\|p_L - e\| \leqslant \sum_M \Big(\prod_{n\in M} \|u_n\|\Big) = \prod_{n\in L} (1 + \|u_n\|) - 1.$$

Now since the series whose general term is $\|u_n\|$ is convergent by hypothesis, the sequence $(1 + \|u_n\|)$ is multipliable in $\mathbf{R}_+^*$ (Chapter IV, § 7, no. 4, Theorem 4). Hence for each $\varepsilon > 0$ there exists a finite subset J_0 of $\mathbf{N}$ such that, for every finite subset L of $\mathbf{N}$ which does not meet J_0, we have $\left|\prod_{n\in L} (1 + \|u_n\|) - 1\right| \leqslant \varepsilon$; hence the result.

COROLLARY. *If the series whose general term is* u_n *is absolutely convergent, and if none of the elements* $e + u_n$ *is a zero divisor in* A, *then the product* $\prod_{n\in \mathbf{N}} (e + u_n)$ *is not a zero divisor in* A.

There is only a finite number of integers n such that $\|u_n\| \geqslant 1$. Let $J = [0, m]$ be an interval of $\mathbf{N}$ containing all these integers. The product of the sequence $(e + u_n)$ is the product of p_J and the element $\prod_{n>m} (e+u_n)$, all of whose factors are units (Chapter IX, § 3, no. 7, Corollary to Proposition 12), and is therefore itself a unit; since p_J is the product of a finite number of non-zero divisors, it is not a zero divisor, and hence $\prod_{n\in \mathbf{N}} (e + u_n)$ is not a zero divisor.

The *sufficient* condition for multipliability given by Theorem 2 is not in general necessary (cf. Exercise 6). However, it is necessary in the important case where A is an algebra of *finite* rank over the field $\mathbf{R}$ (i.e., A is finite-dimensional as a vector space over $\mathbf{R}$); in particular this is the case if A is the division ring of quaternions $\mathbf{H}$, or a matrix algebra $\mathbf{M}_n(\mathbf{R})$:

PROPOSITION 1. *Let* A *be a normed algebra of finite rank over* $\mathbf{R}$. *If* $(e + u_n)$ *is a multipliable sequence in* A, *whose product is a unit of* A, *then the series whose general term is* u_n *is absolutely convergent.*

From Chapter VII, § 3, no. 1, Proposition 2, there exists a number $c > 0$ such that, for every finite family $(x_i)_{i\in I}$ of points of A, we have

$$(1) \qquad \sum_{i\in I} \|x_i\| \leqslant c.\sup_{J\subset I} \Big\|\sum_{i\in J} x_i\Big\|.$$

Let $(a_n)_{n\in \mathbf{N}}$ be an arbitrary sequence of elements of A. For each finite subset I of N, put

$$p_I = \prod_{i\in I} (e + a_i), \quad s_I = \sum_{i\in I} a_i, \quad \sigma_I = \sum_{i\in I} \|a_i\|.$$

Lemma 1. *For each finite subset* I *of* **N**, *let* $\varphi(\mathrm{I}) = \sup_{\mathrm{J} \subset \mathrm{I}} \|p_{\mathrm{J}} - e\|$. *Then for each subset* J *of* I *we have*

$$(2) \qquad \|p_{\mathrm{J}} - e - s_{\mathrm{J}}\| \leqslant \varphi(\mathrm{I})\sigma_{\mathrm{J}}.$$

The lemma is obvious if J is empty; we shall prove it by induction on the number of elements in J. Let $\mathrm{J} = \mathrm{K} \cup \{j\}$, where $\{j\}$ is strictly larger than every $i \in \mathrm{K}$. Then $p_{\mathrm{J}} = p_{\mathrm{K}}(e + a_j)$ and

$$s_{\mathrm{J}} = s_{\mathrm{K}} + a_j,$$

so that

$$p_{\mathrm{J}} - e - s_{\mathrm{J}} = p_{\mathrm{K}} - e - s_{\mathrm{K}} + (p_{\mathrm{K}} - e)a_j,$$

and by the inductive hypothesis and the definition of $\varphi(\mathrm{I})$, we have

$$\|p_{\mathrm{J}} - e - s_{\mathrm{J}}\| \leqslant \varphi(\mathrm{I})\sigma_{\mathrm{K}} + \varphi(\mathrm{I})\|a_j\| = \varphi(\mathrm{I})\sigma_{\mathrm{J}},$$

which proves the lemma.

Lemma 2. *If* I *is a finite subset of* **N** *such that* $\varphi(\mathrm{I}) < 1/c$, *then*

$$\sigma_{\mathrm{I}} \leqslant c\varphi(\mathrm{I})/(1 - c\varphi(\mathrm{I})).$$

For since $\sigma_{\mathrm{J}} \leqslant \sigma_{\mathrm{I}}$ for every subset J of I, we have from (2)

$$\|s_{\mathrm{J}}\| \leqslant \varphi(\mathrm{I})\sigma_{\mathrm{I}} + \|p_{\mathrm{J}} - e\| \leqslant (1 + \sigma_{\mathrm{I}})\varphi(\mathrm{I});$$

and since also $\sigma_{\mathrm{I}} \leqslant c.\sup_{\mathrm{J} \subset \mathrm{I}} \|s_{\mathrm{J}}\|$, from (1), it follows that

$$\sigma_{\mathrm{I}} \leqslant c\varphi(\mathrm{I})\ (1 + \sigma_{\mathrm{I}}),$$

which leads to the result.

Now let $(e + u_n)$ be a multipliable sequence in A, whose product is a unit; by Theorem 1 there exists a finite subset J_0 of **N** such that, for each finite subset H of **N** which does not meet J_0, we have

$$\Big\|\prod_{i \in \mathrm{H}} (e + u_i) - e\Big\| \leqslant 1/2c.$$

By Lemma 2, it follows that $\sum_{i \in \mathrm{H}} \|u_i\| - 1$ for every finite subset H of **N** which does not meet J_0, and hence (Chapter IV, § 7, no. 1, Theorem 1) the family $(\|u_n\|)$ is summable in **R**.

3. INFINITE PRODUCTS

To each sequence (x_n) of points in a normed algebra A, let us make correspond the sequence of *partial products* $p_n = \prod_{k=0}^{n} x_k$; then the *pair* of sequences (x_n) and (p_n) is called the *infinite product* whose general factor is x_n. The infinite product with general factor x_n is said to be *convergent* if the sequence (p_n) is convergent in A; the limit of this sequence is then called the *product* of the sequence (x_n) and is denoted by $\mathop{\mathsf{P}}\limits_{n=0}^{\infty} x_n$.

PROPOSITION 2. *Let (x_n) be a sequence of points in a complete normed algebra* A.

a) *If the infinite product whose general factor is x_n is convergent and if $\mathop{\mathsf{P}}\limits_{n=0}^{\infty} x_n$ is a unit in* A, *then for each $\varepsilon > 0$ there exists an integer n_0 such that*

$$\left\| \prod_{k=m}^{n} x_k - e \right\| \leqslant \varepsilon$$

whenever $n_0 \leqslant m \leqslant n$.

b) *Conversely, if the sequence (x_n) satisfies this condition, the infinite product with general factor x_n is convergent; and if each of the x_n is a unit in* A, *then $\mathop{\mathsf{P}}\limits_{n=0}^{\infty} x_n$ is a unit.*

The proof of this proposition follows step by step the proof of Theorem 1, and is left to the reader (the finite subsets L of **N** in the proof of Theorem 1 are to be replaced by intervals).

COROLLARY 1. *If the infinite product with general factor x_n is convergent, and if $\mathop{\mathsf{P}}\limits_{n=0}^{\infty} x_n$ is a unit, then $\lim\limits_{n\to\infty} x_n = e$.*

COROLLARY 2. *If the infinite product with general factor x_n is convergent, and if $\mathop{\mathsf{P}}\limits_{n=0}^{\infty} x_n$ is a unit, then the infinite product with general factor*

$$y_n = x_{n+h} \qquad (n \geqslant 0)$$

is convergent.

The product of the sequence (y_n) is denoted by $\mathop{\mathsf{P}}\limits_{n=h}^{\infty} x_n$, and is also called the *residue* of index h of the infinite product with general factor x_n.

Still under the assumption that $\mathop{\mathsf{P}}\limits_{n=h}^{\infty} x_n$ is a unit, it follows from Proposition 2 that if (z_n) is a sequence such that $z_n = x_n$ for all but a finite number of indices, then the product with general factor z_n is convergent.

PROPOSITION 3. *Let (k_n) be a strictly increasing sequence of integers $\geqslant 0$, such that $k_0 = 0$; if the infinite product with general factor x_n converges, and* *if we put*

$$u_n = \prod_{p=k_n}^{k_{n+1}-1} x_p,$$

then the infinite product whose general factor is u_n *is convergent and we have*

$$\mathop{\mathsf{P}}_{n=0}^{\infty} u_n = \mathop{\mathsf{P}}_{n=0}^{\infty} x_n.$$

For the sequence of partial products of the sequence (u_n) is a subsequence of the sequence of partial products of the sequence (x_n).

Finally, the same argument as was used for abelian groups (Chapter III, § 5, no. 7) shows that if a sequence (x_n) in a normed algebra A is *multipliable,* then the product whose general factor is x_n is convergent, and

$$\mathop{\mathsf{P}}_{n=0}^{\infty} x_n = \prod_{n\in\mathbf{N}} x_n$$

$\left(\text{which is also written as } \prod_{n=0}^{\infty} x_n\right)$; the converse is of course not true (cf. Exercise 7).

EXERCISES

§ 1

1) A positive real-valued function f on $X \times X$ is a pseudometric on X if and only if $f(x, x) = 0$ for all $x \in X$ and

$$f(x, y) \leqslant f(x, z) + f(y, z)$$

for all x, y, z in X.

2) Let f be a mapping of $X \times X$ into $[0, +\infty]$. Then the family of sets $\overset{-1}{f}([0, a])$, where a runs through the set of all real numbers > 0, forms a fundamental system of entourages of a uniformity on X if and only if f satisfies the following conditions: *a*) $f(x, x) = 0$ for all $x \in X$; *b*) for each $a > 0$ there exists $b > 0$ such that the relation $f(x, y) \leqslant b$ implies $f(y, x) \leqslant a$; *c*) for each $a > 0$ there exists $c > 0$ such that the relations $f(x, z) \leqslant c$ and $f(z, y) \leqslant c$ imply $f(x, y) \leqslant a$.

The conditions *b*) and *c*) are satisfied in particular if there exists a mapping φ of $I = [0, +\infty]$ onto itself which is continuous and zero at the point 0, and a mapping ψ of $I \times I$ into I which is continuous and zero at the point $(0, 0)$, such that we have identically

$$f(y, x) \leqslant \varphi(f(x, y)) \qquad \text{and} \qquad f(x, y) \leqslant \psi(f(x, z), f(z, y)).$$

3) Let X be a topological space, let $\mathfrak{T}$ be the topology of X, let (f_ι) be a *saturated* family (no. 2) of pseudometrics on X, and let $\mathfrak{U}$ be the uniformity defined by the family (f_ι).

a) The topology induced by $\mathfrak{U}$ is *coarser* than $\mathfrak{T}$ if and only if the f_ι are continuous on $X \times X$ (with respect to the product of the topology $\mathfrak{T}$ by itself).

b) The topology induced by $\mathfrak{U}$ is *finer* than $\mathfrak{T}$ if and only if, for each $x_0 \in X$ and each neighbourhood V of x_0 (in the topology $\mathfrak{T}$) there is an index ι and a real number $a > 0$ such that $f_\iota(x_0, x) \geqslant a$ for all $x \in \complement V$.

4) Let X be a non-Hausdorff uniformizable space and let R be the relation "$y \in \overline{\{x\}}$" between two points x, y of X.

a) Show that R is an equivalence relation on X, that every continuous mapping of X into a Hausdorff space is *compatible* with the relation R (*Set Theory*, R, § 5, no. 7), and that the quotient space X/R is completely regular.

b) If $\mathfrak{U}$ is any uniformity compatible with the topology of X, then the Hausdorff uniformity *associated* with $\mathfrak{U}$ (Chapter II, § 3, no. 9) is defined on the quotient space X/R and is compatible with the topology of X/R. The topological space X/R is called the completely regular space *associated* with the uniformizable space X.

5) Let X be a uniformizable space. Show that the family of *all* pseudometrics on X which are *continuous* on X × X defines a uniformity on X compatible with the topology of X. This uniformity $\mathfrak{U}_0$ is called the *universal* uniformity on X; it is the finest of all uniformities compatible with the topology of X. If Y is any uniform space and if $f : X \to Y$ is a continuous mapping, then f is uniformly continuous with respect to the universal uniformity on X; this is not the case for any other uniformity on X compatible with its topology. If there exists a uniformity $\mathfrak{U}$ on X, compatible with its topology and such that X, endowed with this uniformity, is a complete space, then X is also a complete space with respect to any uniformity $\mathfrak{U}'$ which is finer than $\mathfrak{U}$ and coarser than $\mathfrak{U}_0$. (Observe that the Cauchy filters are the same for $\mathfrak{U}'$ as for $\mathfrak{U}_0$.)

¶ 6) *a*) Let X be a completely regular space and let $\mathfrak{U}$ be a uniformity compatible with the topology of X. Let $\mathfrak{U}^*$ be the coarsest uniformity on X which makes uniformly continuous all the mappings of X into [0, 1] which are uniformly continuous with respect to $\mathfrak{U}$. Show that the uniformity $\mathfrak{U}^*$ is Hausdorff and compatible with the topology of X, and that X is *precompact* with respect to $\mathfrak{U}^*$.

b) Let K be a compact space. Show that every mapping $f : X \to K$ which is uniformly continuous with respect to $\mathfrak{U}$ is also uniformly continuous with respect to $\mathfrak{U}^*$ (note that the unique uniformity on K is the coarsest for which all the continuous mappings of K into the interval [0, 1] are uniformly continuous). Hence show that $\mathfrak{U}^*$ is the finest of the uniformities on X which are coarser than $\mathfrak{U}$ and with respect to which X is precompact.

¶ 7) Let X be a completely regular space. If $\mathfrak{U}_0$ denotes the universal uniformity on X (Exercise 5), then $\mathfrak{U}_0^*$ (Exercise 6) is the coarsest uniformity on X for which all continuous mappings of X into [0, 1] are uniformly continuous. Let βX denote the compact space obtained by completing X with respect to the uniformity $\mathfrak{U}_0^*$. βX is called the *Stone-Čech compactification* of X.

a) Show that every continuous mapping of X into a compact space K can be extended to a continuous mapping of βX into K (cf. *Set Theory*, Chapter IV, § 3).

b) Let f be a homeomorphism of X onto a dense subset X′ of a compact space K, and let $\bar{f}$ be the continuous mapping $\beta X \to K$ which extends f. Show that $\bar{f}(\beta X - X) = K - X'$. [Suppose that there is a point $a \in \beta X - X$ such that $\bar{f}(a) = f(b)$, where $b \in X$; consider a continuous mapping $g : \beta X \to [0, 1]$ such that $g(a) = 1$ and $g(b) = 0$, and let P denote the set of all $x \in \beta X$ such that $g(x) \geqslant 1/2$. There is a continuous mapping $h : K \to [0, 1]$ such that $h(f(b)) = 0$ and $h(f(x)) = 1$ for all $x \in P$; show that the relation $h(f(a)) = 0$ contradicts the fact that a is in the closure of P in βX.]

¶ 8) Let X be a completely regular space and let βX be its Stone-Čech compactification.

a) A filter $\mathfrak{F}$ on X is said to be *completely regular* if $\mathfrak{F}$ has a base $\mathfrak{B}$ consisting of open sets such that for each set $A \in \mathfrak{B}$ there exists a set $B \in \mathfrak{B}$ contained in A, and a continuous mapping $f : X \to [0, 1]$ which is equal to 0 on B and 1 on $\complement A$. A completely regular filter is said to be *maximal* if there exists no strictly finer completely regular filter. Show that, if $\mathfrak{F}$ is any completely regular filter on E, there is a maximal completely regular filter finer than $\mathfrak{F}$ (use Zorn's lemma).

b) A completely regular filter $\mathfrak{F}$ is maximal if and only if, for each pair of open sets A, B of X such that $B \subset A$ and such that there is a continuous mapping of X into [0, 1] which is equal to 0 on B and equal to 1 on $\complement A$, we have either $A \in \mathfrak{F}$, or else $A \notin \mathfrak{F}$ there is a set of $\mathfrak{F}$ which does not meet B (if all the sets of $\mathfrak{F}$ meet B, consider the filter generated by the sets of $\mathfrak{F}$ and the sets $\overset{-1}{f}([0, a[)$, where $a \in]0, 1[$).

c) Show that every maximal completely regular filter $\mathfrak{F}$ is a Cauchy filter on X with respect to the uniformity induced by that of βX [show that each continuous mapping $f : X \to [0, 1]$ has a limit with respect to $\mathfrak{F}$, by arguing by *reductio ad absurdum* and using *b*)].

d) Two distinct maximal completely regular filters $\mathfrak{F}, \mathfrak{F}'$ cannot converge to the same point of βX [observe that by *b*) there exists an open set $A \in \mathfrak{F}$ and an open set $A' \in \mathfrak{F}'$ such that $A \cap A' = \emptyset$; then argue by contradiction, using axiom (O_{IV})].

e) The trace on X of the neighbourhood filter of any point $x_0 \in \beta X$ is a maximal completely regular filter $\mathfrak{B}$ (argue by contradiction, using the definitions of a completely regular filter and of the neighbourhood filter of a point in βX).

9) *a*) Let X be a topological space, let $\mathfrak{T}$ be its topology and let $(f_\iota)_{\iota \in I}$ be a family of mappings of X into $K = [0, 1]$ which are continuous in the topology $\mathfrak{T}$. Let $\mathfrak{T}_0$ be the coarsest topology on X for which all the f_ι are continuous. Show that $\mathfrak{T}_0 = \mathfrak{T}$ provided that the family of sets $\overset{-1}{f_\iota}([0, a[)$, where $\iota \in I$ and $a \in]0, 1[$, is a *subbase* of $\mathfrak{T}$ (Chapter I, § 2, no. 3). The space X is then uniformizable, and if it is Hausdorff it is homeomorphic to a subspace of the cube K^I.

b) Suppose that the family (f_ι) is the family of *all* continuous mappings of X into K (with respect to $\mathfrak{T}$). Show that, if Y is any compact space, every mapping of X into Y which is continuous with respect to $\mathfrak{T}$ is continuous with respect to $\mathfrak{T}_0$ (embed Y in a cube). The topology $\mathfrak{T}$ is uniformizable if and only if $\mathfrak{T}_0 = \mathfrak{T}$.

10) Let X be a completely regular space, let K be a compact subset of X, and let V be a neighbourhood of K in X.

a) Show that there is a continuous mapping of X into [0, 1] which is equal to 1 on K and equal to 0 on $\complement V$ [use (O_{IV}), covering K by a finite number of suitably chosen neighbourhoods].

b) Let X′ be the quotient space of X obtained by identifying all the points of K. Show that X′ is completely regular.

¶ 11) Let X be a completely regular space. Two closed sets A, B in X are said to be *completely separated* if there exists a continuous mapping f of X into [0, 1] such that $f(x) = 0$ on A and $f(x) = 1$ on B.

a) Let βX be the Stone-Čech compactification of X. Show that two closed sets A and B in X are completely separated if and only if their closures in βX do not intersect [use Exercises 7 *a*) and 10 *a*)].

b) Let h be a homeomorphism of X onto a dense subset X′ of a compact space K, and let $\bar{h}$ be the continuous mapping $\beta X \to K$ which extends h. Suppose that, for each pair of completely separated closed subsets A, B of X, the closures of $h(A)$ and $h(B)$ in K do not intersect. Show that $\bar{h}$ is a homeomorphism of βX onto K. (Show that $\bar{h}$ is injective: if a, b are distinct points of βX such that $\bar{h}(a) = \bar{h}(b)$, consider a continuous mapping $f: \beta X \to [0, 1]$ such that $f(a) = 0$, $f(b) = 1$, and consider the sets $X \cap \overset{-1}{f}[0, 1/3]$ and $X \cap \overset{-1}{f}[2/3, 1]$.)

¶ 12) Let X be a completely regular space and let βX be its Stone-Čech compactification.

a) Let f be a finite continuous real-valued function on βX, such that $f(x) > 0$ for all $x \in X$, and such that the set $A = \overset{-1}{f}(0) \subset \beta X - X$ is not empty. Then there is a sequence (a_n) of points of X such that the sequence of numbers $\lambda_n = f(a_n) > 0$ is strictly decreasing and such that $\lim_{n \to \infty} \lambda_n = 0$. For each integer $n > 0$, let I_n be an open interval with centre λ_n in $\mathbf{R}_+^*$, such that the closed intervals $\overline{I}_n$ are mutually disjoint, and let $M_n = \overset{-1}{f}(I_n) \cap X$. For each subset H of N, let M_H be the union of the M_n for which $n \in H$; for each filter $\mathfrak{F}$ on N which is finer than the Fréchet filter, let $\mathfrak{F}'$ denote the filter on X which has the $M_H (H \in \mathfrak{F})$ as a base. Show that $\mathfrak{F}'$ is a completely regular filter (Exercise 8), and that if $\mathfrak{F}_1$, $\mathfrak{F}_2$ are two distinct ultrafilters on N, each of which is finer than the Fréchet filter, then there is no filter on X which is finer than each of $\mathfrak{F}_1'$, $\mathfrak{F}_2'$. Deduce that $\text{Card}(A) \geqslant 2^{2^{\text{Card}(\mathbf{N})}}$ [use Exercise 8 *d*) and Chapter I, § 4, Exercise 5].

b) Suppose that X is infinite and *discrete*. Show that the uniformity induced on X by that of βX is the uniformity of finite partitions (Chapter II, § 2, no. 2 and § 4, Exercise 12) [use Exercise 11 *b*)]. Hence show that $\text{Card}(\beta X) = 2^{2^{\text{Card}(\mathbf{N})}}$ (cf. Chapter I, § 4, Exercise 5). Show that there is a continuous real-valued function $f \geqslant 0$ on X such that $\overset{-1}{f}(0)$ is not empty and is contained in $\beta X - X$.

c) Under the same hypotheses as in *b*), show that if A is an infinite closed subset of βX, then $\text{Card}(A) \geqslant 2^{2^{\text{Card}(\mathbf{N})}}$. [Show first that there exists an infinite sequence (a_n) of points of A, and for each n a neighbourhood V_n of a_n in βX, the V_n being mutually disjoint; deduce that every bounded real-valued function f defined on the set D of the a_n can be extended to a continuous function on βX; for this, consider a real-valued function g defined on X and equal to $f(a_n)$ at each point of $X \cap V_n$. Hence conclude that $\overline{D} \subset A$ is homeomorphic to the Stone-Čech compactification of D (cf. § 4, Exercise 17).]

13) Let G be a locally compact, non-compact topological group. Show that the uniformity induced on the product space $H = G \times G$ by the uniformity of its Stone-Čech compactification βH is strictly finer than the uniformity induced by that of $(\beta G) \times (\beta G)$. [Assuming the result false, show, by applying Exercise 7 *a*) to the mapping $(x, y) \to xy^{-1}$ of H into G, that G would be isomorphic to a subgroup of a compact group; now apply Chapter III, § 3, no. 3, Proposition 4, Corollary 1.]

14) If a topological space X is such that every point of X has a *closed* neighbourhood which is a uniformizable subspace of X, show that X is uniformizable.

¶ 15) *a*) Let X be a locally compact space and let Φ be the set of all continuous mappings $g : X \to [0, 1]$ which are *zero on the complement of some compact set* (depending on g). Let $\mathfrak{U}_1$ be the coarsest uniformity on X for which all the mappings $g \in \Phi$ are uniformly continuous. Show that $\mathfrak{U}_1$ is compatible with the topology of X, that the completion $\hat{X}$ of X with respect to $\mathfrak{U}_1$ is compact, and that $X - \hat{X}$ consists of at most one point (show that if X is not compact, the filter of complements of relatively compact subsets of X is a Cauchy filter with respect to $\mathfrak{U}_1$).

b) Show that $\mathfrak{U}_1$ is the *coarsest* uniformity compatible with the topology of X.

c) Conversely, let X be a completely regular space such that the set of uniformities compatible with the topology of X has a coarsest element $\mathfrak{U}_1$. Show that X is locally compact. (Using Exercise 6, show that X is precompact with respect to $\mathfrak{U}_1$; then show that the complement of X in its completion with respect to $\mathfrak{U}_1$ cannot have more than one point.)

d) Suppose that X is locally compact and σ-compact, and therefore the union of an increasing sequence (U_n) of relatively compact open sets such that $\overline{U}_n \subset U_{n+1}$ (Chapter I, § 9, no. 9, Proposition 15). Show that there exists a continuous real-valued function f on X, such that $f(x) \leqslant n$ for $x \in \overline{U}_n$ and $f(x) \geqslant n$ for $x \in \complement U_n$ [cf. Exercise 10 *a*)]. Let $\mathfrak{U}_2$ be the coarsest uniformity on X for which f and all the functions $g \in \Phi$ are uniformly continuous. Show that $\mathfrak{U}_2$ is compatible with the topology of X and that there is an entourage V of $\mathfrak{U}_2$ such that $V(x)$ is relatively compact in X for all $x \in X$ (cf. Chapter II, § 4, Exercise 9).

¶ 16) Let X be a complete Hausdorff uniform space, and let $\mathfrak{U}$ be the uniformity of X.

a) Let A be a subset of X which is the union of a sequence (F_n) of closed sets, and let $x \notin A$. Show that there is a continuous finite real-valued function $f \geqslant 0$ on X, such that $f(x) = 0$ and $f(y) > 0$ for all $y \in A$. (If g_n is a continuous mapping of X into $[0, 1]$ such that $g_n(x) = 0$ and $g_n(y) = 1$ for all $y \in F_n$, consider the function

$$f(x) = \sum_{n=0}^{\infty} g_n(x)/2^n.).$$

b) Let $h \geqslant 0$ be a continuous real-valued function on X, and let V be the open set $\overset{-1}{h}(]0, +\infty[)$. Let $\mathfrak{U}'$ be the uniformity on V which is the least upper bound of the uniformity induced by $\mathcal{U}$ and the uniformity defined by the pseudometric

$$r(x,y) = \left|\frac{1}{h(x)} - \frac{1}{h(y)}\right|.$$

Show that V is complete with respect to $\mathcal{U}'$.

c) Let B be a subset of X which is the intersection of a family of sets (A_λ), each of which is the union of a countable family of closed sets. Show that there exists a uniformity on B which is compatible with the topology induced on B by the topology of X, and such that B is complete with respect to this uniformity [use *b*)].

17) Let X be a topological space such that each $x \in X$ has a fundamental system of neighbourhoods which are both open and closed. Show that X is uniformizable.

18) Let X be a *countable* completely regular space. Show that, for each $x \in X$, the neighbourhoods of x which are both open and closed form a fundamental system of neighbourhoods of x (cf. § 6, no. 4).

19) Let X be a locally compact space. Show that every function $f \geqslant 0$ which is lower semi-continuous on X is the upper envelope of a family of continuous functions $\geqslant 0$, each of which is zero on the complement of a compact set.

20) *a*) Let X be a completely regular space, and let a be a point of X. Then there exists a continuous function $f \geqslant 0$ on X such that $f(a) = 0$ and $f(x) > 0$ whenever $x \neq a$, if and only if there is a sequence (V_n) of neighbourhoods of a in X such that $\bigcap_n V_n = \{a\}$ [cf. Exercise 16 *a*)].

b) Let X be a completely regular space, each point of which has a countable fundamental system of neighbourhoods. Show that, in the Stone-Čech compactification βX, the set X is equal to the set of points which have a countable fundamental system of neighbourhoods in βX [use *a*) and Exercise 12]. Let X′ be another completely regular space, each point of which has a countable fundamental system of neighbourhoods. Show that if the Stone-Čech compactifications βX and βX′ are homeomorphic, then X and X′ are homeomorphic.

¶ 21) *a*) Let X be a topological space. Show that the following two properties are equivalent: α) every finite continuous real-valued function on X is bounded; β) every bounded continuous real-valued function on X attains its bounds. X is said to be *pseudo-compact* if it has these properties. If X is pseudo-compact and if f is any continuous mapping of X into a topological space X′, then $f(X)$ is pseudo-compact.

b) Consider the following two properties of a topological space X: γ) if (U_n) is any countable open covering of X, then X is the union of a finite number of the closures $\overline{U}_n$; δ) every countable filter base on X which is formed of open sets has at least one cluster point. Show that γ) and δ) are equivalent and that they imply that X is pseudo-compact. In particular, every absolutely closed topological space (Chapter I, § 9, Exercise 19) is pseudo-compact. If X has properties γ) and δ), then so does every subspace of X which is the closure of an open set of X [cf. § 4, Exercise 26 *b*)].

c) Consider the following property of a topological space X: ζ) every locally finite open covering of X is finite. Show that ζ) ⟹ γ). [If (U_n) is an increasing sequence of open subsets of X which cover X and are such that $U_{n+1} \not\subset \overline{U}_n$, consider the covering formed by the sets $U_{n+1} \cap \complement\overline{U}_n$ and the complement of a sequence (a_n) such that

$$a_{n+1} \in U_{n+1} \cap \complement\overline{U}_n.]$$

If X is regular, show that γ) ⟹ ζ). [Let (U_α) be an infinite, locally finite, open covering of X. Define by induction a sequence (α_n) of indices, a sequence (x_n) of points of X, and for each x_n two open neighbourhoods V_n and W_n of x_n such that: (i) we have $\overline{V}_n \subset W_n$, $W_n \subset U_{\alpha_n}$, and W_n meets only a finite number of sets U_α; (ii) U_{α_n} meets none of the W_k with indices $k < n$. Now consider the covering formed by the W_n and the complement of the union of the $\overline{V}_n$.]

d) In a completely regular space X the properties α), β), γ), δ) and ζ) are all equivalent, and are equivalent to the following property: θ) X is precompact in any uniformity compatible with its topology. [To show that α) ⟹ ζ), note that if (U_n) is a countably infinite, locally finite, open covering of X and if $a_n \in U_n$, then there is a continuous real-valued function f on X such that $f(a_n) = n$ for each n [use (O_{IV})]. To show that θ) ⟹ α), use Exercise 5 and observe that in a precompact space every uniformly continuous real-valued function is bounded. To show that γ) ⟹ θ), note that if X is not precompact with respect to a uniformity $\mathcal{U}$, then there is a symmetric entourage V of $\mathcal{U}$ and a sequence (x_n) of points of X such that no two of the sets $V(x_n)$ intersect.]

e) If X is a completely regular pseudo-compact space which is complete with respect to some uniformity compatible with its topology, then X is compact.

f) On a completely regular pseudo-compact space X, the universal uniformity (Exercise 5) is induced by the uniformity of the Stone-Čech compactification of X, and is the unique uniformity, compatible with the topology of X, for which all the continuous mappings of X into [0, 1] are uniformly continuous.

¶ 22) Let X be a Hausdorff uniform space, let Y be a closed subset of X, and let f be a bounded, uniformly continuous, real-valued function on Y. Show that f has an extension $\bar{f}$ to X which is bounded and uniformly continuous on X. [We may assume that $f(\mathrm{Y}) \subset [0, 1]$. For each dyadic number $r = k/2^n \in [0, 1]$, let $\mathrm{A}(r)$ be the set of all $x \in \mathrm{Y}$ such that $f(x) \leqslant r$, and let $\mathrm{B}(r) = \mathrm{A}(r) \cup (\mathrm{X} - \mathrm{Y})$. Define by induction * (as in the proof of Theorem 1 of § 4, no. 1) * an open set $\mathrm{U}(r)$ for each dyadic number $r \in [0, 1]$ such that, whenever r and r' are two dyadic numbers and $r < r'$, there is an entourage V of the uniformity for X for which

$$\mathrm{V}(\mathrm{A}(r)) \subset \mathrm{U}(r'), \qquad \mathrm{V}(\mathrm{U}(r)) \subset \mathrm{U}(r'), \qquad \mathrm{V}(\mathrm{U}(r)) \subset \mathrm{B}(r').$$

Then define $\bar{f}(x)$ to be the greatest lower bound of the dyadic numbers r such that $x \in \mathrm{U}(r)$.]

§ 2

1) Let X be a Hausdorff uniform space and let (f_ι) be a family of pseudometrics on X which define the uniformity of X. Let X_ι be the metric space associated with the uniform space obtained by endowing X with the structure defined by the single pseudometric f_ι (no. 1). Show that X is isomorphic to a subspace of the product uniform space $\prod_\iota \mathrm{X}_\iota$.

2) In a connected metric space X for which the metric is not bounded on $\mathrm{X} \times \mathrm{X}$, show that a sphere cannot be empty.

¶ 3) *a*) Let X be a compact metric space. If the closure in X of every open ball is the closed ball of the same centre and radius, then every ball in X is a connected set [if x and y are two points of X, and if S is the closed ball with centre x and radius $d(x, y)$, show that for each $\varepsilon > 0$ the set $\mathrm{A}_{y,\varepsilon}$ of points of S which can be joined to y by a V_ε-chain contained in S (Chapter II, § 4, no. 4) contains points z such that $d(x, z) < d(x, y)$]. Deduce that $x \in \mathrm{A}_{y,\varepsilon}$ for all $\varepsilon > 0$, arguing by contradiction and using Weierstrass's theorem (Chapter IV, § 6, no. 1, Theorem 1).

b) In the space $Y = \mathbf{R}^2$ with the metric

$$d(x, y) = \max(|x_1 - y_1|, |x_2 - y_2|),$$

let X be the compact subspace consisting of all points (x_1, x_2) such that either $x_1 = 0$ and $0 \leqslant x_2 \leqslant 1$, or $0 \leqslant x_1 \leqslant 1$ and $x_2 = 0$. Show that every ball in X is connected but that the closure of an open ball is not necessarily the closed ball with the same centre and radius.

4) A metric space X is said to be an *ultrametric* space if its metric d satisfies the inequality

$$d(x, y) \leqslant \sup(d(x, z), d(y, z))$$

for all x, y, z in X (which implies the triangle inequality) (cf. § 6, Exercise 2).

a) If $d(x, z) \neq d(y, z)$, show that $d(x, y) = \sup(d(x, z), d(y, z))$.

b) Let $V_r(x)$ be the open ball with centre x and radius r. Show that $V_r(x)$ is both open and closed in X (and hence that X is *totally disconnected*) and that for each $y \in V_r(x)$ we have $V_r(y) = V_r(x)$.

c) Show that the closed ball $W_r(x)$ with centre x and radius r is both open and closed in X, and that for each $y \in W_r(x)$ we have

$$W_r(y) = W_r(x).$$

The distinct open balls of radius r contained in $W_r(x)$ form a *partition* of $W_r(x)$, and the distance between any pair of these balls is equal to r.

d) If two balls (open or closed) of X intersect, then one is contained in the other.

e) A sequence (x_n) of points of X is a Cauchy sequence if and only if $d(x_n, x_{n+1})$ tends to 0 as n tends to infinity.

f) If X is compact, show that for each $x_0 \in X$ the set of values of $d(x_0, x)$ in X is a (finite or infinite) countable subset of $[0, +\infty]$, of which all the points, with the possible exception of 0, are *isolated* [for each value r taken by $d(x_0, x)$, consider the least upper bound of $d(x_0, x)$ on the set of points where $d(x_0, x) < r$, and its greatest lower bound on the set of points where $d(x_0, x) > r$].

5) Let X be a metric space, d its metric. For each pair (x, y) of points of X, let $d_0(x, y)$ denote the greatest lower bound of the real numbers $\alpha > 0$ such that x and y can be joined by a V_α-chain (Chapter II, § 4, no. 4). Show that d_0 is a pseudometric on X, and that the

metric space associated with the uniform space defined by the pseudometric d_0 on X (no. 1) is an ultrametric space (Exercise 4).

¶ 6) Let X be a metric space. If A and B are two non-empty subsets of X, put

$$\rho(A, B) = \sup_{x \in A} d(x, B) \qquad \text{and} \qquad \sigma(A, B) = \sup\big(\rho(A, B),\ \rho(B, A)\big);$$

also put $\sigma(\emptyset, \emptyset) = 0$, $\sigma(\emptyset, A) = \sigma(A, \emptyset) = +\infty$ for any non-empty subset A of X. Show that σ is a pseudometric on the set $\mathfrak{P}(X)$ of all subsets of X, and that the uniformity defined by σ coincides with the uniformity constructed from that of X by the procedure of Chapter II, § 1, Exercise 5.

Suppose that X is *bounded*. Then σ is a *metric* on the set $\mathfrak{F}(X)$ of non-empty closed subsets of X. If, moreover, X is *complete*, show that $\mathfrak{F}(X)$ is a complete metric space [let Φ be a Cauchy filter on $\mathfrak{F}(X)$; for each set $\mathfrak{X} \in \Phi$, let $S(\mathfrak{X})$ be the union of the subsets A of X which belong to $\mathfrak{X}$; show that the sets $S(\mathfrak{X})$ form a filter base on X as $\mathfrak{X}$ runs through Φ, that this filter base has a non-empty set C of cluster points, and that Φ converges to C].

¶ 7) Let X be an infinite discrete space. Show that the uniformity of finite partitions (Chapter II, § 2, no. 2) on X, which is compatible with the topology of X, is not metrizable (otherwise there exists a sequence $(\mathfrak{F}_n)$ of finite partitions of X such that every finite partition of X is formed of sets, each of which is a union of sets belonging to one of the partitions $\mathfrak{F}_n$; from this it would follow that the set of finite partitions of X is countable).

8) Let (X_ι) be an uncountable family of Hausdorff topological spaces, each of which has at least two distinct points. Show that, in the product space $\prod_\iota X_\iota$, no point has a countable fundamental system of neighbourhoods.

9) Let X be a topological space, each point of which has a countable fundamental system of neighbourhoods.

a) X is Hausdorff provided that every convergent sequence in X has only one limit.

b) If a is a cluster point of an infinite sequence (x_n) of points of X, then there is an infinite subsequence of (x_n) which converges to a.

c) Let A be a non-empty subset of X, let x_0 be a point of $\overline{A}$, and let f be a mapping of A into a Hausdorff topological space X'. Then a point $a \in X'$ is a limit of f at the point x_0, relative to A, provided that, for every sequence (x_n) of points of A which converges to x_0, the sequence $(f(x_n))$ converges to a.

d) With the notation of *c*), suppose also that every point of X' has a countable fundamental system of neighbourhoods. If $a \in X'$ is a cluster point of f at x_0 relative to A, then there is a sequence (x_n) of points of A which converges to x_0 and is such that the sequence $(f(x_n))$ converges to a.

¶ 10) Let X be a compact space. If there exists a continuous real-valued function f on $X \times X$ such that the relation $f(x, y) = 0$ is *equivalent* to $x = y$, then X is metrizable. (Show that if V runs through a fundamental system of neighbourhoods of 0 in **R**, the sets $\overset{-1}{f}(V)$ form a fundamental system of neighbourhoods of the diagonal Δ in $X \times X$.)

¶ 11) Let X be a compact metric space, with metric d. Show that if f is a mapping of X into X such that, for each pair x, y of points of X, we have $d(f(x), f(y)) \geqslant d(x, y)$, then f is an *isometry* of X *onto* itself. [Let a, b be any two points of X, and put $f^n = f^{n-1} \circ f$, $a_n = f^n(a)$, $b_n = f^n(b)$; show that, for each $\varepsilon > 0$, there exists an index k such that $d(a, a_k) \leqslant \varepsilon$ and $d(b, b_k) \leqslant \varepsilon$, by choosing suitable subsequences of (a_n) and (b_n); hence show that $d(a_1, b_1) = d(a, b)$ and that $f(X)$ is dense in X.]

¶ 12) Let X be a compact metrizable space.

a) Show that there is a continuous mapping of Cantor's triadic set K (Chapter IV, § 2, no. 5) onto X (cf. Chapter IV, § 8, Exercise 11).

b) If in addition X is totally disconnected and has no isolated points, then X is homeomorphic to K. [Argue as in Chapter IV, § 8, Exercise 12, using the fact that every neighbourhood of a point $x \in X$ contains a neighbourhood of x which is both open and closed (Chapter II, § 4, no. 4, Corollary to Proposition 6)].

¶ 13) *a*) On the set **R** of real numbers, consider the topology $\mathfrak{T}$ defined as follows: for each $y > 0$, $U_y(x)$ denotes the union of the intervals $[x, x+y[$ and $]-x-y, -x[$; $\mathfrak{T}$ is the topology for which the $U_y(x)$ form a fundamental system of neighbourhoods of x as y runs through the set of real numbers > 0. Let X be the space obtained by endowing the interval $[-1, +1]$ with the topology induced by $\mathfrak{T}$. Show that X is compact. (Consider a filter $\mathfrak{F}$ on X; if $x \in X$ is a cluster point of $\mathfrak{F}$ with respect to the topology of the real line, show that either x or $-x$ is a cluster point of $\mathfrak{F}$ with respect to $\mathfrak{T}$.)

b) Every point of X has a countable fundamental system of neighbourhoods, and X has a countable dense subset, but X is not metrizable (show that its topology has no countable base).

c) Let A be an open subset of X. Show that A is the union of a countable family (I_n) of open intervals contained in $[0, 1]$, the intervals $-I_n$, a subset of the set of left-hand points of the intervals I_n and $-I_n$, and possibly the point $+1$. Hence show that A is the union of a countable family of closed subsets of X.

d) Let Y be the set $I \times \{1, 2\}$, where I is the interval $[0, 1]$ of $\mathbf{R}$; let X_i denote $I \times \{i\}$ $(i = 1, 2)$ and let f denote the bijection $(x, 1) \to (x, 2)$ of X_1 onto X_2. For each $x \in I$, let $\mathfrak{B}((x, 2))$ denote the set of subsets consisting only of $\{(x, 2)\}$, and let $\mathfrak{B}((x, 1))$ denote the set of subsets of the form $V_1 \cup (f(V_1) - \{x, 2\})$, where $V_1 = V \times \{1\}$ and V runs through a fundamental system of neighbourhoods of x in I. Show that for each $y \in Y$, $\mathfrak{B}(y)$ is a fundamental system of neighbourhoods of y for a topology on Y. Endowed with this topology, Y is compact and every point of Y has a countable fundamental system of neighbourhoods, but Y has no countable dense subset. Every closed subset of Y contained in X_2 is finite; hence show that the compact subset X_1 of Y has no countable fundamental system of neighbourhoods. If R is the equivalence relation on Y whose classes are the set X_1 and the points of $Y - X_1$, R is closed and every equivalence class mod R is compact, but there is a point in Y/R which has no countable fundamental system of neighbourhoods [cf. § 4, Exercise 24 *a*)].

14) *a*) Let X be an accessible topological space (Chapter I, § 8, Exercise 1). Show that the following properties are equivalent:

α) Every sequence of points of X has a cluster point.

β) No infinite discrete subspace of X is closed.

γ) Every countable open covering of X contains a finite open covering X.

δ) Given any infinite open covering $\mathfrak{R}$ of X, there exists an open covering $\mathfrak{S} \subset \mathfrak{R}$ of X, distinct from $\mathfrak{R}$.

A topological space X is said to be *countably compact* if it is Hausdorff and has the above properties.

b) A sequence of points in a countably compact space is convergent if and only if it has a unique cluster point.

c) Every closed subspace of a countably compact space is countably compact. Conversely, if X is Hausdorff and if every point of X has a countable fundamental system of neighbourhoods, then every countably compact subspace of X is closed in X.

d) Let f be a continuous mapping of a countably compact space X into a Hausdorff space X'. Then $f(\mathrm{X})$ is a countably compact subspace of X'.

e) Let X be a countably compact space, every point of which has a countable fundamental system of neighbourhoods. Then every sequence of points of X has a convergent subsequence.

f) Let (X_n) be a countable sequence of topological spaces, in each of which every point has a countable fundamental system of neighbourhoods. The product space $\prod_{n=1}^{\infty} \mathrm{X}_n$ is then countably compact if and only if each of the spaces X_n is countably compact. [Use *e*) and Chapter I, § 6, Exercise 16.]

g) A countably compact space in which every point has a countable fundamental system of neighbourhoods is regular.

h) If a countably compact space has a *countable base* of open sets, it is *compact* and therefore metrizable.

i) Every lower semi-continuous real-valued function on a countably compact space attains its greatest lower bound. In particular, every countably compact space is pseudo-compact (§ 1, Exercise 21) [cf. § 4, Exercise 26 *b*)].

j) Show that the property δ) of *a*) implies the following:

ζ) Every point-finite (§ 4, no. 3) open covering of X contains a finite open covering of X. (Argue by contradiction.) Conversely, a regular space which satisfies ζ) is countably compact [show that ζ) then implies β)].

¶ 15) Let $\mathrm{X} = [a, b[$ be the locally compact space defined in Chapter I, § 9, Exercise 12.

a) A subset of X is relatively compact if and only if it is bounded. Hence show that X is countably compact and therefore (Proposition 15) non-metrizable (observe that every countable subset of X is bounded).

b) Show that every point of X has a metrizable neighbourhood (use Proposition 16).

c) If A and B are two non-compact closed sets in X, their intersection is not empty (form an increasing sequence in which the even-numbered terms are points of A and the odd-numbered terms are points of B). Deduce that every neighbourhood of a non-compact closed set is the complement of a relatively compact set. If A is the set of non-isolated points of X, show that A is closed and is not the intersection of any countable family of open sets.

d) Let X' denote the interval $[a, b]$ endowed with the following topology: for each $x \in \mathrm{X}$ the intervals $]y, x]$, where y runs through the set of elements $< x$, form a fundamental system of neighbourhoods of x; a fundamental system of neighbourhoods of b is formed by the sets V_x where, for each $x \in X$, V_x denotes the union of $\{b\}$ and the set of points of $[x, b]$ which have a predecessor (i.e., which are *isolated* in X). Show that X′ is countably compact but *not regular* [cf. Exercise 14 *g*)] and that the subspace X of X′ is countably compact but *not closed* [cf. Exercise 14 *c*)].

16) Let X and Y be two countably compact spaces. Show that the product $X \times Y$ is countably compact in each of the following two cases: (i) one of X, Y is compact; (ii) one of X, Y is such that each point has a countable fundamental system of neighbourhoods. [In case (i), supposing that Y is compact, consider an increasing sequence (G_n) of open sets whose union is $X \times Y$; for each n, let H_n be the set of all $x \in X$ such that $\{x\} \times Y \subset G_n$; show that H_n is open and that $X = \bigcup_n H_n$.]

¶ 17) Let βN be the Stone-Čech compactification of the discrete space N (§ 1, Exercise 7).

a) Show that there are two countably compact subspaces A, B of βN such that $A \cap B = N$ and $A \cup B = \beta N$. [Let $\aleph_\alpha = 2^{2^{\mathrm{Card}(N)}}$; by virtue of § 1, Exercise 12 *b*), there is a bijection $\xi \to S_\xi$ of the ordinal ω_α (*Set Theory*, Chapter III, § 6, Exercise 10) onto the set of countably infinite subsets of X. By transfinite induction, define two injective mappings $\xi \to x_\xi$, $\xi \to y_\xi$ of ω_α into $\beta N - N$ such that $x_\xi \in \overline{S}_\xi, y_\xi \in \overline{S}_\xi$ for each ξ, and such that if P (resp. Q) is the set of the x_ξ (resp. y_ξ) we have $P \cap Q = \varnothing$, $P \cup Q = \beta N - N$; for this, make use of § 1, Exercise 12 *b*) and *c*). Then show that $A = P \cup N$ and $B = Q \cup N$ satisfy the conditions of the question.]

b) Show that the product space $A \times B$ is not countably compact (note that the intersection of $A \times B$ with the diagonal of $\beta N \times \beta N$ is an infinite discrete closed subspace of $A \times B$).

¶ 18) Let X be a metric space such that, for each $x \in X$, there is an open ball with centre x which, when considered as a subspace of X, has a countable base. Let r_x be the least upper bound of the radii of balls with centre x which have this property.

a) Show, using Zorn's lemma, that there is a *maximal* family (B_α) of open balls in X which are pairwise disjoint and such that, if x_α is the centre of B_α, the radius of B_α is $< r_{x_\alpha}$. Show that the union of the B_α is dense in X, and hence that there is a dense subset M of X such that every open ball with centre at an arbitrary point $x \in X$ and radius $< r_x$ contains only a *countable* infinity of points of M (note that such a ball can meet only a countable infinity of pairwise disjoint open sets, and use Proposition 12).

b) For each point $x \in M$, let S_x denote the open ball with centre x and radius $\frac{1}{3} r_x$. Show that the S_x cover X, and that an arbitrary point of X belongs only to a *countable* infinity of the S_x [note that if $y \in S_x$, we have $d(x, y) \leqslant \frac{1}{2} r_y$].

c) Let R be the following equivalence relation between points x, y of X: there exists a sequence $(z_i)_{1 \leqslant i \leqslant n}$ of points of M such that $x \in S_{z_1}$, $y \in S_{z_n}$ and S_{z_i} meets $S_{z_{i+1}}$ for $1 \leqslant i \leqslant n - 1$. Show that the equivalence classes (mod R) are subspaces of X which are both open and closed in X and which have a countable base; in other words, X is the *topological sum* of metric spaces with countable bases (Chapter I, § 2, no. 4).

¶ 19) In a metric space every relatively compact set is bounded. If X is a topological space, show that the following conditions on X are equivalent: α) there exists a metric on X, compatible with its topology, such that every bounded subset of X (with respect to this metric) is relatively compact; β) X is locally compact and has a countable base. [To show that α) ⟹ β), show that if every bounded set is relatively compact, then X is locally compact and σ-compact. To show that β) ⟹ α), note that X is metrizable by the corollary to Proposition 12; if d is a metric on X compatible with the topology of X, and if f is the function defined in § 1, Exercise 15 *d*), take the uniformity on X defined by the two pseudometrics $d(x, y)$ and $|f(x) - f(y)|$.]

¶ 20) *a*) Give an example of a closed equivalence relation R on a metrizable locally compact space X which has a countable base such that X/R is paracompact but such that there exists a point of X/R which has no countable fundamental system of neighbourhoods [cf. Chapter I, § 10, Exercise 17 and Chapter IX, § 4, Exercise 24 *a*)].

b) Let X be a metrizable space and let R be a closed equivalence relation on X. Show that if every point of X/R has a countable fundamental system of neighbourhoods, then each equivalence class (mod R) has a compact frontier in X. (Show that if the result were false there would exist a sequence in X, all of whose points were distinct, which had no cluster point, and whose image in X/R was a sequence which converged to a point distinct from all the points of the sequence.) For each $z \in X/R$, let C_z be the inverse image of z in X, and let F_z be the frontier of C_z if this frontier is not empty; if C_z is both open and closed, let less F_z be any subset of C_z consisting of a single point. Let Y be the union of the sets F_z as z runs through X/R. Show that Y/R_Y is homeomorphic to X/R.

¶ 21) *a*) Let X be a metrizable space, let d be a bounded metric compatible with the topology of X, let σ be the metric corresponding to d on the set $\mathfrak{F}(X)$ of non-empty closed subsets of X (Exercise 6); and let R be an open and closed equivalence relation on X. Show that the restriction of σ to X/R is a metric compatible with the quotient topology. [Use Exercise 20 *b*) to show that every class mod R is open and compact. Then prove that if a sequence (z_n) in X/R tends to a point a and if $\varphi : X \to X/R$ is the canonical mapping, we have $\sigma(\overset{-1}{\varphi}(a), \overset{-1}{\varphi}(z_n)) \to 0$; argue by contradiction, using the fact that R is open.]

b) On the compact interval $I = [0, 1]$ of $\mathbf{R}$, let R be the equivalence relation whose classes are the points of the Cantor set K (Chapter VI, § 2, no. 5), other than the end-points of the intervals contiguous to K, and the closures of the intervals contiguous to K. Show that the quotient space I/R is homeomorphic to I [Chapter IV, § 8, Exercise 16 *b*)] but that the metric σ is not compatible with the topology of I/R.

¶ 22) Let X be a Hausdorff topological space with a countable base (U_n), and let R be a Hausdorff equivalence relation on X such that every point of X/R has a countable fundamental system of neighbourhoods. Show that the topology of X/R has a countable base. [Let $\varphi : X \to X/R$ be the canonical mapping, and show that the interiors of finite unions of sets $\varphi(U_n)$ form a base of the topology of X/R: if V is a neighbourhood of a point $z \in X/R$, and (W_k) is a sequence of sets belonging to the base (U_n), contained in $\overset{-1}{\varphi}(V)$ and covering $\overset{-1}{\varphi}(z)$, show that there is a finite number of indices k such that the union of the $\overset{-1}{\varphi}(W_k)$ is a neighbourhood of z. Argue by contradiction: if this statement were false we could construct a sequence (y_n) of distinct points of X/R, tending to z and such that y_n did not belong to the union of the $\varphi(W_k)$ for $k \leqslant n$; then show that the union of the $\overset{-1}{\varphi}(y_n)$ would be closed in X.]

Show that the same conclusion is valid if R is assumed to be closed and every equivalence class mod R is compact (similar method).

¶ 23) A topological space is said to be *submetrizable* if its topology is finer than the topology of a metrizable space. A submetrizable space is Hausdorff but not necessarily regular (Chapter I, § 8, Exercise 20).

a) A completely regular space X is submetrizable if and only if there exists a uniformity on X, compatible with the topology of X and defined by a family Φ of pseudometrics which contains at least one metric d (cf. § 1, Exercise 3). Let $\hat{X}$ be the completion of X with respect to this uniformity; the metric d extends to a pseudometric $\bar{d}$ on $\hat{X}$; show that for each $x \in \hat{X}$ there is at most one point $y \in X$ such that $\bar{d}(x, y) = 0$.

b) Show that if X is a completely regular submetrizable space, there exists a uniformity, compatible with the topology of X, and with respect to which X is complete [use *a*) and § 1, Exercise 16 *c*)]. Hence show that if in addition X is pseudo-compact (§ 1, Exercise 21), then X is compact.

c) Show that a locally compact, paracompact, submetrizable space X is metrizable (reduce to the case where X is σ-compact by using Chapter I, § 9, no. 10, Theorem 5; then use the Corollary to Proposition 16) [cf. § 5, Exercise 15].

¶ 24) *a*) Let X be a complete metric space, and let A be a subset of X which is the intersection of a countable family of open sets. Show that there exists a metric on A with respect to which A is a complete metric space, and which defines on A the topology induced by that of X and also defines a uniformity finer than that induced by the uniformity of X [cf. § 1, Exercise 16 *b*) and *c*)].

b) Conversely, let X be a metrizable space and let A be a subset of X such that there exists a metric d on A which is compatible with the topology induced by that of X and for which A is a complete metric space. Show that A is a countable intersection of open sets in X. [For each integer $n > 0$, consider the set G_n of points $x \in \bar{A}$ which have an open neighbourhood U such that the diameter of $A \cap U$ (with respect to d) is $\leqslant 1/n$.]

¶ 25) Let X be a metrizable space, let βX be its Stone-Čech compactification (§ 1, Exercise 7) and let d be a bounded metric compatible with the topology of X. For each $x \in X$, let $f_x(y)$ denote the real-valued function obtained by extending $y \to d(x, y)$ by continuity to βX.

a) Show that $f_x(z) + f_y(z) \geqslant d(x, y)$ whenever x and y are in X and z is in βX. For each $y \in \beta X$ other than $x \in \beta X$, show that $f_x(y) > 0$.

b) Show that if X is complete with respect to the metric d, then X is the intersection of a sequence of open sets in βX. [Consider the set G_n of all $y \in \beta X$ such that $f_x(y) < 1/n$ for at least one point $x \in \varphi X$, and show by using *a*) that $X = \bigcap_n G_n$.]

c) Conversely, show that if X is the intersection of a sequence of open sets G_n in βX, there exists a metric d' on X which is compatible with the topology of X, and with respect to which X is complete. [Considering only the case in which $X \neq \beta X$, note that $\beta X - X$ is the union of a family of compact sets $F_n = \beta X - G_n$; for each n, let $g_n(x) = \inf_{y \in F_n} f_x(y)$, and show by using *a*) that $g_n(x) > 0$ for all $x \in X$, and that g_n is continuous on X. Finish the proof as in Exercise 16 *b*) of § 1.]

§ 3

1) A pseudometric f on a group G (written multiplicatively) is said to be *left-invariant* (resp. *right-invariant*) if $f(zx, zy) = f(x, y)$ [resp. $f(xz, yz) = f(x, y)$] for all x, y, z in G.

a) If f is a left-invariant pseudometric, the real-valued function $g(x) = f(e, x)$ on G (e being the identity element of G) satisfies the following conditions:

(i) $g(x) \geqslant 0$ for all $x \in G$, and $g(e) = 0$;

(ii) $g(x^{-1}) = g(x)$;

(iii) $g(xy) \leqslant g(x) + g(y)$.

Conversely, if g is any real-valued function on G satisfying these conditions, then $f(x, y) = g(x^{-1}y)$ is a left-invariant pseudometric on G.

b) The topology $\mathcal{T}$ defined by a saturated family (f_ι) of left-invariant pseudometrics on a group G is compatible with the group structure of G if and only if, for each $a \in G$, each index ι and each real number $\alpha > 0$, there exists an index $\varkappa$ and a real number $\beta > 0$ such that the relation $f_\varkappa(e, x) \leqslant \beta$ implies $f_\iota(e, axa^{-1}) \leqslant \alpha$. If this condition is satisfied, the uniformity on G defined by the family f_ι is the same as the left uniformity of the topological group obtained by endowing G with the topology $\mathcal{T}$.

c) On every topological group G there exists a family of left-invariant pseudometrics such that the uniformity defined by this family is the same as the left uniformity of G.

2) Let G be a topological group whose left and right uniformities coincide. Show that this unique uniformity can be defined by a family of pseudometrics which are simultaneously left-and right-invariant (using Exercise 3 of Chapter III, § 3, show that if V is any neighbourhood of the identity element e of G, then $V_0 = \bigcap_{x \in G} xVx^{-1}$ is a neighbourhood of e).

3) Let G be a topological group, let (f_ι) be a saturated family of left-invariant pseudometrics on G which define the left uniformity of G, and let $g_\iota(x) = f_\iota(e, x)$. Let H be a normal closed subgroup of G, and for each coset $\dot{x} \in G/H$ let $h_\iota(\dot{x}) = \inf_{x \in \dot{x}} g_\iota(x)$; show that, if $\bar{f}_\iota(\dot{x}, \dot{y}) = h_\iota(\dot{x}^{-1}\dot{y})$, the $\bar{f}_\iota$ form a family of left-invariant pseudometrics on G/H which define the left uniformity of G/H (argue as in Remark 2 of no. 1).

4) Show that the topology of a valued division ring is locally retrobounded (Chapter III, § 6, Exercise 22).

* ¶ 5) Let ω be an absolute value on the field $\mathbf{Q}$ of rational numbers.

a) Show that ω is uniquely determined by its values at the prime numbers.

b) If there is a prime number p such that $\omega(p) \leqslant 1$, show that $\omega(q) \leqslant 1$ for every other prime number q [find an upper bound for $\omega(q^n)$ by writing q^n in the scale of p, and let $n \to \infty$].

c) If there exists a prime number p such that $\omega(p) < 1$, show that $\omega(q) = 1$ for every other prime number q [show that we cannot have $\omega(q) < 1$, by using the fact that for each integer $n > 0$ there exist two rational integers r and s such that $1 = rp^n + sq^n$]. Hence show that ω is then an absolute value equivalent to the *p-adic* absolute value on $\mathbf{Q}$.

d) If $\omega(p) > 1$ for every prime number p, show that if p and q are any two prime numbers, then

$$\frac{\log \omega(p)}{\log p} = \frac{\log \omega(q)}{\log q}$$

[same method as in *b*)]. Hence show that in this case $\omega(x) = |x|^\rho$ for some $\rho \leqslant 1$. *

6) Let E be a left vector space over a division ring K, with a countable base (a_n). Let (r_n) be a decreasing sequence of numbers > 0, tending to 0. For each $x = \sum_k t_k a_k \neq 0$ of E, put $\|x\| = r_h$ where h is the smallest of the indices k such that $t_k \neq 0$, and put $\|0\| = 0$. Show that $\|x - y\|$ is an invariant metric on the additive group E, and that the topology it defines on E is independent of the sequence (r_n) (decreasing and tending to 0) chosen. Hence show that if the definitions of a norm and a normed space (Definitions 5 and 6) are extended to the case where the absolute value on the scalars is improper, Proposition 7 and Theorem 1 are no longer valid.

7) Let E be a normed space over a valued division ring. Show that if every absolutely summable family in E is summable in E, then E is complete. [Let (x_n) be a Cauchy sequence in E, and consider a subsequence (x_{n_k}) of (x_n) such that the series whose general term is $x_{n_{k+1}} - x_n$ is absolutely convergent.]

8) Give an example of a family which is summable but not absolutely summable in the field $\mathbf{Q}_p$ of p-adic numbers. [Show that a family $(x_\iota)_{\iota \in I}$ of points of $\mathbf{Q}_p$ is summable if and only if $\lim x_\iota = 0$ with respect to the filter of complements of finite subsets of I.]

9) A mapping w of a ring A into $\mathbf{R}_+$ is called a *semi-absolute value* on A if it satisfies the conditions: (i) $w(0) = 0$; (ii) $w(x - y) \leqslant w(x) + w(y)$;

(iii) $w(xy) \leqslant w(x)w(y)$. The semi-absolute value w is said to be *Hausdorff* if $w(x) = 0$ implies $x = 0$. If w is a Hausdorff semi-absolute value on A, then $w(x-y)$ is an invariant metric on the additive group of A, and hence defines a topology compatible with the additive group structure of A. Generalize the principal properties of normed algebras, notably Proposition 13, to rings endowed with a semi-absolute value.

If w is a non-Hausdorff semi-absolute value on A, the set $\overset{-1}{w}(0)$ is a two-sided ideal $\mathfrak{a}$ in A; if we endow A with the topology defined by the pseudometric $w(x-y)$, this topology is compatible with the ring structure of A, and the Hausdorff space associated with A is the quotient ring $A/\mathfrak{a}$, on which the function $\overline{w}(\overline{x})$, which for each $\overline{x}$ mod $\mathfrak{a}$ is equal to the common value of $w(x)$ for all $x \in \overline{x}$, is a Hausdorff semi-absolute value said to be *associated* with w; the topology defined by $\overline{w}$ is the quotient by $\mathfrak{a}$ of the topology of A.

¶ 10) *a*) Two semi-absolute values w_1, w_2 on a ring A are said to be *equivalent* if the pseudometrics $w_1(x-y)$, $w_2(x-y)$ are equivalent. Show that, if w is a semi-absolute value on A, then aw and $w^{1/a}$ are semi-absolute values equivalent to w for every real number $a \geqslant 1$.

b) If w_i $(1 \leqslant i \leqslant n)$ are semi-absolute values on a ring A, the functions $w = \sum_i w_i$ and $w' = \sup_i w_i$ are two equivalent semi-absolute values on A. If

$$\mathfrak{a}_i = \overset{-1}{w_i}(0) \qquad \text{and} \qquad \mathfrak{a} = \overset{-1}{w}(0),$$

then $\mathfrak{a} = \bigcap_i \mathfrak{a}_i$. If A_i denotes the completion of the quotient ring $A/\mathfrak{a}_i$ endowed with the Hausdorff semi-absolute value associated with w_i, show that $A/\mathfrak{a}$ endowed with the Hausdorff semi-absolute value associated with w is isomorphic to a subring of the product ring $\prod_i A_i$; show that (assuming that A has an identity element 1) this ring is dense in $\prod_i A_i$ if and only if the w_i satisfy the following condition: for each $\varepsilon > 0$ and for each index i, there is an element $x_i \in A$ such that $w_i(1-x_i) \leqslant \varepsilon$ and $w_k(x_i) \leqslant \varepsilon$ for each index $k \neq i$.

11) Let A be a ring endowed with a Hausdorff semi-absolute value, and suppose that A has an identity element and is complete with respect to the topology defined by the semi-absolute value. Show that every maximal ideal of A is *closed* (use Proposition 13).

¶ 12) Let K be a non-discrete, Hausdorff, topological division ring, and let $\varphi : K \to \mathbf{R}_+$ be a mapping such that $\varphi(0) = 0$, $\varphi(xy) = \varphi(x)\varphi(y)$ and such that if V_n denotes the set of all $x \in K$ such that $\varphi(x) \leqslant 1/n$, the sets V_n form a fundamental system of neighbourhoods of 0 in K.

a) Show that there exists a real number $a > 0$ such that, for all $x \in K$,

$$\varphi(1 + x) \leqslant a(1 + \varphi(x))$$

[if not, there would exist a sequence (x_n) of points of K such that both $(1 + x_n)^{-1}$ and $x_n(1 + x_n)^{-1}$ tended to zero]. Hence show that if $\psi(x) = (\varphi(x))^{\alpha}$ we have

$$\psi(x + y) \leqslant 2 \sup (\psi(x), \psi(y))$$

for α sufficiently small.

b) Show that if $n = 2^p$ we have $\psi\left(\sum_{i=1}^{n} x_i\right) \leqslant n \sup (\psi(x_i))$, and deduce that for every integer $m > 0$ we have $\psi(m) \leqslant 2m$.

c) Deduce from *b*) that for every $n = 2^p$ and every $x \in K$ we have $\psi((1 + x)^{n-1}) \leqslant 2n(1 + \psi(x))^{n-1}$, and hence that ψ is an *absolute value* on K which defines the topology of K.

¶ 13) Let K be a non-discrete Hausdorff topological division ring. Let R be the set of all $x \in K$ such that $\lim_{n \to \infty} x^n = 0$, and let N be the complement of the set $R \cup R^{-1}$. Show that there exists an absolute value on K which defines the topology of K if and only if (i) R is open in K; (ii) for each neighbourhood V of 0 in K, there is a neighbourhood U of 0 in K such that $R.U \subset V$; and (iii) if $x \in R$ and $y \in R \cup N$, then $yx \in R$.

To show that these conditions are sufficient, prove successively that:

a) N is a normal subgroup of the multiplicative group K^* of non-zero elements of K.

b) In the quotient group K^*/N, put $\dot{x} \leqslant \dot{y}$ if there exist $x \in \dot{x}$ and $y \in \dot{y}$ such that $yx^{-1} \in R \cup N$; show that this relation is an ordering compatible with the group structure of K^*/N, and that the ordered group so defined is isomorphic to a subgroup of the additive group **R** (use Exercise 1 of Chapter V, § 3).

c) Complete the proof with the help of Exercise 12.

In particular, the topology of a non-discrete locally compact topological field can be defined by an absolute value.

§ 4

1) *a*) Construct a topological space consisting of four points which satisfies axiom (O_V) but not axiom (O_{III}).

b) A topological space X which satisfies axiom (O_V) also satisfies axiom (O_{III}) if and only if it has the following property: every closed subset of X is the intersection of its neighbourhoods. X is then uniformizable and the completely regular space associated with X (§ 1, Exercise 4) is normal.

c) If a topological space satisfies axioms (C) and (O_{III}), it satisfies (O_V), and the associated completely regular space is compact.

2) Let X be a topological space which satisfies axiom (O_V).

a) Show that the relation $R : \overline{\{x\}} \cap \overline{\{y\}} \neq \emptyset$ between two points x, y of X is equivalent to the relation R′ : " $f(x) = f(y)$ for every real-valued continuous function f on X "; hence show that R is an equivalence relation on X.

b) Show that the quotient space $Y = X/R$ is normal and that, if $\varphi : X \to Y$ is the canonical mapping, every continuous real-valued function f on X can be written in the form $f = g \circ \varphi$, where g is a continuous real-valued function on Y.

3) If X is a topological space, the following conditions are equivalent:

a) Every subspace of X satisfies axiom (O'_V).

b) Every open subspace of X satisfies axiom (O'_V).

c) Given any two subsets A and B of X such that $A \cap \overline{B} = B \cap \overline{A} = \emptyset$, there exist two disjoint open sets U, V such that $A \subset U$ and $B \subset V$.

A topological space X is said to be *completely normal* if it is Hausdorff and satisfies these conditions. Every metrizable space is completely normal. A compact space need not be completely normal.

4) Every linearly ordered set X endowed with either of the topologies $\mathcal{T}_+(X)$, $\mathcal{T}_-(X)$ (Chapter I, § 2, Exercise 5) is completely normal. (If A and B are two subsets of X such that $A \cap \overline{B} = B \cap \overline{A} = \emptyset$, define a neighbourhood V_x of x for each $x \in A$, and a neighbourhood W_y of y for each $y \in B$, such that $V_x \cap W_y = \emptyset$ for all $x \in A$ and all $y \in B$.)

¶ 5) Every linearly ordered set X, endowed with the topology $\mathcal{T}_0(X)$ (Chapter I, § 2, Exercise 5), is completely normal. [Consider first the case where X is *compact*; show with the help of Chapter IV, § 2, Exercise 6 that every open set in X is the union of mutually disjoint open intervals; use this result to prove the proposition, by considering (in the notation of Exercise 3) the complement of the closed set $\overline{A} \cap \overline{B}$, and then the complement of $\overline{B}$; to pass to the general case, in which X is arbitrary, use Chapter IV, § 4, Exercise 7.]

¶ 6) Show that, in a normal space X, every subspace Y of X which is a countable union of closed sets is normal. [Let A, B be two disjoint closed subsets of Y, and suppose that Y is the union of an ascending sequence (Y_n) of closed subsets of X. Define by induction two sequences (U_n), (V_n) of open subsets of X such that: (i) $\overline{U_n \cap Y_n}$ and $\overline{V_n \cap Y_n}$ do not intersect; (ii) $U_n \cap Y_n$ contains $A \cap Y_n$ and all the sets $\overline{U_i \cap Y_i}$ for $1 \leqslant i \leqslant n$; and $V_n \cap Y_n$ contains $B \cap Y_n$ and all $\overline{V_i \cap Y_i}$, $1 \leqslant i \leqslant n$.]

7) A topological space X is said to be *perfectly normal* if it is normal and if each closed subset of X is a countable intersection of open sets (or, equivalently, if each open subset of X is a countable union of closed sets).

a) A Hausdorff space X is perfectly normal if and only if, given any closed subset A of X, there exists a real-valued continuous function f on X such that $\overset{-1}{f}(0) = A$ [use the method of Exercise 16 *a*) of § 1].

b) Show that every perfectly normal space X is completely normal and that every subspace of X is perfectly normal [use Exercises 3 *b*) and 6].

c) Every lower semi-continuous real-valued function on a perfectly normal space is the upper envelope of a sequence of continuous functions (use the method of § 2, no. 7, Proposition 11).

d) The compact space X obtained by adjoining a point at infinity to an uncountable discrete space is completely normal but not perfectly normal, and every subspace of X is paracompact.

¶ 8) Show that the non-metrizable compact space X defined in Exercise 13 *a*) of § 2 is perfectly normal, and that the non-metrizable compact space Y defined in Exercise 13 *d*) of § 2 is completely normal but not perfectly normal.

¶ 9) *a*) Let X be a paracompact space in which every point has a countable fundamental system of neighbourhoods, and let Y be a countably compact normal space (§ 2, Exercise 14). Show that $X \times Y$ is normal. [Let A, B be two disjoint closed subsets of $X \times Y$; for each $x \in X$, show that there is a neighbourhood U_x of x in X and two open sets V_x, W_x in Y such that $\overline{V}_x \cap \overline{W}_x = \varnothing$ and such that for each $z \in U_x$ we have $A(z) \subset V_x$ and $B(z) \subset W_x$; to prove this, argue by contradiction. Let $(T_\lambda)_{\lambda \in L}$ be a locally finite open covering of X, finer than the covering $(U_x)_{x \in X}$, and let (f_λ) be a partition of unity subordinate to (T_λ); for each $\lambda \in L$, let x_λ be such that $T_\lambda \subset U_{x_\lambda}$ and let g_λ be a continuous mapping of Y into $[0, 1]$ which is equal to 0 on $\overline{V}_{x_\lambda}$ and to 1 on $\overline{W}_{x_\lambda}$; consider the function $\sum_\lambda f_\lambda(x)\, g_\lambda(y)$ on $X \times Y$.]

b) Let $Y = [a, b[$ be the locally compact space defined in Chapter I, § 9, Exercise 12, and let $X = [a, b]$ be the compact space obtained by adjoining a point at infinity to Y. Y is normal (Exercise 4) and countably compact (§ 2, Exercise 15), but the product $X \times Y$ is not normal (use Chapter II, § 4, Exercise 4).

¶ 10) Let L be an uncountable set and let X denote the completely regular space N^L (N carrying the discrete topology). Let A (resp. B) be the set of all $x = (x(\lambda))_{\lambda \in L}$ in X such that, for each integer $k \neq 0$ (resp. $k \neq 1$) the set of $\lambda \in L$ such that $x(\lambda) = k$ has at most one element.

a) Show that A and B are disjoint closed subsets of X.

b) Let U, V be two open subsets of X such that $A \subset U$ and $B \subset V$. Let Φ denote the set of all elementary sets in X (Chapter I, § 4, no. 1) whose projections which are distinct from **N** consist of a single element; for each $W \in \Phi$, let $H(W)$ be the set of all $\lambda \in L$ such that $pr_\lambda(W)$ consists of a single element. Show that there is a sequence (x_n) of points of A, a sequence (U_n) of sets of Φ, a sequence (λ_n) of distinct elements of L and a strictly increasing sequence $(m(n))_{n \in N}$ of integers with the following properties: (i) $U_n \subset U$ is a neighbourhood of x_n; (ii) $H(U_n)$ is the set of all λ_k such that $k \leqslant m(n)$; (iii) $x_0(\lambda) = 0$ for each $\lambda \in L$, and for each $n > 0$, $x_n(\lambda_k) = k$ if $k \leqslant m(n-1)$ and $x_n(\lambda) = 0$ for all other $\lambda \in L$.

c) Let $y \in B$ be the point such that $y(\lambda_k) = k$ for each integer k and $y(\lambda) = 1$ for all $\lambda \in L$ other than the λ_k. Let $V_0 \subset V$ be a set of Φ which contains y, and let n be an integer such that $\lambda_k \in L - H(V_0)$ for all $k > m(n)$. Show that $U_{n+1} \cap V_0 \neq \emptyset$, and deduce that X is not normal.

11) Deduce from Exercise 10 that:

a) If a product $\prod_{\iota \in I} X_\iota$ of non-empty Hausdorff spaces is normal, then X_ι is countably compact (§ 2, Exercise 14) for all save a countable set of indices.

b) A product $\prod_{\iota \in I} X_\iota$ of non-empty metrizable spaces is normal if and only if X_ι is compact for all but a countable set of indices; and the product space $\prod_{\iota \in I} X_\iota$ is then paracompact.

12) Let X, Y be two Hausdorff spaces.

a) Suppose that there is a closed set A in X which is not a countable intersection of open sets, and that there is a countably infinite subset B of Y which is not closed. Let $b \in \overline{B} - B$, and consider the subsets $C = A \times B$, $D = (X - A) \times \{b\}$ in $X \times Y$. Show that $C \cap \overline{D} = \overline{C} \cap D = \emptyset$, but that there is no pair U, V of open subsets of $X \times Y$ such that $C \subset U$, $D \subset V$ and $U \cap V = \emptyset$.

b) If $X \times Y$ is completely normal then either (i) one of the spaces X, Y is perfectly normal, or (ii) one of the spaces X, Y has the property that every countably infinite subset is closed (cf. § 5, Exercise 16).

c) Let $X = [a, b]$ be an uncountable well-ordered set such that $[a, x[$ is countable for each $x < b$. Endow X with the topology in which $\{x\}$ is open for each $x < b$ and the intervals $[x, b]$ $(x < b)$ form a fundamental system of neighbourhoods of b. Show that $X \times X$ is completely normal but that X is not perfectly normal.

d) Show that if X is a compact space such that $X \times X \times X$ is completely normal, then X is metrizable (use Theorem 1 of § 2, no. 4).

¶ 13) Let $(X_\iota)_{\iota \in I}$ be an *uncountable* family of Hausdorff topological spaces, each containing more than one point. For each $\iota \in I$, let a_ι and b_ι be two distinct points of X_ι. Let X be the product space $\prod_{\iota \in I} X_\iota$, and let Y be the subspace of X consisting of these points $(x_\iota)_{\iota \in I}$ such that $x_\iota = a_\iota$ for all but a *countable* set of indices ι; and let b be the point b_ι of X.

In the product space $X \times Y$, let A be the diagonal of $Y \times Y$ and let B be the set $\{b\} \times Y$. Show that A and B are closed sets and that there is no pair of open sets U, V in $X \times Y$ such that $A \subset U$, $B \subset V$, and $U \cap V = \emptyset$. [Let U, V be two open sets such that $A \subset U$ and $B \subset V$; show that there is an increasing sequence (H_n) of countable subsets of I, such that if x_n denotes the point of Y for which each coordinate with index $\iota \in H_n$ is equal to b_ι, and each coordinate with index $\iota \notin H_n$ is equal to a_ι, then the point (x_{n+1}, x_n) belongs to V for each integer n; prove that the sequence (x_{n+1}, x_n) tends to a point of A, and deduce that $U \cap V \neq \emptyset$.]

Hence construct an example of a connected topological ring which is not normal.

Deduce also that the product of an uncountable family of Hausdorff topological spaces, each of which has at least two distinct points, cannot be completely normal (use the result above when each X_ι is a discrete space of two points).

14) Let X be a non-normal Hausdorff space, and let A, B be two disjoint closed subsets of X such that there is no pair of disjoint open sets U, V in X satisfying the relations $A \subset U$ and $B \subset V$. Let R be the equivalence relation on X whose equivalence classes are the set A, the set B, and the sets $\{x\}$ where $x \in \complement(A \cup B)$. Show that the graph of R is closed in $X \times X$ and that the relation R is closed, but that the quotient space X/R is not Hausdorff (cf. Chapter I, § 8, no. 3, Proposition 8).

15) *a*) Let X be a normal space and let R be a closed equivalence relation on X. Show that the quotient space X/R is normal (use Proposition 10 of Chapter I, § 5, no. 4).

b) Let X be a locally compact σ-compact space, and let R be an equivalence relation on X whose graph is closed in X×X. Show that X/R is normal (cf. Chapter I, § 10, Exercise 19).

c) Let X be the complement in **R** of the set of points of the form $1/n$, where n is any integer except 0 or ± 1. Consider the equivalence relation R on the metrizable space X for which the equivalence class of each $x \in X$, which is not an integer, consists of the points x and $1/x$, and the equivalence class of each integer consists of that integer alone. Show that R is open and that the graph of R is closed in X×X, but that X/R is not regular.

16) For each covering $\mathfrak{R}$ of a topological space X, let $V_{\mathfrak{R}}$ denote the union of all the sets $U \times U$ such that $U \in \mathfrak{R}$.. A covering $\mathfrak{R}$ of X is said to be *even* if there is a neighbourhood W of the diagonal Δ of X×X such that the covering $(W(x))_{x \in X}$ is finer than $\mathfrak{R}$. $\mathfrak{R}$ is said to be *divisible* if there is a neighbourhood W of Δ in X×X such that $\overset{2}{W} \subset V_{\mathfrak{R}}$.

a) Show that every even open covering of X is divisible [cf. Exercise 19 *b*)].

b) Let $\mathfrak{R}$ be an open covering of X such that there exists a locally finite closed covering $\mathfrak{S}$ of X which is finer than $\mathfrak{R}$. Show that $\mathfrak{R}$ is even. (For every set $A \in \mathfrak{S}$, let U_A be a set of $\mathfrak{R}$ which contains A, and let V_A be the union in X×X of $U_A \times U_A$ and $(X - A) \times (X - A)$; consider the set $W = \bigcap_{A \in \mathfrak{S}} V_A$.)

¶ 17) *a*) If every finite open covering of a topological space X is divisible (Exercise 16), then X satisfies Axiom (O'_V) (consider a *binary* open covering of X, i.e., one formed by two subsets of X). Conversely, if X satisfies (O'_V), then every finite open covering of X is even [use Theorem 3 of no. 3, which is valid provided that X satisfies (O'_V), to reduce to the case of a binary covering, and then again to deal with this case]. As $\mathfrak{R}$ runs through the set of finite open coverings of X, the sets $V_{\mathfrak{R}}$ form a fundamental system of entourages of a uniformity on X, compatible with the topology of X; this uniformity is called the " uniformity of finite open coverings ".

b) Suppose that X is normal. Show that the uniformity $\mathfrak{U}$ of finite open coverings is the same as the coarsest uniformity $\mathfrak{U}'$ for which all continuous mappings of X into [0, 1] are uniformly continuous [i.e., the uniformity induced on X by that of its Stone-Čech compactification (§ 1, Exercise 7)]. [To show that $\mathfrak{U}$ is coarser than $\mathfrak{U}'$, use Theorem 3 of no. 3 and Axiom (O_V) in order to construct a finite family of pseudometrics on X of the form $(x, y) \to |f(x) - f(y)|$, in such a way that an entourage of the uniformity defined by these pseudometrics is contained in an entourage of the uniformity $\mathfrak{U}$].

c) Let X be a normal space and let βX be its Stone-Čech compactification. For each closed subset Y of X, let $\overline{Y}$ be the closure of Y in βX. Show that the continuous mapping of βY into $\overline{Y}$ which extends the identity mapping of Y (βY being the Stone-Čech compactification of Y) is a homeomorphism [either use *b*) or else Theorem 2 and the universal property of βY]. If Y and Z are two closed subsets of X, show that $\overline{Y \cap Z} = \overline{Y} \cap \overline{Z}$ in βX (if $x_0 \notin \overline{Y \cap Z}$ and $x_0 \in \overline{Y}$, consider a closed neighbourhood V of x_0 in βX, such that $V \cap Y$ and Z are disjoint closed subsets of X).

¶ 18) A family (A_α) of subsets of a topological space X is said to be *discrete* if, for each $x \in X$, there is a neighbourhood of x which meets at most one set A_α of the family. A Hausdorff space X is said to be *collectively normal* if, for each discrete family (A_α) of closed subsets of X, there is a family (U_α) of mutually disjoint open subsets of X such that $A_\alpha \subset U_\alpha$ for each index α. Every collectively normal space is normal.

a) Every open covering of a completely regular space X is divisible (Exercise 16) if and only if the set of neighbourhoods of the diagonal Δ in $X \times X$ is the filter of entourages of the universal uniformity of X (§ 1, Exercise 5). Show that if this condition is satisfied, then X is collectively normal [if (A_α) is a discrete family of closed subsets of X, consider the open covering (V_α), where $V_\alpha = X - \bigcup_{\beta \neq \alpha} A_\beta$ for each α].

b) Let $Y = [a, b[$ be the locally compact space defined in Chapter I, § 9, Exercise 12; let Y_0 be the set Y endowed with the discrete topology, and let Z be the set $Y_0 \cup \{b\}$, where the points of Y_0 are open sets, and the sets $]x, b[$, where x runs through Y_0, from a fondamental system of neighbourhoods of b. Show that the space $X = Y \times Z$ is collectively normal (recall that no infinite family of subsets of Y can be discrete, and use Exercise 4). Let $\mathfrak{R}$ be the open covering of X consisting of $Y \times Y_0$ and the products $[a, x] \times]x, b]$ (where $x < b$); show that $\mathfrak{R}$ is not divisible and that consequently the set of neighbourhoods of the diagonal Δ in $X \times X$ is not the filter of entourages of a uniformity on X [use Exercise 12 *a*) of Chapter I, § 9].

¶ 19) *a*) A regular space X is paracompact if and only if every open covering of X is even (Exercise 16). [To prove necessity, use Lemma 5 and Exercise 16 *b*); for sufficiency, use Exercise 16 *a*) of § 4, Proposition 2 of § 1, no. 4, and Theorem 4 of § 4, no. 5].

b) Give an example of a divisible non-even covering of a collectively normal, non-paracompact space [use *a*) and Chapter II, § 4, Exercise 4].

c) Show that a paracompact space X is complete with respect to its universal uniformity (§ 1, Exercise 5). [If a Cauchy filter $\mathfrak{F}$ with respect to this uniformity has no cluster point, then every point $x \in X$ has an open neighbourhood V_x which does not meet at least one set of $\mathfrak{F}$; use *a*) and Exercise 18 *a*).]

20) *a*) A regular space X is paracompact if, for each open covering $\mathfrak{R}$ of X, there is a sequence $(\mathfrak{S}_n)$ such that $\mathfrak{S} = \bigcup_n \mathfrak{S}_n$ is an open covering of X which refines $\mathfrak{R}$ and such that each $\mathfrak{S}_n$ is a locally finite family (cf. Lemmas 4, 5 and 6).

b) Deduce from *a*) that in a paracompact space every countable union of closed sets is a paracompact subspace (cf. § 5, Exercise 15).

c) In a regular space X, let $\mathfrak{F}$ be a locally finite family of closed sets each of which is a paracompact subspace of X. Show that the union of the sets of $\mathfrak{F}$ is a paracompact subspace of X (use Lemma 6 of no. 5; cf. § 5, Exercise 15).

d) Show that if X is a paracompact space and Y is a regular space which is a countable union of compact sets, then $X \times Y$ is paracompact [embed Y in its Stone-Čech compactification and use *b*); cf. § 5, Exercise 16].

¶ 21) *a*) Let X be a paracompact space and let R be a closed equivalence relation on X such that every equivalence class with respect to R is compact. Show that X/R is paracompact [use Exercise 15 *a*), Theorem 3 of no. 3, and Lemma 6 of no. 5; cf. Exercise 15 *c*)].

b) Let X be a Hausdorff space and let R be a closed equivalence relation on X such that every equivalence class with respect to R is compact. Show that if X/R is paracompact, then so is X (use the argument of Chapter I, § 9, no. 10, Proposition 17).

c) Let Y be a locally compact but not paracompact space (cf. Chapter I, § 9, Exercise 12), and let Y_0 be the one-point compactification of Y. Let X be the topological sum of Y and Y_0, and let R be the equivalence relation on X which identifies each point of Y with its canonical image in Y_0. The relation R is open, every equivalence class mod R contains at most two points, and X/R is homeomorphic to Y_0, hence is compact.

¶ 22) A regular space X is metrizable if and only if there exists a sequence $(\mathfrak{B}_n)$ of locally finite families of open subsets of X such that $\mathfrak{B} = \bigcup_n \mathfrak{B}_n$ is a base of the topology of X. [To show that the condition is necessary, use Lemma 3 of no. 5. To show that it is sufficient, show first that X is paracompact by using Lemmas 4, 5 and 6. For each pair of integers m, n and each $U \in \mathfrak{B}_m$, let U' be the union of the sets $V \in \mathfrak{B}_n$ such that $\overline{V} \subset U$. Show that $\overline{U'} \subset U$. Let $f_U : X \to [0, 1]$ be a continuous mapping which is equal to 0 on $X - U$ and to 1 on U', and let $d_{mn}(x, y) = \sum_{U \in \mathfrak{B}_m} |f_U(x) - f_U(y)|$; show that the pseudometrics d_{mn} define the topology of X.] ("*Theorem of Nagata-Smirnov.*") In particular, every regular space with a countable base is metrizable.

23) *a*) Every regular Lindelöf space (Chapter I, § 9, Exercise 15) is paracompact [use Exercise 20 *a*)].

b) The product of a Lindelöf space and a compact space is a Lindelöf space (argue as in the proof of Proposition 17 of Chapter I, § 9, no. 10; cf. § 5, Exercise 16).

¶ 24) *a*) Let X be a metrizable space and let R be a closed equivalence relation on X, all of whose equivalence classes are compact. Show that X/R is metrizable. [Apply the Nagata-Smirnov theorem (Exercise 22) by proving the analogue of Lemma 3 of no. 5 for open coverings of X formed of sets which are *saturated* with respect to R; let $F_{n\alpha}$ denote the largest saturated open set contained in the set of all $x \in X$ such that $d(x, X - U_\alpha) > 2^{-n}$, let $F'_{n\alpha}$ denote the saturation of the set of all $x \in X$ such that $d(x, X - U_\alpha) \geqslant 2^{-n}$ and let $G_{n\alpha}$ denote the set of all $x \in F_{n\alpha}$ such that $x \notin F'_{n+1, \beta}$ for all $\beta < \alpha$.]

b) Extend the result of *a*) to the case in which the relation R is closed and every point of X/R has a countable fundamental system of neighbourhoods [use Exercise 20 *b*) of § 2].

c) Let X be a topological space and let (F_α) be a locally finite closed covering of X. Show that if each of the subspaces F_α is metrizable then X is metrizable (consider X as a quotient space of the topological sum of the F_α).

d) A paracompact space, in which every point has a metrizable neighbourhood, is metrizable [apply *c*); cf. Chapter I, § 9, Exercise 12 *b*), also Chapter IX, § 4, Exercise 25 *d*), § 2, Exercise 15 and § 5, Exercise 15].

25) A Hausdorff space X is said to be *metacompact* if, given any open covering R of X, there is a point-finite open covering of X which refines R.

a) Show that every closed subspace of a metacompact space is metacompact, and that if every open subspace of a metacompact space X is metacompact, then every subspace of X is metacompact.

b) Let X be a Hausdorff space and let R be a closed equivalence relation on X, all of whose equivalence classes are compact. Show that if X/R is metacompact, then X is metacompact [argue as in Exercise 21 *b*)].

c) Show that a topological space which is both metacompact and countably compact (§ 2, Exercise 14) is compact. [Use Zorn's lemma to show that every point-finite open covering contains a minimal open covering, and use § 2, Exercise 14 *a*).]

d) The locally compact, completely normal space $X = [a, b[$ defined in Chapter I, § 9, Exercise 12 is not metacompact (*).

e) Let Y be a metrizable space and let A be a subset of Y such that both A and its complement are dense in Y. Let X denote the non-regular space obtained by endowing Y with the topology generated by the open sets of Y and the set A. Show that X is metacompact. (For each $x \in X$, consider a neighbourhood V_x of x contained in a set of a given open covering $\mathfrak{R}$ of X, such that $V_x \subset A$ if $x \in A$, and such that V_x is a neighbourhood of x in Y if $x \notin A$. Note that the subspace A of X is paracompact, and that the union of the V_x for $x \notin A$ is also paracompact.)

26) *a*) Show that every pseudo-compact normal space X (§ 1, Exercise 22) is countably compact (§ 2, Exercise 14). [Show that if (x_n) is a sequence of distinct points of X with no cluster point, then there is a continuous finite real-valued function f on X such that $f(x_n) = n$ for all n.]

b) Let $Y = [a, b[$ be the locally compact space defined in Chapter I, § 9, Exercise 12, and let Y_0 be the compact space obtained by adjoining to Y a point at infinity (which may be identified with b). Let c be the smallest element of Y such that $[a, c[$ is infinite, and let Z be the compact subspace $[a, c]$ of Y. Let X be the complement of the point (b, c) in the compact product space $Y_0 \times Z$. Show that X is locally compact but not normal [cf. Exercise 12 *a*)]; that X is pseudo-compact but not countably compact. Give an example of a closed subspace of X which is not pseudo-compact. Also, give an example of a lower semi-continuous function on X which does not attain its greatest lower bound [cf. § 2, Exercise 14 *i*)].

27) A topological space X is said to be *locally paracompact* if every point of X has a *closed* neighbourhood which is a paracompact subspace of X.

(*) There are spaces which are perfectly normal and collectively normal but not metacompact, and spaces which are perfectly normal and metacompact but not collectively normal (and therefore not paracompact); cf. E. Michael, *Can. Journ. of Math.* 7 (1955), pp. 275-279.

a) Show that a locally paracompact space is completely regular.

b) In the compact space $Y_0 \times Z$ defined in Exercise 26 *b*), let H be the complement of the set of points (b, y) where $y < c$. Show that H is normal but not locally paracompact [cf. Chapter I, § 9, Exercise 12 *b*)].

c) Let X be a topological space and let Φ be a covering of X by closed sets which are paracompact subspaces of X, such that each $A \in \Phi$ has a neighbourhood in X which belongs to Φ. Let X′ be the set which is the sum of X and a point ω; define a topology on X′ by taking as a fundamental system of neighbourhoods of $x \in X$ in X′ the set of all neighbourhoods of x in X, and as a fundamental system of neighbourhoods of ω the filter base generated by the complements in X′ of the sets of Φ [cf. Exercise 20 *c*)]. Show that X′ is paracompact.

¶ 28) Let X be a metric space, let d be the metric on X, let A be a closed subset of X and let $f: A \to [1, 2]$ be a continuous real-valued function. Show that the real-valued function g on X which is such that $g(x) = f(x)$ if $x \in A$, and

$$g(x) = \frac{1}{d(x, A)} \inf_{y \in A} (f(y) d(x, y))$$

if $x \in \complement A$, is continuous on X. Hence give a proof of Theorem 2 for metric spaces.

29) Let X be a Hausdorff topological space. Then X is normal if and only if, for each closed subset A of X and each real-valued function f defined on A and continuous with respect to the induced topology, there exists a continuous pseudometric d on X with respect to which f is uniformly continuous. (To show that the condition is sufficient, show that every real-valued function f defined on a closed subset A, and continuous with respect to the induced topology, can be extended to a continuous function on X; for this, consider the Hausdorff space $\overline{X}$ associated with the uniformity defined on X by the pseudometric d.)

30) Let X be a normal space and let g (resp. f) be an upper (resp. lower) semi-continuous function on X, such that $g \leqslant f$. Show that there is a continuous real-valued function h on X such that $g \leqslant h \leqslant f$. [Reduce to the case where f and g take their values in $[0, 1]$. For each dyadic number $r \in [0, 1]$, let $F(r)$ [resp. $G(r)$] be the set of all $x \in X$ such that $f(x) \leqslant r$ (resp. $g(x) < r$). Define by induction a family of open sets $U(r)$ (where r runs through the set of dyadic numbers contained in $[0, 1]$) such that when $r < r'$ we have $F(r) \subset U(r')$, $\overline{U(r)} \subset U(r')$ and $\overline{U(r)} \subset G(r')$, by imitating the proof of Theorem 1 of no. 1; then complete the proof as in Theorem 1.]

§ 5

1) *a*) If A is a subset of a topological space X, show that the following statements are equivalent: α) the frontier of A is nowhere dense; β) A is the union of an open set and a nowhere dense set; γ) A is the difference between a closed set and a nowhere dense set.

b) Show that if a subset A of X is such that, for each point $x \in A$, there is a neighbourhood V of x in X such that $V \cap A$ is nowhere dense in X, then A is nowhere dense in X.

2) A subset A of a topological space X is said to be a *thin* set if, for each perfect set $P \subset X$, $P \cap A$ is nowhere dense in P.

a) Every finite union of thin sets is thin.

b) The frontier of a thin set is nowhere dense.

c) There is a largest perfect set in X whose complement is thin.

¶ 3) Let A be a subset of a topological space X, and let D(A) denote the set of all $x \in X$ such that, for each neighbourhood V of x, the set $V \cap A$ is not meagre. D(A) is contained in $\overline{A}$.

a) If B is a subset of X such that $D(B) = \emptyset$, then

$$D(A \cup B) = D(A)$$

for every subset A of X.

b) $D(A) = \emptyset$ if and only if A is meagre. [Begin by showing that if (U_ι) is a family of pairwise disjoint open subsets of X and if, for each ι, V_ι is a nowhere dense set (relative to X) contained in U_ι, then $\bigcup_\iota V_\iota$ is nowhere dense. Then consider a *maximal* set $\mathfrak{M}$ of open subsets of X, each pair of which are disjoint, and such that $A \cup U$ is meagre for each $U \in \mathfrak{M}$; the existence of such a maximal set is established by Zorn's lemma. Show that if $D(A) = \emptyset$, then $A \cap \complement W$ is nowhere dense, where W is the union of the sets of $\mathfrak{M}$, and hence show that A is meagre.]

c) Show that D(A) is closed and that $A \cap \complement D(A)$ is meagre for all $A \subset X$ [show, with the help of *b*), that $A \cap \complement D(A)$ is meagre relative to the open subspace $\complement D(A)$].

d) Show that D(A) is equal to the closure of its interior. [If D'(A) is the closure of the interior of D(A), show that $A \cap \complement D'(A)$ is meagre, by observing that this set is the union of

$$A \cap \complement D(A) \quad \text{and} \quad A \cap D(A) \cap \complement D'(A).]$$

4) Let X and Y be two topological spaces and let A (resp. B) be a subset of X (resp. Y).

a) $A \times B$ is nowhere dense (resp. perfect) in $X \times Y$ if and only if one of the two sets A, B is nowhere dense (resp. A and B are closed and one of them is perfect).

b) $A \times B$ is thin in $X \times Y$ if and only if A and B are both thin sets.

¶ 5) *a*) Let X and Y be two topological spaces and let A be a nowhere dense subset of $X \times Y$. If the topology of Y has a countable base, show that the set of all $x \in X$ such that the section $A \cap (\{x\} \times Y)$ of A at x is not nowhere dense relative to $\{x\} \times Y$ is a meagre set in X (if U is any non-empty open set in Y, show that the set of all $x \in X$ such that $\{x\} \times U$ is contained in the closure of the section of A at x is a nowhere dense set in X).

b) Show by an example that the result of *a*) can be false if the topology of Y does not have a countable base (take X to be a Hausdorff space with no isolated points and Y to be the set X with the discrete topology; and take A to be the diagonal of $X \times Y$).

c) If B, C are subsets of X, Y respectively and if one of B, C is meagre (in X, Y respectively), then $B \times C$ is meagre in $X \times Y$. Conversely, if $B \times C$ is meagre in $X \times Y$ and if the topology of either X or Y has a countable base, then one of the sets B, C is meagre.

d) Deduce from *c*) that if the topology of one of the spaces X, Y has a countable base, then $D(B \times C) = D(B) \times D(C)$ in the notation of Exercise 3.

6) A set A in a topological space X is said to be *almost open* if there is an open set U in X such that both $U \cap \complement A$ and $A \cap \complement U$ are meagre sets.

a) The complement of an almost open set is almost open.

b) Any countable union of almost open sets is almost open.

c) Show that the following statements are equivalent: α) A is almost open; β) there is a meagre subset M of X such that $A \cap \complement M$ is both open and closed in the subspace $X - M$; γ) there is a set $G \subset A$ which is a countable intersection of open sets in X, such that $A - G$ is meagre in X; δ) there is a set $F \supset A$ which is a countable union of closed sets in X, such that $F - A$ is meagre in X; ζ) the set $D(A) \cap D(X - A)$ (Exercise 3) is nowhere dense in X; θ) the set $D(A) \cap \complement A$ is meagre in X. [Use Exercises 3 *a*) and 3 *b*) to show that $\alpha) \Rightarrow \zeta) \Rightarrow \theta) \Rightarrow \alpha)$.]

d) Let X, Y be two topological spaces and suppose that the topology of either X or Y has a countable base. Then a subset of $X \times Y$

of the form $A \times B$ is almost open if and only if either both A and B are almost open or one of them is meagre [use ζ) of *c*)].

7) A topological space X is said to be *non-meagre* if it is not meagre relative to itself.

a) X is non-meagre if and only if every countable family of dense open sets in X has a non-empty intersection.

b) In a non-meagre space, the complement of every meagre set is a non-meagre subspace.

8) Let X be a topological space in which there is a non-empty open subset A which is a non-meagre subspace. Show that X is non-meagre and that A is not a meagre set relative to X.

9) Let X be the subspace of $\mathbf{R}^2$ which is the union of $\mathbf{Q}^2$ and the line $y = 0$. Show that X is meagre relative to itself. Deduce that a topological space can have non-meagre subspaces without being non-meagre itself.

10) Show that every completely regular pseudo-compact space (§ 1, Exercise 21) is a Baire space (argue as in Theorem 1).

11) If X is a Baire space with no isolated points, show that every point of X is a point of condensation of X (Chapter I, § 9, Exercise 17).

¶ 12) A topological space X is said to be *totally non-meagre* if every non-empty *closed* subspace of X is non-meagre. A locally compact space is totally non-meagre, and so is a complete metric space.

a) Show that a totally non-meagre regular space is a Baire space.

b) In an accessible totally non-meagre space, a non-empty countable closed set is thin (Exercise 2) and hence has at least one isolated point.

c) In a totally non-meagre space X, every subspace A which is the intersection of a countable intersection of open sets in X is totally non-meagre (if F is a subset of A which is closed in A, let $\overline{F}$ be the closure of F in X, and show that $\overline{F} \cap \complement F$ is meagre relative to $\overline{F}$).

d) If every point of a topological space X has a neighbourhood which is a totally non-meagre subspace, then X is totally non-meagre.

¶ 13) Let X be a totally non-meagre space which is connected and locally connected. Show that X cannot be the union of an infinite sequence (F_n) of non-empty mutually disjoint closed sets. (Let H be the union of the frontiers H_n of the F_n in X; show that H is closed in X and that each set H_n is nowhere dense in H; to establish the latter point, consider a fundamental system of connected neighbourhoods of a point of H_n, and argue by contradiction, using Chapter I, § 11, Proposition 3.)

14) Let X be the subspace of $\mathbf{R}^2$ consisting of the points $(r, 0)$, where r runs through the set $\mathbf{Q}$ of rational numbers, and the points $(k/n, 1/n)$, where n runs through the set of integers $\geqslant 1$ and k runs through the set of all rational integers. Show that X is a Baire space but is not totally non-meagre.

¶ 15) Let X be the subset of the plane $\mathbf{R}^2$ consisting of the line $D = \{0\} \times \mathbf{R}$ and the points $(1/n, k/n^2)$, where n runs through the set of integers > 0 and k runs through the set of all rational integers.

a) For each point $(0, y)$ of D and each integer $n > 0$, let $T_n(y)$ be the set of points $(u, v) \in X$ such that $u \leqslant 1/n$ and $|v - y| \leqslant u$. Show that if we take as a fundamental system of neighbourhoods of each point $(0, y)$ the set of sets $T_n(y)$, and as a fundamental system of neighbourhoods of each other point of X the set consisting of this point alone, then we have defined a topology $\mathcal{T}$ on X, which is finer than the topology induced by that of $\mathbf{R}^2$, and with respect to which X is locally compact.

b) Show that X is the union of a countable family of metrizable closed subspaces and that every point of X has a metrizable compact neighbourhood, but that X *is not normal.* [Let A be the set of all points $(0, y)$ where y is rational, and let $B = D - A$; show that every neighbourhood U of A in X meets every neighbourhood V of B in X; for this, consider for each n the set B_n of points $y \in B$ such that $T_n(y) \subset V$, and note that the set B is not meagre in $\mathbf{R}$.]

¶ 16) Let $\mathbf{R}_-$ denote the topological space obtained by endowing the linearly ordered set $\mathbf{R}$ of real numbers with the topology $\mathcal{T}_-(\mathbf{R})$ (Chapter I, § 2, Exercise 5).

a) Show that every open set in $\mathbf{R}_-$ is the union of a countable family of mutually disjoint intervals which are either open or half-open on the left. Deduce that $\mathbf{R}_-$ is perfectly normal (§ 4, Exercise 8) and totally non-meagre (Exercise 12).

b) Show that the space $\mathbf{R}_- \times \mathbf{R}_-$ is not normal, and hence that $\mathbf{R}_-$ is not metrizable, although $\mathbf{R}_-$ contains a countable dense subset. (Considering the set of points (x, y) in $\mathbf{R}_- \times \mathbf{R}_-$ such that $x + y = 1$, show that $\mathbf{R}_- \times \mathbf{R}_-$ has a closed subspace which is homeomorphic to the space X defined in Exercise 15.)

c) Show that $\mathbf{R}_-$ is a Lindelöf space (Chapter I, § 9, Exercise 15) and therefore is paracompact (§ 4, Exercise 23). (For each $x \in \mathbf{R}_-$, let U_x be a semi-open interval $]y(x), x]$ with x as right-hand end-point. Show that the set of $z \in \mathbf{R}_-$ which do not belong to any open interval $]y(x), x[$ is countable; use Chapter IV, § 2, Exercise 1.)

d) Show that every compact subset of $\mathbf{R}_-$ is countable [if A is a compact set in $\mathbf{R}_-$, the topologies induced on A by $\mathfrak{T}_-(\mathbf{R})$ and $\mathfrak{T}_0(\mathbf{R})$ are identical; deduce that every point of A is the left-hand end-point of an interval contiguous to A].

17) *a*) Show that every product $X = \prod_{\iota \in I} X_\iota$ of complete metric spaces is a Baire space. [Argue as in the proof of Theorem 1, taking G_n to be an elementary set (Chapter I, § 4, no. 1) all of whose projections are either the whole space or have a diameter $\leqslant 1/n$; then note that there is a countable subset J of I such that each G_n is of the form $H_n \times \prod_{\iota \notin J} X_\iota$, where $H_n \subset \prod_{\iota \in J} X_\iota$.]

b) Show by an example that if I is uncountable then X need not be totally non-meagre (Exercise 12). [Use Exercises 1 and 23 *b*) of § 2.]

¶ 18) *a*) Show that in a complete metric space, every perfect set contains a subset homeomorphic to Cantor's triadic set (Chapter IV, § 2, no. 5) (use Exercise 11 of Chapter IV, § 8).

b) Deduce that, in a complete metric space with a countable base, every perfect set has the power of the continuum and that every closed set and every open set is either countable or has the power of the continuum [use Exercise 11 *b*)].

c) Show that in an uncountable complete metric space with a countable base, the set of perfect subsets has the power of the continuum. [Prove the assertion first for the compact space $\{0, 1\}^{\mathbf{N}}$, then use *a*) and § 2, Exercise 12 *b*).]

d) Let X be an uncountable complete metric space with a countable base. Show that there is a subset Z of X such that both Z and X — Z have the power of the continuum and such that neither Z nor X — Z contains a non-empty perfect set. [Use *b*) and *c*) and the method described in *Set Theory*, Chapter III, § 6, Exercise 24 *a*).] If, furthermore, X has no isolated point, show that neither Z nor X — Z is meagre. (If A is a countable intersection of dense open subsets of X, show that A contains non-empty perfect sets by using *c*) and § 2, Exercise 24.) Deduce that in this case Z is not an almost open set in X (Exercise 6; note that the reasoning above shows that if U is any non-empty open subset of X, $U \cap Z$ is not meagre).

19) Let A be a dense subspace of a topological space X, and let f be a continuous mapping of A into a complete metric space Y. Show that the set of points of X where f has no limit (relative to A) is a meagre set in X [consider, for each $n > 0$, the set of points $x \in E$ at which the oscillation $\omega(x; f)$ of f is $< 1/n$].

¶ 20) Let f be a homeomorphism of a subset A of a complete metric space X onto a subset A′ of a complete metric space X′. Show that there exists an extension $\bar{f}$ of f to a set B which is a countable intersection of open sets in X, such that f is a homeomorphism of B onto a set B′ which is a countable intersection of open sets in X′ (apply Exercise 19 to f and to the inverse homeomorphism g).

¶ 21) *a*) Let (f_n) be a sequence of continuous mappings of a topological space X into a perfectly normal space Y (§ 4, Exercise 8) such that, for each $x \in \mathrm{X}$, the sequence $(f_n(x))$ has a limit $f(x)$ in Y. Show that, for every closed subset S of Y, $\mathrm{A} = \overset{-1}{f}(\mathrm{S})$ is a countable intersection of open sets. [There exists a decreasing sequence (U_n) of open sets in Y such that $\mathrm{S} = \bigcap_n \overline{\mathrm{U}}_n$; note that $f(x) \in \mathrm{S}$ if and only if, for each integer $n > 0$, there exists an integer $k \geqslant 0$ such that $f_{n+k}(x) \in \mathrm{U}_n$.] Deduce that $\overline{\mathrm{A}} - \mathrm{A}$ is meagre in X.
b) Deduce from *a*) that if we suppose in addition that Y is a metrizable space of countable type, then the set of all $x \in \mathrm{X}$ such that f is not continuous at x is meagre. [If (V_n) is a base of the topology of Y and if $\mathrm{S}_n = \mathrm{Y} - \mathrm{V}_n$ and $\mathrm{A}_n = \overset{-1}{f}(\mathrm{S}_n)$, show that x must belong to one of the sets $\overline{\mathrm{A}}_n - \mathrm{A}_n$.]

22) *a*) Let X be a Baire space, Y a metric space, (f_n) a sequence of continuous mappings of X into Y such that, for each $x \in \mathrm{X}$, the sequence $(f_n(x))$ has a limit $f(x)$ in Y. Then the set of all points $x \in \mathrm{X}$ at which f is not continuous is meagre. [Prove that, for each integer $n > 0$, the set of all points $x \in \mathrm{X}$ at which the oscillation $\omega(x; f)$ of f is $\leqslant 1/n$ contains a dense open set; for this, if G_p denotes the set of $x \in \mathrm{X}$ for which the distance from $f_p(x)$ to $f_q(x)$ is $\leqslant 1/2n$ for *all* $q \geqslant p$, show that the union of the interiors of the sets G_p is a dense open set.]
b) Let K be the compact interval $[0, 1]$ in **R**. Give an example of a sequence of continuous mappings f_n of K into the compact space K^K such that $\lim_{n\to\infty} f_n(x) = f(x)$ exists for each $x \in \mathrm{K}$, but such that f is not continuous at any point of K.

¶ 23) Let X be a topological space, and let Y, Z be two metric spaces, with metrics d, d' respectively. Let $f\colon \mathrm{X} \times \mathrm{Y} \to \mathrm{Z}$ be a mapping such that $y \to f(x_0, y)$ is continuous on Y for each $x_0 \in \mathrm{X}$ and such that $x \to f(x, y_0)$ is continuous on X for each $y_0 \in \mathrm{Y}$.
a) For each real number $\varepsilon > 0$, each $b \in \mathrm{Y}$ and each $x \in \mathrm{X}$, let $g(x; b, \varepsilon)$ denote the least upper bound of the numbers $\alpha > 0$ such that

the relation $d(b, y) < \alpha$ implies $d'(f(x, b), f(x, y)) \leqslant \varepsilon$. Show that the function $x \to g(x; b, \varepsilon)$ is upper semi-continuous.

b) If X is a Baire space, deduce from *a*) and Theorem 2 that for each $b \in Y$ there is a set S_b in X, whose complement is meagre, such that f is continuous at (a, b) for each $a \in S_b$.

24) Let X, Y be two topological spaces. A mapping $f: X \to Y$ is said to be *almost open* if, for every closed subset S of Y, $\overset{-1}{f}(S)$ is almost open in X (Exercise 6).

a) Let $g: X \to Y$ be a mapping with the property that there exists a meagre set M in X such that $g|\complement M$ is continuous. Show that g is almost open. Consider the converse, when the topology of Y has a countable base.

b) Suppose that Y is perfectly normal (§ 4, Exercise 8). Let (f_n) be a sequence of almost open mappings of X into Y such that

$$\lim_{n \to \infty} f_n(x) = f(x)$$

exists for each $x \in X$. Show that f is almost open [argue as in Exercise 21 *a*)].

25) Let R be an open equivalence relation on a topological space X. Show that if X is non-meagre (Exercise 7) [resp. a Baire space, resp. totally non-meagre (Exercise 12)], then so is X/R.

¶ 26) Let X be a locally compact metrizable space, let d be a metric compatible with the topology of X, and let R be a closed equivalence relation on X, all of whose equivalence classes are compact. Then X/R is locally compact (Chapter I, § 10, Proposition 9) and metrizable (§ 4, Exercise 24).

a) For each pair of points u, v of X/R, set

$$\rho(u, v) = \sup_{\varphi(x) = u} d(x, \overset{-1}{\varphi}(v)),$$

where $\varphi : X \to X/R$ is the canonical mapping, and for each $u \in X/R$ let

$$\lambda(u) = \limsup_{v \to u,\ v \neq u} \rho(u, v).$$

Show that λ is upper semi-continuous on X/R and that, for each $\alpha > 0$, the set of points $u \in X/R$ such that $\lambda(u) \geqslant \alpha$ is nowhere dense. [Show that otherwise there would exist a compact subset K of X and an infinite sequence (x_n) of points of K such that $d(x_m, x_n) > \frac{1}{2}\alpha$ for $m \neq n$.]

b) Deduce from *a*) that there is a dense subset M of X/R which is a countable intersection of open sets in X/R and is such that for each $x \in \overset{-1}{\varphi}(M)$ the image under φ of every neighbourhood of x in X is a neighbourhood of $\varphi(x)$ in X/R (use Theorem 1).

¶ 27) *a*) Let G be a topological group, and let A be an almost open set (Exercise 6) in G. Show that if A is not meagre then AA^{-1} is a neighbourhood of the identity element in G. [Note first that G is a Baire space by virtue of Exercise 3. Let A^* be the union of all open sets $U \subset G$ such that $U \cap \complement A$ is meagre. Show that A^* is not empty and that if $x \in G$ is such that xA^* meets A^*, then xA meets A. Hence show that $AA^{-1} \supset A^*(A^*)^{-1}$.]

b) Deduce from *a*) that an almost open subgroup of a topological group G is either meagre or else open (and closed). In particular, a countable topological group which is a Hausdorff Baire space is discrete.

c) Show that there exist subgroups H of the topological group **R** such that **R**/H is countably infinite (use a Hamel base); such a subgroup is dense but not meagre nor almost open.

d) Let G be a metrizable topological group whose left and right uniformities coincide. Show that if there exists a metric on G, compatible with the topology of G, and with respect to which G is a complete metric space, then G is a complete metrizable group [use *b*); § 3, Exercise 2; and § 2, Exercise 24].

¶ 28) Let G, G′ be two topological groups. Show that every continuous homomorphism f of G *onto* G′ is a *strict morphism* in each of the following two cases:

a) G is locally compact and σ-compact, and G′ is non-meagre [show that there is a relatively compact open set U in G such that the interior of $f(\overline{U})$ is not empty; deduce that for each compact neighbourhood V of the identity element of G, the interior of $f(V)$ is not empty, and then apply Chapter III, § 2, Exercise 18].

b) G is complete, the topology of G has a countable base, and G′ is non-meagre. [By using the fact that, for each neighbourhood U of the identity element e of G, G is the union of a sequence of sets of the form x_nU, show that for each open set A in G there is an open set A′ in G′ which contains $f(A)$ and is such that $f(A)$ is dense in A′. Then consider a fundamental system (U_n) of symmetric neighbourhoods of e in G, such that $\overline{U}^2_{n+1} \subset U_n$, and show that $f(U_p)$ contains the open set U'_{p+1}; to do this, take any point $a' \in U'_{p+1}$ and then construct by induction a sequence (b_n) of points of G such that $b_n \in U_{p+n}$ and such that $f(b_1b_2\ldots b_n)$ tends to a'; note that, if $x' \in U'_k$, then there exists $y \in U_k$ such that $f(y) \in x'U'_{k+1}$.] The hypothesis that G has a countable base is essential, as is shown by the example where G′ is the group **R** with the topology of the real line, G is the group **R** with the discrete topology and $f : G \to G'$ is the identity mapping.

29) Let G be a locally compact, σ-compact, topological group, or else a complete metrizable group whose topology has a countable base. Let H be a closed normal subgroup of G and let A be a closed subgroup of G. Then the quotient group $A/(A \cap H)$ is isomorphic to AH/H if and only if AH is a closed subgroup of G [to show that the condition is necessary, note that $A/(A \cap H)$ is complete, by Chapter III, § 4, no. 6, Proposition 13, and Chapter IX, § 3, no. 1, Proposition 4; to show that the condition is sufficient, use Exercise 28].

¶ 30) Let G be a group and let d be a metric on G such that, with respect to the topology $\mathcal{T}$ defined by d on G, the mappings $y \to x_0 y$ and $y \to y x_0$ of G into G are continuous for each $x_0 \in G$.

a) Show that if G is complete with respect to the topology $\mathcal{T}$, then the mapping $(x, y) \to xy$ is continuous on $G \times G$ (use Exercise 23).

b) Suppose in addition that $\mathcal{T}$ has a countable base. Show that the mapping $x \to x^{-1}$ is then continuous on G, i.e., that the topology $\mathcal{T}$ is *compatible* with the group structure of G. [In the group $G \times G$, endowed with the topology $\mathcal{T} \times \mathcal{T}$, consider the set F of all points (x, x^{-1}) where $x \in G$; F is closed in $G \times G$, and the law of composition $(x, x^{-1})(y, y^{-1}) = (xy, y^{-1}x^{-1})$ defines a group structure on F; show that the topology induced on F by that of $G \times G$ is compatible with this group structure by using *a*). Then prove, by arguing as in Exercise 28 *b*), that the projection pr_1 of F onto G is bicontinuous.]

§ 6

1) *a*) Every zero-dimensional space is completely regular. The non-normal locally compact spaces defined in § 4, Exercise 26 *b*) and § 5, Exercise 15 are zero-dimensional.

b) A topological space X is said to be *strongly zero-dimensional* if for each closed subset A of X and each neighbourhood U of A, there is an open and closed neighbourhood of A contained in U. Every strongly zero-dimensional space is normal. A normal space is strongly zero-dimensional if and only if its Stone-Čech compactification (§ 1, Exercise 7) is totally disconnected [use Exercise 17 *c*) of § 4].

c) Let X be a strongly zero-dimensional Hausdorff space and let (U_n) be an increasing sequence of non-empty open sets in X. Show that there exists a sequence (G_n) of mutually disjoint sets which are both open and closed in X, such that $G_n \subset U_n$ for each n, and $\bigcup_n G_n = \bigcup_n U_n$ (define the G_n by induction). If, moreover, X is perfectly normal (§ 4, Exercise 8), every open set in X is the union of a countable family of mutually disjoint sets which are both open and closed in X.

d) Let X be a normal space which is the union of a sequence (A_n) of strongly zero-dimensional subspaces. Show that X is strongly zero-dimensional. [Let B, C be two disjoint closed subsets of X; define by induction two increasing sequences (G_n), (H_n) of open sets in X such that $\overline{G}_n \cap \overline{H}_n \neq \emptyset$, $A_n \subset G_n \cup H_n$, $B \cap A_n \subset G_n$, $C \cap A_n \subset H_n$.]

¶ 2) *a*) Let X be a metrizable space. Show that the following properties are equivalent:

α) There exists a metric d compatible with the topology of X, with respect to which X is an ultrametric space (§ 2, Exercise 4).

β) X is strongly zero-dimensional.

γ) There exists a sequence $(\mathfrak{B}_n)$ of locally finite families of open and closed subsets of X such that $\mathfrak{B} = \bigcup_n \mathfrak{B}_n$ is a base for the topology of X.

[To show that *α*) implies *β*), note that if F is a closed set in an ultrametric space X, then the set of all $x \in X$ such that $d(x, F) = \rho > 0$ is both open and closed in X. To show that *β*) implies *γ*), use Lemma 3 of § 4, no. 5 and Exercise 1 *c*) of § 6. Finally, to establish that *γ*) implies *α*), reduce to the case where $\mathfrak{B}_n = (U_{n,\lambda})_{\lambda \in L}$, and if $f_{n,\lambda}$ is the characteristic function of $U_{n,\lambda}$, consider the mapping $x \rightarrow f(x) = f_{n,\lambda}(x)$ of X into the product group $G^{N \times L}$, where $G = \mathbf{Z}/2\mathbf{Z}$ (identified with the set $\{0, 1\}$); for each $n \in N$, let G_n be the subgroup of G consisting of the elements whose (k, λ) coordinates are zero for all $\lambda \in L$ and all $k < n$; show that if G is endowed with the group topology for which (G_n) is a fundamental system of neighbourhoods of 0, then f is a homeomorphism of X onto a subspace of G; and complete the proof by noting that the topology of G can be defined by an invariant metric with respect to which G is an ultrametric space.]

b) Deduce from *a*) that every zero-dimensional Hausdorff space X which has a countable base is a strongly zero-dimensional metrizable space. (Show by using § 2, no. 8, Proposition 13 that there exists a countable base formed of sets which are both open and closed.) Furthermore, X is then homeomorphic to a subspace of Cantor's triadic set K (cf. § 2, Exercise 12 (*).)

(*) It is not known whether every zero-dimensional metrizable space is strongly zero-dimensional.

¶ 3) *a*) Show that every totally disconnected subspace of $\mathbf{R}$ is zero-dimensional.

b) Let K be Cantor's triadic set; every $x \in \mathrm{K}$ can be written uniquely in the form $x = \sum_k 2/3^{n_k}$, where $(n_k)_{k \geqslant 1}$ is a strictly increasing (finite or infinite) sequence of integers > 0 (Chapter IV, § 8, Exercise 9); and define $f(x)$ to be the number $\sum_k (-1)^{n_k}/2^k$. Let G be the graph of the function f in $\mathrm{K} \times \mathbf{R}$. Show that the metrizable space G is totally disconnected but is not zero-dimensional. [Note that the intersection of $\overline{\mathrm{G}}$ and the line $\{0\} \times \mathbf{R}$ contains an open interval containing the point $(0, 0)$.]

4) Let X be a metrizable space.

a) Show that the set of Borel subsets of X is the smallest subset $\mathfrak{F}$ of $\mathfrak{P}(\mathrm{X})$ which contains all closed subsets of X and is such that countable unions and countable intersections of sets of $\mathfrak{F}$ belong to $\mathfrak{F}$. (Consider the subset $\mathfrak{G}$ of $\mathfrak{F}$ consisting of all $\mathrm{A} \in \mathfrak{F}$ such that $\mathrm{X} - \mathrm{A} \in \mathfrak{F}$.)

b) Show that the set of Borel subsets of X is the smallest subset $\mathfrak{F}$ of $\mathfrak{P}(\mathrm{X})$ which contains all open subsets of X and is such that every countable intersection of sets of $\mathfrak{F}$ belongs to $\mathfrak{F}$, and every union of a sequence of mutually disjoint sets of $\mathfrak{F}$ belongs to $\mathfrak{F}$ (same method).

c) Every ordinal α can be written uniquely in the form $\alpha = \omega\beta + n$, where $n < \omega$ (*Set Theory*, Chapter III, § 2, Exercise 15); α is said to be *even* or *odd* according as n is even or odd. For every countable ordinal α, we define by transfinite induction sets of subsets $\mathfrak{F}_\alpha(\mathrm{X})$, $\mathfrak{G}_\alpha(\mathrm{X})$ (or simply $\mathfrak{F}_\alpha$, $\mathfrak{G}_\alpha$) as follows: $\mathfrak{F}_0$ (resp. $\mathfrak{G}_0$) is the set of all closed (resp. open) subsets of X; if α is even and > 0, $\mathfrak{F}_\alpha$ (resp. $\mathfrak{G}_\alpha$) is the set of all subsets of X which are countable intersections (resp. countable unions) of sets belonging to $\bigcup_{\xi<\alpha} \mathfrak{F}_\xi$ (resp. $\bigcup_{\xi<\alpha} \mathfrak{G}_\xi$); if α is odd, then $\mathfrak{F}_\alpha$ (resp. $\mathfrak{G}_\alpha$) is the set of all subsets of X which are countable unions (resp. countable intersections) of sets belonging to $\bigcup_{\xi<\alpha} \mathfrak{F}_\xi$ (resp. $\bigcup_{\xi<\alpha} \mathfrak{G}_\xi$). Show that the union of the $\mathfrak{F}_\alpha$ (resp. $\mathfrak{G}_\alpha$) is the set of Borel subsets of X [use *a*)]. Deduce that if X is of countable type, the set of Borel subsets of X has power at most equal to the power of the continuum.

d) The relation $\mathrm{A} \in \mathfrak{F}_\alpha$ is equivalent to $\complement \mathrm{A} \in \mathfrak{G}_\alpha$. Every *finite* union (resp. intersection) of sets of $\mathfrak{F}_\alpha$ (resp. $\mathfrak{G}_\alpha$) belongs to $\mathfrak{F}_\alpha$ (resp. $\mathfrak{G}_\alpha$). We have $\mathfrak{F}_\alpha \subset \mathfrak{F}_{\alpha+1} \cap \mathfrak{G}_{\alpha+1}$ and $\mathfrak{G}_\alpha \subset \mathfrak{F}_{\alpha+1} \cap \mathfrak{G}_{\alpha+1}$ (use transfinite induction).

e) Let f be a continuous mapping of X into a metrizable space Y. Show that $\overset{-1}{f}(\mathfrak{F}_\alpha(Y)) \subset \mathfrak{F}_\alpha(X)$ and $\overset{-1}{f}(\mathfrak{G}_\alpha(Y)) \subset \mathfrak{G}_\alpha(X)$.

f) If X and Y are two metrizable spaces and $A \in \mathfrak{F}_\alpha(X)$, $B \in \mathfrak{F}_\alpha(Y)$, [resp. $A \in \mathfrak{G}_\alpha(X)$, $B \in \mathfrak{G}_\alpha(Y)$], then $A \times B \in \mathfrak{F}_\alpha(X \times Y)$ [resp. $A \times B \in \mathfrak{G}_\alpha(X \times Y)$].

g) Let (X_n) be a sequence of metrizable spaces. Show that if A_n is a Borel set in X_n for each n, then $\prod_n A_n$ is a Borel set in $\prod_n X_n$.

h) Let $\alpha > 0$ be an even (resp. odd) countable ordinal and let (A_n) be a countable covering of X formed of sets belonging to $\mathfrak{G}_\alpha$ (resp. $\mathfrak{F}_\alpha$). Show that there is a partition (B_n) of X such that $B_n \subset A_n$ and $B_n \in \mathfrak{F}_\alpha \cap \mathfrak{G}_\alpha$ for each n [use *d*)].

¶ 5) *a*) Let J be the Polish space N^N (N carrying the discrete topology) and let X be a metrizable space of countable type. Show that for each countable ordinal α there is a subset G_α of $J \times X$ such that (i) $G_\alpha \in \mathfrak{G}_\alpha(J \times X)$ (Exercise 4) and (ii) for every subset $U \in \mathfrak{G}_\alpha(X)$ there exists $z \in J$ such that $G_\alpha(z) = U$. [We may proceed as follows by transfinite induction: the spaces J and J^N being homeomorphic, let f be a homeomorphism of J onto J^N, and for each integer n let $f_n = pr_n \circ f$. If (A_n) is a countable base of X containing the empty set, take G_0 such that $G_0(z) = \bigcup_n A_{f_n(z)}$ for each $z \in J$. For each countable ordinal α, take $G_{\alpha+1}$ to be such that $G_{\alpha+1}(z) = \bigcap_n G_\alpha(f_n(z))$ if α is even, and $G_{\alpha+1}(z) = \bigcup_n G_\alpha(f_n(z))$ if α is odd. Finally, if α has no predecessor and if (λ_n) is an increasing sequence of ordinals such that $\alpha = \sup_n \lambda_n$, take G_α to be such that $G_\alpha(z) = \bigcup_n G_{\lambda_n}(f_n(z))$. Use the fact that f_n is continuous on J.]

b) In particular take $X = J$. Show that the set $pr_1(G_\alpha \cap \Delta)$, where Δ is the diagonal of $J \times J$, belongs to $\mathfrak{G}_\alpha(J)$ but not to $\mathfrak{F}_\alpha(J)$ [argue by contradiction, using *a*)].

6) *a*) Let J be the Polish space N^N. Show that if X is any Souslin space, there is a continuous surjection of J onto X [reduce to the case where X is Polish and consider a sifting of X by a sieve (C_n) where all the C_n are infinite].

b) Let X be a metrizable space. Show that if A is any Souslin subspace of X, there is a closed subset F of $J \times X$ such that $A = pr_2(F)$ [use *a*)].

c) Let X be a metrizable space of countable type; let $L = J \times X$ and let F be a closed subset of $J \times L$ such that for each closed subset

M of L there exists $z \in J$ such that $F(z) = M$ [Exercise 5 *a*)]. Show that, for each Souslin subspace S of X, there exists $z \in J$ such that $pr_2(F(z)) = S$. Deduce that, if $X = J$, the set T of all $z \in J$ such that $z \in pr_2(F(z))$ is a Souslin set, but that $J - T$ is not a Souslin set [same reasoning as in Exercise 5 *b*)] (*).

7) *a*) Show that every zero-dimensional Polish space is homeomorphic to a closed subspace of the product $J = \mathbf{N}^{\mathbf{N}}$ (consider a strict sifting of X by sets which are both open and closed).

b) Let X be a zero-dimensional Polish space and let Y be a dense subspace of X which has no interior point and which is a countable intersection of open sets in X. Show that Y is homeomorphic to J. (Note that, in a metrizable zero-dimensional space, an open but not closed set is the union of an *infinite* countable sequence of pairwise disjoint sets which are both open and closed; use Theorem 1 of no. 1.) Deduce that every dense subspace of $\mathbf{R}$, which has no interior point and is a countable intersection of open sets, is homeomorphic to J (note that such a set is contained in the complement of a countable dense subset D of $\mathbf{R}$, and that $\complement D$ is zero-dimensional).

c) If X is any uncountable zero-dimensional Polish space, there exists a partition of X into a countable set and a subspace homeomorphic to J [use *b*) and Exercise 11 of § 5].

8) Let X be a Souslin space and let f be a continuous mapping of X into a Hausdorff space Y. Show that if $f(X)$ is uncountable, there exists a subspace A of X homeomorphic to Cantor's triadic set K (Chapter IV, § 2, no. 5) such that $f|A$ is injective. [Reduce to the case where X is a complete metric space: show that there exists a sieve $C = (C_n, p_n)$ such that, for each n, C_n has 2^n elements, and that there exists for each n a mapping φ_n of C_n into the set of all non-empty closed subsets of diameter $\leqslant 2^{-n}$ in X, such that (i) $\varphi_{n+1}(c) \subset \varphi_n(p_n(c))$ for all $c \in C_{n+1}$; (ii) whenever c and c' are distinct elements of C_n, $f(\varphi_n(c))$ and $f(\varphi_n(c'))$ are disjoint. Use § 5, Exercise 11.] In particular, every uncountable Souslin space contains a subspace homeomorphic to K and therefore has the power of the continuum.

¶ 9) *a*) Let X be a Polish space, and let R be an equivalence relation on X. Show that there exists a subset of X which is a countable inter-

(*) * It can be shown that in the space $\mathcal{C}(I;\mathbf{R})$ of continuous real-valued functions on a compact interval $I \subset \mathbf{R}$, endowed with the topology of uniform convergence (which makes it into a Polish space), the set of differentiable functions is not a Souslin set but its complement is a Souslin set [cf. S. Mazurkiewicz, *Fund. Math.* **27** (1936), p. 244]. *

section of open sets and which meets each equivalence class in exactly one point, provided one or the other of the following two hypotheses is satisfied:

α) R is closed;

β) X is locally compact and the graph of R is closed in $X \times X$.

[For α), follow the method of the proof of Theorem 4 of no. 8, making use of the following remark: If (G_n) is a sequence of sets, each of which is a countable intersection of open sets, and if for each n there exists a neighbourhood V_n of G_n such that the family (V_n) is locally finite, then $\bigcup_n G_n$ is a countable intersection of open sets. Same method for β); observe that the saturation of a compact set is now closed (cf. Chapter I, § 10, Exercise 16).]

b) Let X be a Polish space, let R be an equivalence relation on X which is both open and closed, and let $\varphi : X \to X/R$ be the canonical mapping. Show that there exists a subset A of X which is a countable intersection of open sets in X and is such that the restriction of φ to A is a homeomorphism of A onto $\varphi(A)$ and such that $\varphi(A)$ is dense in X/R and is a countable intersection of open sets. [With the notation of the proof of Theorem 4 of no. 8, show that we may assume that for each n the $\varphi_n(c)$ have been defined in such a way that the image under φ of the union of the interiors of the $h_n(c)$, as c runs through C_n, is dense in X/R; use *a*) and Theorem 1 of no. 1.]

c) Show that the conclusion of *b*) remains valid when X is locally compact and has a countable base and R is a closed equivalence relation on X all of whose equivalence classes are compact. [Reduce to case *b*) by the use of § 5, Exercise 26.]

¶ 10) *a*) Let X be a metrizable space and let S be a Souslin subspace of X. Show that S is almost open in X (§ 5, Exercise 6). [Let P be a Polish space, g a continuous mapping of P onto S, (C_n, p_n, φ_n) a sifting of P, and f the corresponding continuous mapping of L(C) onto P (in the notation of no. 5); let $h = g \circ f$, and for each $c \in C_n$ let $q_n(c)$ denote the subspace of L(C) consisting of sequences (c_k) such that $c_n = c$. Let $F_n(c) = h(q_n(c))$ and let $U_n(c) = g(\varphi_n(c)) \supset F_n(c)$ for each $c \in C_n$; let $W_n(c)$ denote the union of $D(U_n(c))$ (§ 5, Exercise 3) and a meagre set in X, such that $F_n(c) \subset W_n(c) \subset \overline{U_n(c)}$, so that $W_n(c)$ is almost open in X. Show that

$$V_n(c) = W_n(c) \cap \complement\Big(\bigcup_{p_n(c')=c} W_{n+1}(c')\Big)$$

is meagre in X by noting that $F_n(c) = \bigcup_{p_n(c')=c} F_{n+1}(c')$; show also that the set $W_n(c) - F_n(c)$ is contained in the union of the sets $V_m(d)$ where $m \geqslant n$ and, for each m, d runs through the set of elements of C_m such that $p_{nm}(d) = c$. Hence show that $F_n(c)$ is almost open.]

b) Give an example of a continuous mapping f of $[0, 1]$ into itself and an almost open set $B \subset I$ such that $f(B)$ is not almost open. [Use Exercise 16 *b*) of Chapter IV, § 8, and show that we may take B to be a subset of the Cantor set K such that $f(B) = Z$ has the property described in § 5, Exercise 18 *d*).]

¶ 11) Let $X = \mathbf{R}^n$ and let M be a closed subset of X. A point $x \in M$ is said to be *linearly accessible* if there exists a point $y \in X - M$ such that the open segment with extremities x, y is contained in $X - M$. Show that the set $L \subset M$ of linearly accessible points of M is a Souslin set. [Let $d(x, y)$ be the Euclidean metric in X, and for each $z \in X$ let $f_z(x, y) = d(x, z) + d(z, y) - d(x, y)$. Note that the set of all $(x, y, z) \in X \times X \times X$ such that $x \in M, y \in X - M, z \in M, z \neq x$ and $f_z(x, y) = 0$ is a countable union of compact sets, and consequently so is its projection on the product of the first two factors.]

12) Let X be a metrizable space, and let f be a mapping of the set $\mathfrak{K}(X)$ of compact subsets of X into $\overline{\mathbf{R}}$ which satisfies the conditions (CA_I) (for two compact subsets of X) and (CA_{III}). For each subset A of X, let $f_*(A)$ denote the least upper bound of the numbers $f(K)$ for all compact subsets K of A, and let $f^*(A)$ denote the greatest lower bound of the $f_*(U)$ for all open sets U containing A. A is said to be *admissible* with respect to f if $f_*(A) = f_*(A)$.

a) Show that for every compact subset L of X and every $\varepsilon > 0$, there exists an open set $U \supset L$ such that, for every compact subset K of X for which $L \subset K \subset U$, we have $f(K) \leqslant f(L) + \varepsilon$. (Argue by contradiction.) Deduce that every compact subset and every open subset of X is admissible with respect to f.

b) Suppose that f satisfies the following condition:

(AL) For each pair of compact subsets K, K′ of X, we have

$$f(K \cup K') + f(K \cap K') \leqslant f(K) + f(K').$$

Show that if (K_i) is a finite family of compact sets with the property that

$$f\left(\bigcup_i K_i\right) < +\infty,$$

and if (K'_i) is another finite family of compact sets indexed by the same set, such that $K'_i \subset K_i$ and $f(K'_i) > -\infty$ for each i, then we have

$$f\left(\bigcup_i K_i\right) - f\left(\bigcup_i K'_i\right) \leqslant \sum_i (f(K_i) - f(K'_i)).$$

Let (A_i), (B_i) be two finite families of subsets of X, with the same index set, such that $B_i \subset A_i$ for each i, $f^*\left(\bigcup_i A_i\right) < +\infty$ and $f^*(B_i) > -\infty$ for each i. Show that if f satisfies (AL), we have

$$f^*\left(\bigcup_i A_i\right) - f^*\left(\bigcup_i B_i\right) \leqslant \sum_i (f^*(A_i) - f^*(B_i)).$$

[Reduce to the case of an index set with two elements. Consider first the case where A_i and B_i are open sets, and use the following lemma: if U_1, U_2 are two open sets and K is a compact set contained in $U_1 \cup U_2$, then there exist compact sets $K_i \subset U_i$ $(i = 1, 2)$ such that $K \subset K_1 \cup K_2$.]

c) Suppose that f satisfies condition (AL). Show that for every increasing sequence (A_n) of subsets of X such that $f^*(A_n) > -\infty$, we have $f^*\left(\bigcup_n A_n\right) = \sup_n f^*(A_n)$ [use *b*)]. Deduce that if f does not take the value $-\infty$ on $\mathfrak{K}(X)$, then f^* is a capacity on X.

d) Suppose that f satisfies the condition (AL). Show that the union of any sequence (A_n) of admissible sets such that $f^*(A_n) > -\infty$ for each n is admissible. [Use *b*) and *c*).]

* 13) Let K be a compact subset of $\mathbf{R}^2$. For each real number y, let $\delta_1(K, y)$ [resp. $\delta_2(K, y)$] denote the diameter of $K \cap ([0, +\infty[\times \{y\})$ [resp. $K \cap (]-\infty, 0] \times \{y\})$]. Show that the functions $y \to \delta_1(K, y)$ and $y \to \delta_2(K, y)$ are upper semi-continuous. Let φ be an increasing continuous mapping of $[0, +\infty]$ into itself such that $\varphi(0) = 1$ and $\varphi(+\infty) = 2$, and let

$$\psi(K, y) = \varphi(\delta_1(K, y)\delta_2(K, y)) \qquad \text{and} \qquad f(K) = \int_{pr_2(K)} \psi(K, y)\, dy.$$

Show that f satisfies the conditions (CA_I) and (CA_{III}) and that we have $f(K \cup K') \leqslant f(K) + f(K')$ for each pair of compact subsets K, K′ of $\mathbf{R}^2$. But if A is the closed set defined by $x \geqslant 0$ and $0 \leqslant y \leqslant 1$, show that $f_*(A) = 1$ and $f^*(A) = 2$. *

14) Let X be a metrizable space and let f be a capacity on X such that $f(A \cup B) \leqslant f(A) + f(B)$.

a) With the notation of Exercise 12, show that $f^*(A \cup B) \leqslant f^*(A) + f^*(B)$ and $f_*(A \cup B) \leqslant f_*(A) + f^*(B)$ for any two subsets A, B of X. [If K is a compact set contained in $A \cup B$, and U is an open set containing B, write K in the form $K = (K \cap U) \cup (K \cap \complement U)$.]

b) Let K be a compact subset of X having the power of the continuum, and let (A, B) be a partition of K into two sets such that every compact subset of A or of B is countable [cf. § 5, Exercise 18 *d*)]. Show that if $f(\{x\}) = 0$ for all $x \in X$ and if $f(K) > 0$, then neither A nor B is capacitable with respect to f.

* 15) Let μ be the Lebesgue measure on $\mathbf{R}$; define a capacity f on $\mathbf{R}^2$ satisfying the condition (AL) of Exercise 12 by setting $f(A) = \mu^*(\mathrm{pr}_1 A)$ (Proposition 15).

a) Let A be a non-capacitable bounded subset of $\mathbf{R}^2$ (Exercise 14), let B_0 be a circle $\|x\| = r$ such that A is contained in the open disc $\|x\| < r$, and let B_1 be a circle $\|x\| = r'$ of radius $r' > r$. Show that $A \cup B_0$ and $A \cup B_1$ are capacitable, although their intersection A is not.

b) For each n, let C_n be the set of all $x \in \mathbf{R}^2$ such that

$$r < \|x\| < r + \frac{1}{n}.$$

Show that each of the sets $A \cup C_n$ is capacitable, although their intersection is not, and that we have $\inf_n f(C_n) \neq f\left(\bigcap_n C_n\right)$. *

¶ 16) Let X, X′ be two metrizable spaces. A mapping $f: X \to X'$ is said to be a *Borel* mapping if, for each closed subset F′ of X′, $\overset{-1}{f}(F')$ is a Borel set in X. For each even (resp. odd) countable ordinal α, f is said to be *of class* α if, for each closed subset F′ of X′, $\overset{-1}{f}(F')$ belongs to $\mathfrak{F}_\alpha(X)$ [resp. $\mathfrak{G}_\alpha(X)$] (Exercise 4); then, for each open set G′ in X′, $\overset{-1}{f}(G')$ belongs to $\mathfrak{G}_\alpha(X)$ [resp. $\mathfrak{F}_\alpha(X)$]. The Borel mappings of class 0 are precisely the continuous mappings.

a) The characteristic function φ_A of a subset A of X is of class α if and only if $A \in \mathfrak{F}_\alpha \cap \mathfrak{G}_\alpha$.

b) Let $f: X \to X'$ be a mapping of class α. Show that for every subset $B' \in \mathfrak{F}_\beta(X')$ [resp. $B' \in \mathfrak{G}_\beta(X')$], $\overset{-1}{f}(B')$ belongs to $\mathfrak{F}_{\alpha+\beta}(X)$ [resp. $\mathfrak{G}_{\alpha+\beta}(X)$]. (Use transfinite induction on β.)

c) Let $f: X \to X'$ be a mapping of class α and let $g: X' \to X''$ be a mapping of class β, where X″ is a third metrizable space. Show that $g \circ f$ is a Borel mapping of class $\alpha + \beta$.

d) For each even (resp. odd) countable ordinal α, let (A_n) be a sequence of subsets of X which belong to $\mathfrak{G}_\alpha$ (resp. $\mathfrak{F}_\alpha$) and which cover X. If a mapping $f: X \to X'$ is such that $f|A_n$ is of class α for each n, then f is of class α. Give the analogous result for a *finite* sequence of subsets of X which belong to $\mathfrak{F}_\alpha$ (resp. $\mathfrak{G}_\alpha$) and cover X.

17) *a*) Let X, X′ be two metrizable spaces, X′ being of countable type, and let (U'_n) be a countable base for the topology of X′. Let α

be an even (resp. odd) countable ordinal. If a mapping $f: X \to X'$ is such that $\overset{-1}{f}(U'_n)$ belongs to $\mathfrak{G}_\alpha(X)$ [resp. $\mathfrak{F}_\alpha(X)$] for each n, then f is a Borel mapping of class α.

b) Let (X'_n) be a sequence of metrizable spaces of countable type. A mapping $f = (f_n) : X \to \prod_n X'_n$ is of class α if and only if each f_n is of class α [use *a*)]. Deduce that, if X is of countable type, the finite real-valued functions of class α on X form a ring.

c) Let X, X′ be two metrizable spaces of countable type, and let α be an even (resp. odd) countable ordinal. If a mapping $f: X \to X'$ is of class α, show that its graph belongs to $\mathfrak{F}_\alpha(X \times X')$ [resp. $\mathfrak{G}_\alpha(X \times X')$] [use *b*) and Exercise 16 *b*)].

18) Let X be a Souslin space, X′ a metrizable space of countable type and let f be a mapping of X into X′.

a) Show that if the graph G of f is a Souslin set in $X \times X'$, then f is a Borel function and therefore [Exercise 17 *c*)] G is a Borel set (use Theorem 2, Corollary).

b) Show that if f is an injective Borel mapping and if X′ is a Souslin space, then the image under f of any Souslin set in X is a Souslin set in X′ (use Exercise 17 *c*)). If f is bijective, the inverse mapping is also a Borel mapping.

19) Let X, X′ be two metrizable spaces and let f be an almost open mapping (§ 5, Exercise 24) of X into X′. Show that if B′ is any Borel set in X′, then $\overset{-1}{f}(B')$ is almost open in X (cf. § 5, Exercise 6). Deduce that if g is a Borel mapping of X′ into a metrizable space X″, then $g \circ f$ is almost open. Give an example of a continuous mapping f and an almost open mapping g such that $g \circ f$ is not almost open [cf. Exercise 10 *b*)].

APPENDIX

1) Let X be a Hausdorff topological space on which there is defined an associative law of composition written multiplicatively. Generalize Definition 1 to the case of a family $(x_\iota)_{\iota \in I}$ of elements of X, whose index set I is *linearly ordered*. For each finite subset J of I, let p_J denote the product $\prod_{\iota \in J} x_\iota$ of the sequence $(x_\iota)_{\iota \in J}$.

If X is a *complete topological group*, then a family $(x_\iota)_{\iota \in I}$ of points of X is multipliable if and only if, given any neighbourhood V of the identity element e of X, there exists a finite subset J_0 of I such that, for every finite subset J of I containing J_0, we have $p_J(p_{J_0})^{-1} \in V$ [or $(p_{J_0})^{-1}p_J \in V$].

¶ 2) In a linearly ordered set I, we shall say that a non-empty subset J is a *slice* of I if, whenever α and β are elements of J such that $\alpha < \beta$, then $[\alpha, \beta] \subset J$.

a) Let $(x_\iota)_{\iota \in J}$ be a multipliable family in a *complete group* G. Show that, for every slice J of I, the family $(x_\iota)_{\iota \in J}$ is multipliable (consider first the particular case in which the set of lower bounds of J which do not belong to J is empty).

b) A partition $(J_\lambda)_{\lambda \in L}$ of I is said to be an *ordered partition* if L is a linearly ordered set and if the relations $\lambda < \mu$, $\alpha \in J_\lambda$, $\beta \in J_\mu$ imply $\alpha < \beta$; the J_λ are then slices of I. If $(x_\iota)_{\iota \in I}$ is a multipliable family in a complete group G, show that if we put $p_\lambda = \prod_{\iota \in J_\lambda} x_\iota$, then the family $(P_\lambda)_{\lambda \in L}$ is multipliable and has the same product as the family $(x_\iota)_{\iota \in I}$.

c) Conversely, if $(J_\lambda)_{\lambda \in L}$ is a *finite* ordered partition of I, and if $(x_\iota)_{\iota \in I}$ is such that each of the families $(x_\iota)_{\iota \in J_\lambda}$ is multipliable, then the family $(x_\iota)_{\iota \in I}$ is multipliable.

¶ 3) Let G be a Hausdorff topological group which has a neighbourhood V_0 of the identity element e on which $x \to x^{-1}$ is uniformly continuous (as a mapping $G_d \to G_d$: cf. Chapter III, § 3, Exercises 7 and 8). Let

$(x_\iota)_{\iota \in I}$ be a multipliable family of points of G; show that, for each neighbourhood U of e, there exists a finite subset J_0 of I such that, for every finite subset K of I contained in a slice of I which does not meet J_0, we have $p_K \in U$. [Show first that there exists a finite subset H_0 of I and a finite number of points $a_1, a_2, \ldots, a_q$ of G such that, for every finite subset L of I contained in a slice of I which does not meet H_0, we have $p_L \in a_k V'_0 a_k^{-1}$ for at least one index k, where V'_0 is a neighbourhood of e such that $V'_0 V'_0 \subset V_0$. Consequently we may restrict ourselves to considering finite subsets contained in one of the open intervals whose end-points are two consecutive indices of H_0; use the argument of Exercise 2 *a*) and the uniform continuity of x^{-1} on each of the neighbourhoods $a_k V_0 a_k^{-1}$.]

Deduce that under the same conditions we have $\lim x_\iota = e$ with respect to the filter of complements of finite subsets of I (if I is infinite).

4) Let G be a locally compact group. If (x_ι) is a multipliable family of points of G, show that the set of finite partial products p_J of this family is relatively compact.

5) *a*) Let G be a complete group such that every neighbourhood of the identity element e in G contains an open subgroup of G. Show that a sequence (x_n) of points of G is multipliable if and only if $\lim_{n \to \infty} x_n = e$.

b) Let G be a complete group in which every neighbourhood of e contains a *normal* open subgroup of G. Show that a family $(x_\iota)_{\iota \in I}$ of points of G is multipliable if and only if $\lim x_\iota = e$ with respect to the filter of complements of finite subsets of I (use Exercise 3).

6) Let A be the algebra (over **R**) of bounded real-valued functions defined on the interval $X = [1, +\infty[$ of **R**, with the norm $\|f\| = \sup_{x \in X} |f(x)|$. For each integer $n > 0$, let u_n be the function which is equal to $1/n$ for $n \leqslant x \leqslant n + 1$ and equal to 0 elsewhere. Show that the family $(1 + u_n)$ is multipliable in A but that the series whose general term is u_n is not absolutely summable in A.

7) Let $(\boldsymbol{x}_n)$ be a sequence of points of a normed algebra A, such that for every permutation σ of **N**, the infinite product with general factor $\boldsymbol{x}_{\sigma(n)}$ is convergent and $\displaystyle\mathop{\mathrm{P}}_{n=0}^{\infty} \boldsymbol{x}_{\sigma(n)}$ is a unit. Show that each of the sequences $(\boldsymbol{x}_{\sigma(n)})$ is multipliable (same argument as in Chapter III, § 5, no. 7, Proposition 9).

¶ 8) Let A be a complete normed algebra. Show that if $(\boldsymbol{x}_n)$ is a sequence of points of A such that, for every permutation σ of **N**, the sequence $(\boldsymbol{x}_{\sigma(n)})$ is multipliable and its product is a unit, then each

x_n is a unit. (First establish the following algebraic lemma: in a ring with an identity element, if two elements x, y are such that xy is a unit and yx is not a zero divisor, then x and y are both units.)

DIAGRAM OF THE PRINCIPAL TYPES OF TOPOLOGICAL SPACES

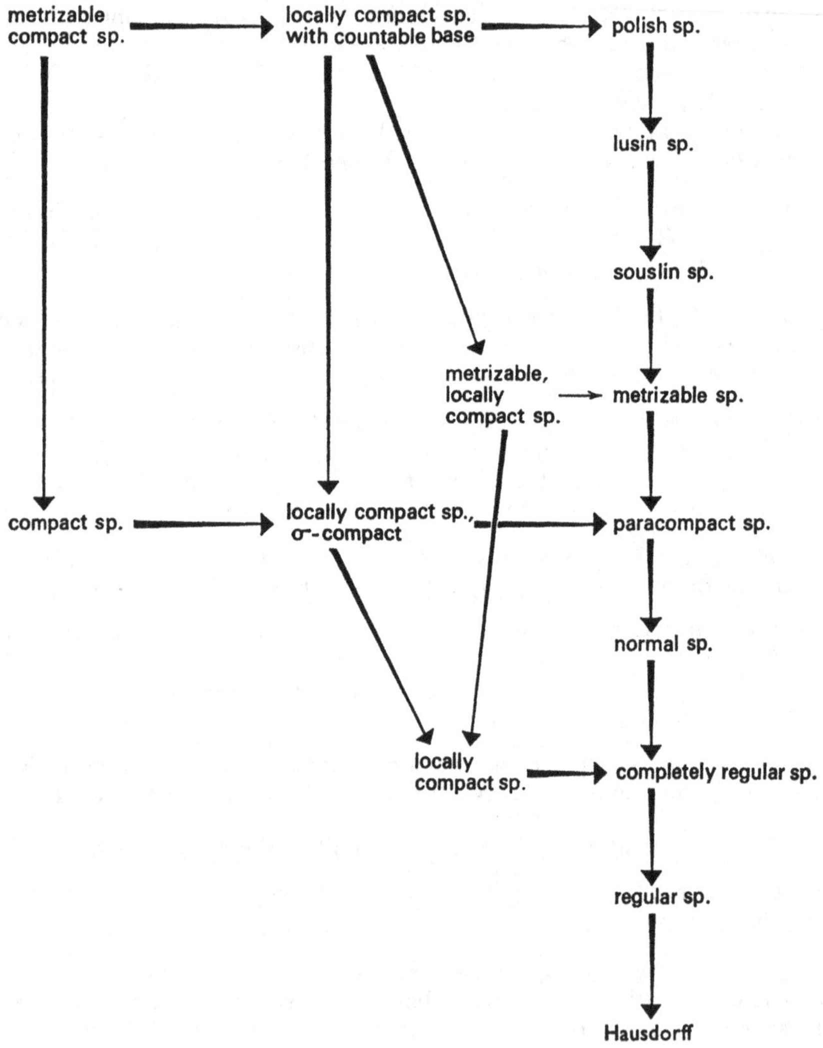

HISTORICAL NOTE

(Numbers in brackets refer to the bibliography at the end of this Note.)

As we remarked in the Historical Note to Chapter II, the notion of a metric space was introduced in 1906 by M. Fréchet, and developed some years later by F. Hausdorff in his "*Mengenlehre*". It acquired great importance after 1920, partly as a consequence of the fundamental work of S. Banach and his school on normed spaces and their applications to functional analysis, and also because of the significance of the notion of absolute value in the theory of numbers and algebraic geometry (where in particular the process of completion with respect to an absolute value has proved to be a powerful instrument).

The decade 1920-1930 saw a whole series of investigations into the properties of metric spaces. These studies, undertaken by the Moscow school, aimed especially at getting necessary and sufficient conditions for the metrizability of a given topology. This movement of ideas brought out the significance of the notion of a normal space, which had been defined by Tietze in 1923 but whose important role was recognized only as a consequence of the work of Urysohn [6] on the extension of continuous real-valued functions. Except for the trivial case of functions of a real variable, the problem of extending to the whole space a continuous real-valued function defined on a closed set was first considered (for the case of the plane) by H. Lebesgue [3]; before Urysohn's definitive result, it was solved for metric spaces by H. Tietze [4]. The generalization of this problem to include functions with values in an arbitrary topological space has acquired considerable importance in algebraic topology in the last few years. Recent work has also shown that in questions of this sort the notion of a normal space is not easily handled, since it allows too many possibilities of "pathology"; it often has to be replaced by the more restrictive concept of paracompactness, introduced in 1944 by J. Dieudonné [9]. In this theory, the most remarkable result is the theorem of A. H. Stone [10] which states that every metrizable space is paracompact (*).

We have already noted (Historical Note to Chapter IV) the important work at the end of the nineteenth century and the beginning of the twentieth (E. Borel, Baire, Lebesgue, Osgood, W. H. Young) on the classification of point-sets in $\mathbf{R}^n$, and on the classification and characterization of real-valued functions obtained from continuous functions by iterating the process of passing to the limit (with respect to sequences of functions). It was quickly realized that metric spaces, whose development after 1910 is largely the work of the Russian and Polish schools, formed a natural domain for investigations of this nature. Among other things, the study of metric spaces has shown clearly the fundamental role played in modern analysis by the notion of a meagre set and by the theorem on the countable intersection of dense open sets in a complete metric space (§ 5, no. 3, Theorem 1), which was first proved (independently) by Osgood [1] for the real line and by Baire [2] for the spaces $\mathbf{R}^n$.

On the other hand, Souslin in 1917 [5], correcting an error of Lebesgue's, showed that the continuous image of a Borel set is not necessarily a Borel set; this led him to the definition and study of a larger class of sets, since called "analytic sets" or "Souslin sets". After Souslin's premature death this work was carried forward particularly by N. Lusin (whose ideas had inspired Souslin's work) and the Polish mathematicians (see [7] and [8]). The importance of these sets nowadays lies in their applications to the theory of integration (where, thanks to their special properties, they allow constructions which would be impossible for arbitrary measurable sets) and to modern potential theory, in which the fundamental theorem on the capacitability of Souslin sets, proved recently by G. Choquet [11], has already shown itself to be rich in diverse applications.

(*) This theorem permitted a more satisfactory solution to the metrization problem than the criteria obtained around 1930 by the Russo-Polish school (the "Nagata-Smirnov criterion", cf. § 4, Exercise 22). But it should be noted that hitherto these criteria have had hardly any applications; as so often in the history of mathematics, it appears that the importance of the problem of metrization has lain less in its solution than in the new ideas it has stimulated.

BIBLIOGRAPHY

[1] W. Osgood, Non-uniform convergence and the integration of series term by term, *Amer. Journ. of Math.*, **19** (1897), pp. 155-190.

[2] R. Baire, Sur les fonctions de variables réelles, *Ann. di Mat.* (3), **3** (1899), p. 1.

[3] H. Lebesgue, Sur le problème des Dirichlet, *Rend. Circ. mat. di Palermo*, **24** (1907), pp. 371-402.

[4] H. Tietze, Über Funktionen, die auf einer abgeschlossenen Menge stetig sind, *J. de Crelle*, **145** (1915), pp. 9-14.

[5] M. Souslin, Sur une définition des ensembles mesurables B sans nombres transfinis, *C. R. Acad. Sci.*, **164** (1917), pp. 88-91.

[6] P. Urysohn, Über die Mächtigkeit der zusammenhängenden Mengen, *Math. Ann.*, **94** (1925), p. 262.

[7] N. Lusin, *Leçons sur les ensembles analytiques et leurs applications*, Paris (Gauthier-Villars), 1930.

[8] K. Kuratowski, *Topologie* I, 2nd edition, Warszawa-Vrocaw, 1948.

[9] J. Dieudonné, Une généralisation des espaces compacts, *Journ. de Math.* (9), **23** (1944), pp. 65-76.

[10] A. H. Stone, Paracompactness and product spaces, *Bull. Amer. Math. Soc.*, **54** (1948), pp. 977-982.

[11] G. Choquet, Theory of capacities, *Ann. Inst. Fourier*, **5** (1953-1954) pp. 131-295.

CHAPTER X

Function spaces

1. THE UNIFORMITY OF $\mathfrak{S}$-CONVERGENCE

Notation. If X and Y are any two sets, we recall that the set of all mappings of X into Y is denoted by $\mathscr{F}(X; Y)$, and may be identified with the product set Y^X (*Set Theory*, Chapter II, § 5, no. 2). For each subset H of $\mathscr{F}(X; Y)$ and each $x \in X$, we shall denote by $H(x)$ the set of elements $u(x) \in Y$ as u runs through H. If Φ is a filter base on $\mathscr{F}(X; Y)$, we denote by $\Phi(x)$ the filter base on Y formed by the sets $H(x)$ as H runs through Φ. Finally, we recall that, for each $u \in \mathscr{F}(X; Y)$ and each subset A of X, $u|A$ denotes the *restriction* of u to A, which is a mapping of A into Y; if H is a subset of $\mathscr{F}(X; Y)$, $H|A$ will denote the set of restrictions $u|A$ of functions $u \in H$.

1. THE UNIFORMITY OF UNIFORM CONVERGENCE

Let X be a set and let Y be a *uniform space*. For each entourage V of Y, let $W(V)$ denote the set of all pairs (u, v) of mappings of X into Y such that $(u(x), v(x)) \in V$ for all $x \in X$. As V runs through the set of entourages of Y, the sets $W(V)$ form a *fundamental system of entourages* of a uniformity on $\mathscr{F}(X; Y)$. For they clearly satisfy Axiom (U'_I) (Chapter II, § 1, no. 1); if V, V′ are two entourages of Y such that $V \subset V'$, we have $W(V) \subset W(V')$, and therefore the sets $W(V)$ satisfy (B_I) (Chapter I, § 6, no. 3); we have $\overset{-1}{\widehat{W(V)}} = W(\overset{-1}{V})$ so that (U'_{II}) is satisfied; finally, the relations "$(u(x), v(x)) \in V$ for all $x \in X$" and "$(v(x), w(x)) \in V$ for all $x \in X$" imply the relation "$(u(x), w(x)) \in \overset{2}{V}$ for all $x \in X$"; in other words, we have $\overset{2}{\widehat{W(V)}} \subset W(\overset{2}{V})$, which proves (U'_{III}).

DEFINITION 1. *The uniformity on the set* $\mathcal{F}(X; Y)$ *which has as a fundamental system of entourages the set of subsets* $\boldsymbol{W}(V)$, *where* V *runs through the set of entourages of* Y, *is called the uniformity of uniform convergence. The topology it induces is called the topology of uniform convergence. If a filter* Φ *on* $\mathcal{F}(X; Y)$ *converges to an element* u_0 *with respect to this topology,* Φ *is said to converge uniformly to* u_0.

Note that the *topology* of uniform convergence on $\mathcal{F}(X; Y)$ depends on the uniform structure of Y and not merely on the topology of Y (Exercise 4).

The uniform space obtained by endowing $\mathcal{F}(X; Y)$ with the uniformity of uniform convergence is denoted by $\mathcal{F}_u(X; Y)$.

2. $\mathfrak{S}$-CONVERGENCE

DEFINITION 2. *Let* X *be a set,* Y *a uniform space,* $\mathfrak{S}$ *a set of subsets of* X. *The uniformity of uniform convergence in the sets of* $\mathfrak{S}$, *or simply the uniformity of* $\mathfrak{S}$*-convergence, is the coarsest uniformity on* $\mathcal{F}(X; Y)$ *which makes uniformly continuous the restriction mappings* $u \to u|A$ *of* $\mathcal{F}(X, Y)$ *into the uniform spaces* $\mathcal{F}_u(A; Y)$, *where* A *runs through* $\mathfrak{S}$. *The uniform space obtained by endowing* $\mathcal{F}(X; Y)$ *with the uniformity of* $\mathfrak{S}$*-convergence is denoted by* $\mathcal{F}_{\mathfrak{S}}(X; Y)$.

The topology induced by the uniformity of $\mathfrak{S}$-convergence is called the *topology of* $\mathfrak{S}$*-convergence*; it is the coarsest for which all the mappings $u \to u|A$ of $\mathcal{F}(X; Y)$ into $\mathcal{F}_u(A; Y)$ $(A \in \mathfrak{S})$ are continuous (Chapter II, § 2, no. 3, Proposition 4, Corollary).

A filter Φ on $\mathcal{F}(X; Y)$ converges to u_0 with respect to the topology of $\mathfrak{S}$-convergence if and only if $u|A$ *converges uniformly to* $u_0|A$ with respect to Φ for all $A \in \mathfrak{S}$ (Chapter I, § 7, no. 6, Proposition 10), and Φ is therefore said to *converge uniformly to* u_0 *on the sets of* $\mathfrak{S}$.

A filter base Φ on $\mathcal{F}_{\mathfrak{S}}(X; Y)$ is a Cauchy filter base if and only if, for each $A \in \mathfrak{S}$, the image of Φ under the mapping $u \to u|A$ is a Cauchy filter base on $\mathcal{F}_u(A; Y)$ (Chapter II, § 3, no. 1, Proposition 4).

Let f be a mapping of a topological (resp. uniform) space Z into $\mathcal{F}_{\mathfrak{S}}(X; Y)$. Then f is continuous (resp. uniformly continuous) if and only if, for each $A \in \mathfrak{S}$, the mapping $z \to f(z)|A$ of Z into $\mathcal{F}_u(A; Y)$ is continuous (resp. uniformly continuous) (Chapter I, § 2, no. 3, Proposition 4; Chapter II, § 2, no. 3, Proposition 4).

Finally, let M be a subset of $\mathcal{F}_{\mathfrak{S}}(X; Y)$; then M is precompact if and only if, for each $A \in \mathfrak{S}$, the set of restrictions $u|A$ for $u \in M$ is a precompact subset of $\mathcal{F}_u(A; Y)$ (Chapter II, § 4, no. 2, Proposition 3).

Remarks. 1) The general definition of the entourages of an initial uniformity (Chapter II, § 2, no. 3, Proposition 4) shows that a fundamental system of entourages of $\mathcal{F}_{\mathfrak{S}}(X; Y)$ may be obtained as follows: for each $A \in \mathfrak{S}$ and each entourage V of a fundamental system of entourages $\mathfrak{B}$ of Y, let $W(A, V)$ be the set of all pairs of mappings (u, v) of X into Y such that $(u(x), v(x)) \in V$ for each $x \in A$; as A runs through $\mathfrak{S}$ and V runs through $\mathfrak{B}$, the *finite intersections of the sets* $W(A, V)$ form a fundamental system of entourages of $\mathcal{F}_{\mathfrak{S}}(X; Y)$.

This description shows immediately that if $\mathfrak{S}$, $\mathfrak{S}'$ are two sets of subsets of X such that $\mathfrak{S} \subset \mathfrak{S}'$, then the uniformity of $\mathfrak{S}'$-convergence is *finer* than that of $\mathfrak{S}$-convergence.

2) However, the uniformity of $\mathfrak{S}$-convergence is unaltered by replacing $\mathfrak{S}$ by the set $\mathfrak{S}'$ of all subsets of X *which are contained in finite unions of sets of* $\mathfrak{S}$. In the study of $\mathfrak{S}$-convergence we may therefore always restrict ourselves to the case where the set $\mathfrak{S}$ satisfies the following two conditions:

(F'_I) *Every subset of a set of* $\mathfrak{S}$ *belongs to* $\mathfrak{S}$.

(F'_{II}) *Every finite union of sets of* $\mathfrak{S}$ *belongs to* $\mathfrak{S}$.

If (F'_{II}) is satisfied, we obtain a fundamental system of entourages of $\mathcal{F}_{\mathfrak{S}}(X; Y)$ by taking all the sets $W(A, V)$, where A runs through $\mathfrak{S}$ and V runs through a fundamental system of entourages of Y.

3) The uniformity of $\mathfrak{S}$-convergence is the inverse image, under the mapping $u \to (u|A)_{A\in\mathfrak{S}}$ of $\mathcal{F}(X; Y)$ into $\prod_{A\in\mathfrak{S}} \mathcal{F}_u(A; Y)$, of the uniformity of this product space (Chapter II, § 2, no. 6, Proposition 8). If $\mathfrak{S}$ is a *covering* of X, this mapping is *injective* and $\mathcal{F}_{\mathfrak{S}}(X; Y)$ is therefore isomorphic to the uniform subspace of $\prod_{A\in\mathfrak{S}} \mathcal{F}_u(A; Y)$ which is the image of this mapping.

Proposition 1. *If* Y *is Hausdorff and* $\mathfrak{S}$ *is a covering of* X, *then the space* $\mathcal{F}_{\mathfrak{S}}(X; Y)$ *is Hausdorff.*

Let u, v be two elements of $\mathcal{F}_{\mathfrak{S}}(X; Y)$ such that $(u, v) \in W(A, V)$ for every entourage V of Y and every $A \in \mathfrak{S}$. Since Y is Hausdorff it follows that u and v coincide on every set $A \in \mathfrak{S}$, and since $\mathfrak{S}$ covers X we must have $u = v$.

Remarks. 4) Let H be a subset of $\mathcal{F}(X; Y)$. By abuse of language, the uniformity (resp. topology) induced on H by the uniformity (resp. topology) of $\mathfrak{S}$-convergence on $\mathcal{F}(X; Y)$ is called the uniformity (resp. topology) of $\mathfrak{S}$-convergence on the set H.

5) Let L be a set filtered by a filter $\mathfrak{G}$, and let $\lambda \to u_\lambda$ be a mapping of L into $\mathcal{F}_{\mathfrak{S}}(X; Y)$ which has a limit v with respect to $\mathfrak{G}$; we say then that, *with respect to the filter* $\mathfrak{G}$, *the mappings* u_λ *of* X *into* Y *converge uniformly to* v [or that *the family* (u_λ) *is uniformly convergent to* v] *in every set of* $\mathfrak{S}$. If $L = \mathbf{N}$ and $\mathfrak{G}$ is the Fréchet filter, we omit mention of $\mathfrak{G}$ in this statement.

More particularly, suppose that there is a commutative and associative law of composition (written additively) defined on Y. If (u_n) is any sequence of mappings of X into Y, let v_n be the mapping defined by

$$v_n(x) = \sum_{k=0}^{n} u_k(x) \quad (n \in \mathbf{N});$$

we say that *the series whose general term is* u_n *is uniformly convergent in every set of* $\mathfrak{S}$ if the sequence (v_n) is uniformly convergent in every set of $\mathfrak{S}$. Likewise we define a *uniformly summable family* $(u_\lambda)_{\lambda \in L}$ of mappings of X into Y by considering the mappings $x \to \sum_{\lambda \in J} u_\lambda(x)$ for all finite subsets J of L and the limit of these mappings in $\mathscr{F}_\mathfrak{S}(X; Y)$ with respect to the directed set of finite subsets of L (Chapter III, § 5, no. 1).

6) It follows immediately from Definitions 1 and 2 that, for every $x \in \bigcup_{A \in \mathfrak{S}} A$, the mapping $u \to u(x)$ of $\mathscr{F}_\mathfrak{S}(X; Y)$ into Y is *uniformly continuous*. Hence, in particular, if $\overline{H}$ denotes the closure of a subset H of $\mathscr{F}_\mathfrak{S}(X; Y)$, we have $\overline{H}(x) \subset \overline{H(x)}$ for all $x \in \bigcup_{A \in \mathfrak{S}} A$ (Chapter I, § 2, no. 1, Theorem 1).

3. EXAMPLES OF $\mathfrak{S}$-CONVERGENCE

I. *Uniform convergence in a subset of* X. Let A be a subset of X and take $\mathfrak{S} = \{A\}$. The uniformity (resp. topology) of $\mathfrak{S}$-convergence is then called the *uniformity* (resp. *topology*) *of uniform convergence in* A; if a filter Φ on $\mathscr{F}_\mathfrak{S}(X; Y)$ converges to u_0, it is said to converge to u_0 *uniformly in* A. When $A = X$ we recover the structure of uniform convergence defined in no. 1.

II. *Pointwise convergence in a subset of* X. Let A be a subset of X, and take $\mathfrak{S}$ to be the set of all subsets of X which consist of a single point belonging to A (by Remark 2 of no. 2 it comes to the same thing if we take $\mathfrak{S}$ to be the set of all finite subsets of A). The uniformity (resp. topology) of $\mathfrak{S}$-convergence is then called the *uniformity* (resp. *topology*) *of pointwise convergence in* A; if a filter Φ on $\mathscr{F}_\mathfrak{S}(X; Y)$ converges to u_0, it is said to converge to u_0 *pointwise in* A. This will be the case if and only if, for each $x \in A$, $u_0(x)$ is a limit of $u(x)$ with respect to the filter Φ.

In particular, when $A = X$, the uniformity (resp. topology) of pointwise convergence in X is called simply the *uniformity* (resp. *topology*) *of pointwise convergence*; the uniform space obtained by endowing $\mathscr{F}(X; Y)$ with this structure is denoted by $\mathscr{F}_s(X; Y)$. Note that the topology of pointwise convergence is just the *product* topology on Y^X and therefore depends only on the topology of Y, and not on its uniform structure.

III. *Compact convergence.* Suppose that X is a *topological space*, and take $\mathfrak{S}$ to be the set of all *compact* subsets of X. The uniformity (resp. the topology) of $\mathfrak{S}$-convergence is then called the *uniformity* (resp. the *topology*) *of compact convergence*, and the uniform space obtained by endowing $\mathcal{F}(X; Y)$ with this uniformity is denoted by $\mathcal{F}_c(X; Y)$. The structure of compact convergence is coarser than that of uniform convergence, and the two coincide if X is compact; also it is finer than the structure of pointwise convergence, and these two coincide if X is discrete.

If X is a *uniform space* we can define on $\mathcal{F}(X; Y)$ the uniformity of *precompact convergence* by taking $\mathfrak{S}$ to be the set of all *precompact* subsets of X. Again, if X is a *metric space* we may take $\mathfrak{S}$ to be the set of all *bounded* subsets of X; the uniformity of $\mathfrak{S}$-convergence is then called the uniformity of *bounded convergence.*

4. PROPERTIES OF THE SPACES $\mathcal{F}_{\mathfrak{S}}(X; Y)$

PROPOSITION 2. *Let* X_1, X_2 *be two sets, let* Y *be a uniform space and let* $\mathfrak{S}_i$ *be a set of subsets of* X_i $(i = 1, 2)$ *and* $\mathfrak{S}_1 \times \mathfrak{S}_2$ *the set of subsets of* $X_1 \times X_2$ *of the form* $A_1 \times A_2$, *where* $A_i \in \mathfrak{S}_i$, $i = 1, 2$. *Then the canonical bijection*

$$\mathcal{F}(X_1 \times X_2; Y) \to \mathcal{F}(X_1; \mathcal{F}(X_2; Y))$$

(*Set Theory,* R, § 4, no. 14) *is an isomorphism of the uniform space*

$$\mathcal{F}_{\mathfrak{S}_1 \times \mathfrak{S}_2}(X_1 \times X_2; Y)$$

onto $\mathcal{F}_{\mathfrak{S}_1}(X_1; \mathcal{F}_{\mathfrak{S}_2}(X_2; Y))$.

Let V be an entourage of Y and let $A_i \in \mathfrak{S}_i$ $(i = 1, 2)$; then it follows immediately from the definitions that $W(A_1 \times A_2, V)$ is identified with $W(A_1, W(A_2, V))$ by the canonical bijection, and the result is immediate.

PROPOSITION 3. a) *Let* X *be a set; let* $\mathfrak{S}$ *be a set of subsets of* X; *let* Y, Y' *be two uniform spaces; and let* $f : Y \to Y'$ *be a uniformly continuous mapping. Then the mapping* $u \to f \circ u$ *of* $\mathcal{F}_{\mathfrak{S}}(X; Y)$ *into* $\mathcal{F}_{\mathfrak{S}}(X; Y')$ *is uniformly continuous.*

b) *Let* X, X' *be two sets; let* $\mathfrak{S}$ (resp. $\mathfrak{S}'$) *be a set of subsets of* X (resp. X'); *let* Y *be a uniform space; and let* $g : X' \to X$ *be a mapping such that, for each* $A' \in \mathfrak{S}'$, $g(A')$ *is contained in a finite union of sets of* $\mathfrak{S}$. *Then the mapping* $u \to u \circ g$ *of* $\mathcal{F}_{\mathfrak{S}}(X, Y)$ *into* $\mathcal{F}_{\mathfrak{S}'}(X'; Y)$ *is uniformly continuous.*

PROPOSITION 4. *Let* X, Y *be two sets, let* $(X_\lambda)_{\lambda \in L}$ *be a family of sets and let* $(Y_\mu)_{\mu \in M}$ *be a family of uniform spaces. For each* $\lambda \in L$, *let* $\mathfrak{S}_\lambda$ *be a set of subsets of* X_λ, *let* g_λ *be a mapping of* X_λ *into* X, *and let* $\mathfrak{S}$ *be the set of subsets of* X *which is the union of the sets* $g_\lambda(\mathfrak{S}_\lambda)$. *For each* $\mu \in M$, *let* f_μ

be a mapping of Y *into* Y_μ, *and endow* Y *with the coarsest uniformity for which the* f_μ *are uniformly continuous. Then the uniformity of* $\mathfrak{S}$*-convergence on* $\mathscr{F}(X; Y)$ *is the coarsest uniformity which makes uniformly continuous the mappings* $u \to f_\mu \circ u \circ g_\lambda$ *of* $\mathscr{F}(X; Y)$ *into* $\mathscr{F}_{\mathfrak{S}_\lambda}(X_\lambda, Y_\mu)$.

These propositions are immediate consequences of the description of a fundamental system of entourages for the uniformity of $\mathfrak{S}$-convergence given in no. 2, Remark 1; the details of the proofs are left to the reader. Proposition 4 implies, in particular:

COROLLARY. *Let* X *be a set, let* $(Y_\iota)_{\iota \in I}$ *be a family of uniform spaces and let* $\mathfrak{S}$ *be a set of subsets of* X. *If we endow* $\prod_{\iota \in I} Y_\iota$ *with the product uniformity, the canonical bijection of the uniform space* $\mathscr{F}_{\mathfrak{S}}\left(X, \prod_{\iota \in I} Y_\iota\right)$ *onto the product uniform space* $\prod_{\iota \in I} \mathscr{F}_{\mathfrak{S}}(X; Y_\iota)$ (*Set Theory*, R, § 4, no. 13) *is an isomorphism.*

5. COMPLETE SUBSETS OF $\mathscr{F}_{\mathfrak{S}}(X; Y)$

PROPOSITION 5. *Let* Φ *be a set,* Y *a uniform space and* $\mathfrak{S}$ *a set of subsets of* X. *Then a filter* Φ *on* $\mathscr{F}_{\mathfrak{S}}(X; Y)$ *converges to* u_0 *if and only if* Φ *is a Cauchy filter with respect to the uniformity of* $\mathfrak{S}$*-convergence and converges pointwise to* u_0 *in* $B = \bigcup_{A \in \mathfrak{S}} A$.

Since the structure of pointwise convergence in B is coarser than that of $\mathfrak{S}$-convergence, it is enough to show that for each $A \in \mathfrak{S}$ and each *closed* entourage V of Y, $W(A, V)$ is *closed* in B with respect to the topology of pointwise convergence (Chapter II, § 3, no. 3, Proposition 7). Now $W(A, V)$ is the intersection of the inverse images of V under the mappings $(u, v) \to (u(x), v(x))$ as x runs through A; these mappings are continuous with respect to the topology of pointwise convergence (no. 2, Remark 6), and the result follows.

COROLLARY 1. *A subspace* H *of* $\mathscr{F}_{\mathfrak{S}}(X; Y)$ *is complete if and only if, for each Cauchy filter* Φ *on* H, *there exists* $u_0 \in H$ *such that* Φ *converges pointwise to* u_0 *in* $B = \bigcup_{A \in \mathfrak{S}} A$.

This follows immediately from Proposition 5.

COROLLARY 2. *Let* $\mathfrak{S}_1, \mathfrak{S}_2$ *be two sets of subsets of* X, *whose union is the same and which are such that* $\mathfrak{S}_1 \subset \mathfrak{S}_2$, *and let* H *be a subset of* $\mathscr{F}(X; Y)$. *Then if* H *is complete with respect to* $\mathfrak{S}_1$*-convergence, it is complete with respect to* $\mathfrak{S}_2$*-convergence.*

For every Cauchy filter with respect to $\mathfrak{S}_2$-convergence is also a Cauchy filter with respect to $\mathfrak{S}_1$-convergence, and we may apply Corollary 1.

COROLLARY 3. *Let* H *be a subset of* $\mathcal{F}(X; Y)$ *such that, for each*

$$x \in B = \bigcup_{A \in \mathfrak{S}} A,$$

the closure of $H(x)$ *in* Y *is a complete subspace of* Y. *Then the closure* $\overline{H}$ *of* H *in* $\mathcal{F}_{\mathfrak{S}}(X; Y)$ *is a complete subspace.*

Let Φ be a Cauchy filter on $\overline{H}$, and define a mapping $v : X \to Y$ as follows. If $x \in B$, $\Phi(x)$ is a Cauchy filter on $\overline{H(x)}$ (no. 2, Remark 6), hence by hypothesis it has at least one limit point; take $v(x)$ to be one of these limits. If $x \notin B$, take $v(x)$ to be any point of Y. With this definition of v, it is clear that Φ converges pointwise to v in B, and v is therefore a limit of Φ in $\mathcal{F}_{\mathfrak{S}}(X; Y)$ by Proposition 5.

In particular, if Y is complete, the hypothesis of Corollary 3 of Proposition 5 is satisfied for every $H \subset \mathcal{F}(X; Y)$; hence:

THEOREM 1. *Let* X *be a set, let* $\mathfrak{S}$ *be a set of subsets of* X, *and let* Y *be a complete uniform space. Then the uniform space* $\mathcal{F}_{\mathfrak{S}}(X; Y)$ *is complete.*

6. $\mathfrak{S}$-CONVERGENCE IN SPACES OF CONTINUOUS MAPPINGS

Let X, Y be two topological spaces, and let $\mathcal{C}(X; Y)$ denote the set of all *continuous mappings of* X *into* Y. If $\mathfrak{S}$ is a set of subsets of X and if Y is a uniform space, we denote by $\mathcal{C}_{\mathfrak{S}}(X; Y)$ the set $\mathcal{C}(X; Y)$ endowed with the uniformity of $\mathfrak{S}$-convergence. In particular $\mathcal{C}_s(X; Y)$, $\mathcal{C}_c(X; Y)$ and $\mathcal{C}_u(X; Y)$ denote the set $\mathcal{C}(X; Y)$ endowed respectively with the uniformity of pointwise convergence, compact convergence and uniform convergence.

PROPOSITION 6. *Let* X *be a topological space,* Y *a uniform space and* $\mathfrak{S}$ *a set of subsets of* X. *For each* $A \in \mathfrak{S}$ *and each closed entourage* V *of* Y, *the traces on* $\mathcal{C}(X; Y) \times \mathcal{C}(X; Y)$ *of* $W(A, V)$ *and* $W(\overline{A}, V)$ *are the same.*

For if u, v are continuous mappings of X into Y, the mapping $x \to (u(x), v(x))$ of X into $Y \times Y$ is continuous, and the hypothesis that $(u(x), v(x)) \in V$ for all $x \in A$ therefore implies that $(u(x), v(x)) \in \overline{V} = V$ for all $x \in \overline{A}$ (Chapter I, § 2, no. 1, Theorem 1).

If $\overline{\mathfrak{S}}$ denotes the set of closures in X of the sets of $\mathfrak{S}$, Proposition 6 shows that, *on* $\mathcal{C}(X; Y)$, the structures of $\mathfrak{S}$-convergence and $\overline{\mathfrak{S}}$-convergence are identical.

COROLLARY. *Let* B *be a dense subset of* X. *On* $\mathcal{C}$ (X; Y), *the structure of uniform convergence is identical with the structure of uniform convergence in* B.

PROPOSITION 7. *Let* X *be a topological space, let* $\mathfrak{S}$ *be a set of subsets of* X *and let* Y *be a uniform space. If* Y *is Hausdorff and if the union* B *of the sets of* $\mathfrak{S}$ *is dense in* X, *then* $\mathcal{C}_{\mathfrak{S}}$ (X; Y) *is Hausdorff.*

For if (u, v) belongs to all the sets $W(A, V)$, where $A \in \mathfrak{S}$ and V is an entourage of Y, the hypothesis that Y is Hausdorff tells us that $u(x) = v(x)$ for all $x \in B$; if u and v are continuous, then $u = v$ by the principle of extension of identities (Chapter I, § 8, no. 1, Proposition 2, Corollary 1).

In particular, on $\mathcal{C}$ (X; Y), the topology of pointwise convergence in a *dense* subset of X is Hausdorff.

PROPOSITION 8. *Let* X *be a set,* $\mathfrak{F}$ *a filter on* X, *and let* Y *be a uniform space. Then the set* H *of mappings* $u : X \to Y$ *such that* $u(\mathfrak{F})$ *is a Cauchy filter base on* Y *is closed in* $\mathcal{F}_u$ (X; Y).

Let $u_0 : X \to Y$ lie in the closure of H in $\mathcal{F}_u$ (X; Y). For each symmetric entourage V of Y, there is a mapping $u \in H$ such that $(u_0(x), u(x)) \in V$ for all $x \in X$; on the other hand, by hypothesis there is a set $M \in \mathfrak{F}$ such that $(u(x), u(x')) \in V$ whenever x and x' are in M. Since $(u_0(x), u(x)) \in V$ and $(u_0(x'), u(x')) \in V$, it follows that $(u_0(x), u_0(x')) \in \overset{3}{V}$ whenever x and x' are in M, and therefore $u_0(\mathfrak{F})$ is a Cauchy filter base on Y.

COROLLARY 1. *Let* X *be a topological space and* Y *a uniform space. The set of mappings of* X *into* Y *which are continuous at a point* $x_0 \in X$ *is closed in* $\mathcal{F}_u$ (X; U).

If V is the neighbourhood filter of x_0 in X, $u(x_0)$ is a cluster point of $u(V)$; hence u is continuous at x_0 if and only if $u(V)$ is a Cauchy filter base on Y (Chapter II, § 3, no. 2, Proposition 5, Corollary 2).

COROLLARY 2. *Let* X, L *be two sets filtered by filters* $\mathfrak{F}$, $\mathfrak{G}$ *respectively, and let* Y *be a complete uniform space. For each* $\lambda \in L$, *let* u_λ *be a mapping of* X *into* Y. *Suppose that* (i) *the family* $(u_\lambda)_{\lambda \in L}$ *converges uniformly in* X *(with respect to the filter* $\mathfrak{G}$*) to a mapping* $v : X \to Y$; (ii) *for each* $\lambda \in L$, u_λ *has a limit* y_λ *with respect to the filter* $\mathfrak{F}$. *Under these conditions,* v *has a limit with respect to* $\mathfrak{F}$, *and every limit of* v *with respect to* $\mathfrak{F}$ *is a limit of the family* $(y_\lambda)_{\lambda \in L}$ *with respect to* $\mathfrak{G}$.

For v lies in the closure of the set of the u_λ in $\mathcal{F}_u$ (X; Y), and therefore $v(\mathfrak{F})$ is a Cauchy filter base on Y by virtue of Proposition 8; this shows that v has a limit y with respect to $\mathfrak{F}$ because Y is complete. Let

$X' = X \cup \{\omega\}$ be the topological space associated with the filter $\mathfrak{F}$ (Chapter I, § 6, no. 5), and extend u_λ (resp. v) to a mapping $\bar{u}_\lambda$ (resp. $\bar{v}$) of X' into Y by putting $\bar{u}_\lambda(\omega) = y_\lambda$ [resp. $\bar{v}(\omega) = y$]. Then the mappings $\bar{u}_\lambda$, $\bar{v}$ are continuous on X', and $\bar{u}_\lambda$ converges uniformly *in* X to $\bar{v}$ with respect to $\mathfrak{G}$; since X is dense in X', the Corollary to Proposition 6 shows that $\bar{u}_\lambda$ converges uniformly *in* X' to $\bar{v}$, and in particular that $y = \lim_{\mathfrak{G}} y_\lambda$.

THEOREM 2. *Let* X *be a topological space,* Y *a uniform space. Then the set* $\mathcal{C}(X; Y)$ *of continuous mappings of* X *into* Y *is a closed subset of the space* $\mathcal{F}(X; Y)$ *endowed with the topology of uniform convergence.*

For each $x \in X$, the set of mappings of X into Y which are continuous at x is closed in $\mathcal{F}_u(X; Y)$ (Proposition 8, Corollary 1); hence the intersection $\mathcal{C}(X; Y)$ of these sets is also closed.

This result may be expressed in the form that the *uniform limit of continuous functions is continuous.*

COROLLARY 1. *If* Y *is a complete uniform space, then* $\mathcal{C}_u(X; Y)$ *is complete.*

For, by Theorem 2, $\mathcal{C}_u(X; Y)$ is a *closed* uniform subspace of the uniform space $\mathcal{F}_u(X; Y)$, which is complete by Theorem 1 of no. 5.

COROLLARY 2. *Let* X *be a topological space,* $\mathfrak{S}$ *a set of subsets of* X, *and* Y *a uniform space. Let* $\tilde{\mathcal{C}}_{\mathfrak{S}}(X; Y)$ *denote the set of all mappings of* X *into* Y *whose restriction to each set of* $\mathfrak{S}$ *is continuous. Then* $\tilde{\mathcal{C}}_{\mathfrak{S}}(X; Y)$ *is a closed subspace of the uniform space* $\mathcal{F}_{\mathfrak{S}}(X; Y)$ *and is complete if* Y *is complete.*

Suppose that u lies in the closure of $\tilde{\mathcal{C}}_{\mathfrak{S}}(X; Y)$ in $\mathcal{F}_{\mathfrak{S}}(X; Y)$; then (no. 2), for each $A \in \mathfrak{S}$, $u|A$ lies in the closure of $\mathcal{C}(A; Y)$ in $\mathcal{F}_u(A; Y)$, and is therefore continuous by Theorem 2.

COROLLARY 3. *Let* X *be a topological space which is either metrizable or locally compact, and let* Y *be a uniform space. Then* $\mathcal{C}(X; Y)$ *is closed in the uniform space* $\mathcal{F}_c(X; Y)$; *if in addition* Y *is complete, the uniform space* $\mathcal{C}_c(X; Y)$ *is complete.*

By virtue of Corollary 2 it is enough to show that, if we take $\mathfrak{S}$ to be the set of compact subsets of X, we have $\tilde{\mathcal{C}}_{\mathfrak{S}}(X; Y) = \mathcal{C}(X; Y)$ in both cases under consideration. This is clear if X is locally compact. If X is metrizable, and $u : X \to Y$ is a mapping whose restriction to every compact subset of X is continuous, then for each $x \in X$ and each sequence (z_n) of points of X which converges to x, we have $u(x) = \lim_{n \to \infty} u(z_n)$, and therefore u is continuous at x (Chapter IX, § 2, no. 6, Proposition 10).

Note that the argument above applies whenever every point of x has a *countable* fundamental system of neighbourhoods.

Remarks. 1) In general, the set $\mathcal{C}(\mathrm{X}; \mathrm{Y})$ is not closed in $\mathcal{F}(\mathrm{X}; \mathrm{Y})$ with respect to the topology of *pointwise* convergence: in other words, a pointwise limit of continuous functions is not necessarily continuous [Exercise 5 *a*)].

2) A filter on $\mathcal{C}(\mathrm{X}; \mathrm{Y})$ can converge *pointwise* to a *continuous* function without converging uniformly to this function.

> For example, on the interval $\mathrm{I} = [0, 1]$, let u_n be the real-valued function which is equal to 0 for $x = 0$ and $2/n \leqslant x \leqslant 1$, equal to 1 for $x = 1/n$, and linear in each of the intervals $[0, 1/n]$ and $[1/n, 2/n]$. The sequence (u_n) converges pointwise to 0, but does not converge uniformly to 0 in I (cf. Exercise 6).

3) If X is a uniform space, a proof analogous to that of Proposition 8 shows that the set of *uniformly continuous* mappings of X into Y is *closed* in $\mathcal{F}_u(\mathrm{X}; \mathrm{Y})$.

4) Suppose that the uniform space Y carries a commutative and associative law of composition, written additively, such that the mapping $(y, y') \rightarrow y + y'$ is continuous on $\mathrm{Y} \times \mathrm{Y}$. Then, if (u_n) is a sequence of continuous mappings of X into Y such that the series whose general term is u_n is *uniformly convergent* in X, the sum of the series is continuous on X.

We leave it to the reader to state the corresponding result for *uniformly summable* families (no. 1, Remark 5) of continuous mappings.

PROPOSITION 9. *Let* X *be a topological space,* Y *a uniform space. Then the mapping* $(f, x) \rightarrow f(x)$ *of* $\mathcal{C}_u(\mathrm{X}; \mathrm{Y}) \times \mathrm{X}$ *into* Y *is continuous.*

Let $f_0 : \mathrm{X} \rightarrow \mathrm{Y}$ be a continuous mapping, let x_0 be a point of X and let V be an entourage of Y. The set T of continuous mappings $f : \mathrm{X} \rightarrow \mathrm{Y}$ such that $(f(x), f(x_0)) \in \mathrm{V}$ for all $x \in \mathrm{X}$ is a neighbourhood of f_0 in $\mathcal{C}_u(\mathrm{X}; \mathrm{Y})$. On the other hand, since f_0 is continuous, there is a neighbourhood U of x_0 in X such that $(f_0(x), f_0(x_0)) \in \mathrm{V}$ for all $x \in \mathrm{U}$. Consequently we have $(f(x), f(x_0)) \in \overset{2}{\mathrm{V}}$ whenever $(f, x) \in \mathrm{T} \times \mathrm{U}$, and the result is proved.

2. EQUICONTINUOUS SETS

1. DEFINITION AND GENERAL CRITERIA

DEFINITION 1. *Let* X *be a topological space and* Y *a uniform space. A subset* H *of* $\mathcal{F}(\mathrm{X}; \mathrm{Y})$ *is said to be equicontinuous at a point* $x_0 \in \mathrm{X}$ *if, for each entourage* V *of* Y, *there is a neighbourhood* U *of* x_0 *in* X *such that*

$(f(x_0), f(x)) \in V$ for all $x \in U$ and all $f \in H$. H is said to be equicontinuous if it is equicontinuous at every point of X.

Definition 2. *Let X and Y be two uniform spaces. A subset H of $\mathcal{F}(X; Y)$ is said to be uniformly equicontinuous if, for each entourage V of Y, there exists an entourage U of X such that we have $(f(x), f(x')) \in V$ whenever $(x, x') \in U$ and $f \in H$.*

A family $(f_\iota)_{\iota \in I}$ of mappings of X into Y is said to be equicontinuous at a point x_0 (resp. equicontinuous, uniformly equicontinuous) if the set of the f_ι is equicontinuous at x_0 (resp. equicontinuous, uniformly equicontinuous).

It is clear that if $H \subset \mathcal{F}(X; Y)$ is equicontinuous at x_0, then each $f \in H$ is continuous at x_0; if H is equicontinuous, then each $f \in H$ is continuous on X, i.e. $H \subset \mathcal{C}(X; Y)$. Likewise, if H is uniformly equicontinuous (X being a uniform space), every $f \in H$ is uniformly continuous on X. It is also clear that a uniformly equicontinuous set of mappings is equicontinuous; but a set of uniformly continuous mappings can be equicontinuous without being uniformly equicontinuous (see Exercise 1; Corollary 2 to Proposition 1; and no. 2, Proposition 4).

Examples. 1) Let X be a topological space (resp. a uniform space) and Y a uniform space. Every *finite* set of continuous (resp. uniformly continuous) mappings of X into Y is equicontinuous (resp. uniformly equicontinuous).

2) Let X, Y be two metric spaces, d (resp. d') the metric on X (resp. Y), and let k, α be two real numbers > 0. Then the set of all mappings f: $X \to Y$ such that

$$d'(f(x), f(x')) \leqslant k(d(x, x'))^\alpha$$

for each pair of points x, x' of X, is uniformly equicontinuous. For example, the set of all *isometries* (Chapter IX, § 2, no. 2) of X onto a subset of Y is uniformly equicontinuous.

* Let H be a set of real-valued functions defined on an interval $I \subset \mathbf{R}$, which are *differentiable* on I and are such that $|f'(x)| \leqslant k$ for all $x \in I$ and all $f \in H$. Then H is *uniformly* equicontinuous, because if x_1, x_2 are any two points of I we have $|f(x_1) - f(x_2)| \leqslant k|x_1 - x_2|$ for each $f \in H$ by the mean value theorem. *

3) Let G be a topological group, let Y be a uniform space and let $f: G \to Y$ be a uniformly continuous mapping [G being endowed with its left uniformity (Chapter III, § 3, no. 1)]. For each $s \in G$, let f_s be the mapping $x \to f(sx)$ of G into Y. Then the set of mappings $f_s (s \in G)$ is uniformly equicontinuous, since the relation $x^{-1}x' \in V$ is equivalent to $(sx)^{-1}(sx') \in V$.

Proposition 1. *Let T be a set, let $\mathfrak{S}$ be a set of subsets of T, let Y be a uniform space,* X *a topological* (resp. *uniform*) *space, and let f be a mapping* *of* $T \times X$ *into* Y. *For each* $A \in \mathfrak{S}$, *let* $H_A \subset \mathcal{F}(X; Y)$ *be the set of all mappings* $x \to f(t, x)$ *as* t *runs through* A. *Then the mapping* $x \to f(., x)$ *of* X *into* $\mathcal{F}_{\mathfrak{S}}(T; Y)$ *is continuous at a point* $x_0 \in X$ (resp. *uniformly continuous*) *if and only if the set* H_A *is equicontinuous at* x_0 (resp. *uniformly equicontinuous*) *for all* $A \in \mathfrak{S}$.

Consider first the particular case where $\mathfrak{S} = \{T\}$, i.e. $\mathcal{F}_{\mathfrak{S}}(T; Y) = \mathcal{F}_u(T; Y)$. For each entourage V of Y, the condition $(f(., x), f(., x')) \in \mathcal{W}(V)$ signifies that $(f(t, x), f(t, x')) \in V$ for all $t \in T$. To say that $x \to f(., x)$ is continuous at x_0 (resp. is uniformly continuous) is therefore equivalent to saying that, for each entourage V of Y, there is a neighbourhood U of x_0 in X (resp. an entourage M of X) such that the relation $x \in U$ [resp. $(x, x') \in M$] implies $(f(t, x), f(t, x_0)) \in V$ [resp. $(f(t, x), f(t, x')) \in V$] for all $t \in T$, and the proposition follows from Definitions 1 and 2. In the general case, we have to express that, for each $A \in \mathfrak{S}$, the mapping $x \to f(., x)|A$ of X into $\mathcal{F}_u(A; Y)$ is continuous at x_0 (resp. uniformly continuous), by virtue of § 1, no. 2; from what has been said, this is equivalent to saying that, for each $A \in \mathfrak{S}$, H_A is equicontinuous at x_0 (resp. uniformly equicontinuous).

Proposition 1 allows us to translate Definitions 1 and 2 into forms which are sometimes useful, by applying it to the case where $T = H$ and f is the mapping $(h, x) \to h(x)$ of $H \times X$ into Y; since $f(., x)$ is the mapping $h \to h(x)$ of H into Y, we see that:

Corollary 1. *Let* X *be a topological* (resp. *uniform*) *space,* Y *a uniform space and* H *a subset of* $\mathcal{F}(X; Y)$. *For each* $x \in X$, *let* $\tilde{x}$ *denote the mapping* $h \to h(x)$ *of* H *into* Y. *Then* H *is equicontinuous at* x_0 (resp. *uniformly equicontinuous*) *if and only if the mapping* $x \to \tilde{x}$ *of* X *into the uniform space* $\mathcal{F}_u(H; Y)$ *is continuous at* x_0 (resp. *uniformly continuous*).

In particular, if X is compact, every continuous mapping of X into $\mathcal{F}_u(H; Y)$ is uniformly continuous (Chapter II, § 4, no. 1, Theorem 2). Therefore:

Corollary 2. *Let* X *be a compact space,* Y *a uniform space. Then every equicontinuous subset of* $\mathcal{F}(X; Y)$ *is uniformly equicontinuous.*

Now suppose we have a set T, a topological space X, a uniform space Y and a mapping $f: T \times X \to Y$. Let $\tilde{f}$ denote the mapping $x \to f(., x)$ of X into $\mathcal{F}_u(T; Y)$, and let us consider the canonical mapping $\theta: (t, g) \to g(t)$ of $T \times \mathcal{F}_u(T; Y)$ into Y. It is clear that the diagram

$$\begin{array}{ccc} T \times X & \xrightarrow{f} & Y \\ & {\scriptstyle \iota_T \times \tilde{f}} \searrow \quad \nearrow {\scriptstyle \theta} & \\ & T \times \mathcal{F}_u(T; Y) & \end{array}$$

(where ι_T is the identity mapping of T) is commutative. Suppose now that T is endowed with a topology and that, for each $x \in X$, the mapping $f(., x) : t \to f(t., x)$ is continuous; we can then replace $\mathcal{F}_u(T; Y)$ by $\mathcal{C}_u(T; Y)$ in the above diagram. But we know that θ is continuous from § 1, no. 6, Proposition 9; hence if $\tilde{f}$ is continuous it follows that f is continuous. Since the continuity of $\tilde{f}$ can be expressed with the help of Proposition 1, we obtain the following result:

COROLLARY 3. *Let* T, X *be topological spaces, let* Y *be a uniform space and let* f *be a mapping of* $T \times X$ *into* Y. *Then* f *is continuous, provided that the following conditions are satisfied:*

1) *For each* $x \in X$, *the partial mapping* $t \to f(t, x)$ *is continuous.*

2) *As* t *runs through* T, *the partial mappings* $x \to f(t, x)$ *form an equicontinuous subset of* $\mathcal{F}(X; Y)$.

In particular, take T to be a subset H of $\mathcal{F}(X; Y)$ and f to be the canonical mapping $(h, x) \to h(x)$ of $H \times X$ into Y; condition 1) of Corollary 3 means that H is endowed with a topology finer than that of pointwise convergence, and condition 2) means that H is equicontinuous. Hence:

COROLLARY 4. *Let* X *be a topological space,* Y *a uniform space,* H *an equicontinuous set of mappings of* X *into* Y. *If* H *is endowed with the topology of pointwise convergence, then the mapping* $(h, x) \to h(x)$ *of* $H \times X$ *into* Y *is continuous.*

More intuitively, this expresses the fact that if $h \in H$ converges *pointwise* to $h_0 \in H$ and if $x \in X$ converges to x_0, then $h(x)$ converges to $h_0(x_0)$.

COROLLARY 5. *Let* X *be a topological space, let* Y, Z *be two uniform spaces and let* H *be an equicontinuous set of mappings of* Y *into* Z. *If* H, $\mathcal{C}(X; Y)$ *and* $\mathcal{C}(X; Z)$ *are endowed with the topology of pointwise convergence, then the mapping* $(u, v) \to u \circ v$ *of* $H \times \mathcal{C}(X; Y)$ *into* $\mathcal{C}(X; Z)$ *is continuous.*

We have to show that, for each $x \in X$, the mapping $(u, v) \to u(v(x))$ of $H \times \mathcal{C}(X; Y)$ into Z is continuous. Now $v \to v(x)$ is continuous on H (§ 1, no. 2, Remark 6), and it follows from Corollary 4 that $(u, y) \to u(y)$ is a continuous mapping of $H \times Y$ into Z; since $(u, v) \to u(v(x))$ is the composition of $(u, y) \to u(y)$ and $(u, v) \to (u, v(x))$, the result is proved.

The following proposition and its corollary are the analogues of Corollaries 3 and 4 of Proposition 1 for uniformly equicontinuous sets of mappings:

PROPOSITION 2. *Let* T, X, Y *be uniform spaces and let* f *be a mapping of* $T \times X$ *into* Y. *Then* f *is uniformly continuous if and only if the following two conditions are satisfied:*

1) *The mappings* $x \to f(t, x)$ $(t \in T)$ *form a uniformly equicontinuous subset of* $\mathcal{F}(X; Y)$.

2) *The mappings* $t \to f(t, x)$ $(x \in X)$ *form a uniformly equicontinuous subset of* $\mathcal{F}(T; Y)$.

It is easily seen that the conditions are necessary. Conversely, suppose that they are satisfied. Let W be an entourage of Y; then there exists an entourage U of T and an entourage V of X such that:

1) $(t', t'') \in U$ implies that, for each $x \in X$,

$$(f(t', x), f(t'', x)) \in W.$$

2) $(x', x'') \in V$ implies that, for each $t \in T$,

$$(f(t, x'), f(t, x'')) \in W.$$

It is now clear that the relation "$(t', t'') \in U$ and $(x', x'') \in V$" implies that $(f(t', x'), f(t'', x'')) \in \overset{2}{W}$, whence the result.

In particular, take T to be a subset H of $\mathcal{F}(X; Y)$, endowed with the uniformity of uniform convergence, and take f to be the canonical mapping $(h, x) \to h(x)$; then condition 2) of Proposition 2 is automatically satisfied because, for each entourage W of Y, the set of pairs (h', h'') such that $(h'(x), h''(x)) \in W$ for all $x \in X$ is by definition an entourage of the uniform structure of H. Hence only condition 1) has to be expressed; in other words:

COROLLARY. *Let* X, Y *be two uniform spaces and let* H *be a subset of* $\mathcal{F}(X; Y)$. *Then* H *is uniformly equicontinuous if and only if the mapping* $(h, x) \to h(x)$ *of* $H \times X$ *into* Y *is uniformly continuous,* H *being endowed with the uniformity of uniform convergence.*

2. SPECIAL CRITERIA FOR EQUICONTINUITY

It is clear that every subset of an equicontinuous (resp. uniformly equicontinuous) set is equicontinuous (resp. uniformly equicontinuous). Again, if X is a topological (resp. uniform) space and Y is a uniform space, every *finite* union of equicontinuous (resp. uniformly equicontinuous) subsets of $\mathcal{F}(X; Y)$ is equicontinuous (resp. uniformly equicontinuous).

Let X, X' be two topological (resp. uniform) spaces, let Y, Y' be two uniform spaces, let $f: X \to X'$ be a continuous (resp. uniformly continuous) mapping and let $g: Y \to Y'$ be a uniformly continuous mapping.

It follows immediately from the definitions that the mapping $u \to g \circ u \circ f$ of $\mathscr{F}(X; Y)$ into $\mathscr{F}(X'; Y')$ transforms equicontinuous (resp. uniformly equicontinuous) sets into equicontinuous (resp. uniformly equicontinuous) sets.

Proposition 3. *Let* X *be a topological* (resp. *uniform*) *space, let* $(Y_\iota)_{\iota \in I}$ *be a family of uniform spaces, let* Y *be a set, and for each* $\iota \in I$, *let* f_ι *be a mapping of* Y *into* Y_ι. *Let* Y *be endowed with the coarsest uniformity for which all the* f_ι *are uniformly continuous. For a subset* H *of* $\mathscr{F}(X; Y)$ *to be equicontinuous* (resp. *uniformly equicontinuous*) *it is necessary and sufficient that, for each* $\iota \in I$, *the image of* H *under the mapping* $u \to f_\iota \circ u$ *be an equicontinuous* (resp. *uniformly equicontinuous*) *subset of* $\mathscr{F}(X; Y_\iota)$.

This is an immediate consequence of Definitions 1 and 2 and the definition of the entourages of Y.

Proposition 4. *Let* X, Y *be two uniform spaces and let* H *be a set of uniformly continuous mappings of* X *into* Y. *Let* $\hat{X}, \hat{Y}$ *be the Hausdorff completions of* X, Y *respectively, and let* $\tilde{H}$ *denote the set of mappings* $\hat{u} : \hat{X} \to \hat{Y}$ *as* u *runs through* H (Chapter II, § 3, no. 7, Proposition 15). *Then* H *is uniformly equicontinuous if and only if* $\tilde{H}$ *is uniformly equicontinuous.*

We recall that the diagram

$$\begin{array}{ccc} X & \xrightarrow{u} & Y \\ {\scriptstyle i}\downarrow & & \downarrow{\scriptstyle j} \\ \hat{X} & \xrightarrow{\hat{u}} & \hat{Y} \end{array} \tag{1}$$

is commutative, where i and j are the canonical mappings; moreover, the uniformity of X (resp. Y) is the inverse image under i (resp. j) of that of $\hat{X}$ (resp. $\hat{Y}$). Hence H is uniformly equicontinuous if and only if its image under the mapping $u \to j \circ u$ is uniformly equicontinuous (Proposition 3), and we may already restrict ourselves to the case where Y is Hausdorff and complete; moreover, if $\tilde{H}$ is uniformly equicontinuous then so is H, because it is the image of $\tilde{H}$ under the mapping $\hat{u} \to \hat{u} \circ i$; thus it remains to prove the converse when $Y = \hat{Y}$. Let V be a closed entourage of Y; by hypothesis there is an entourage U of X such that the relations $(x, x') \in U$ and $u \in H$ imply that $(u(x), u(x')) \in V$. Now, if U' is the image of U under $i \times i$, the closure $\overline{U'}$ of U' in $\hat{X} \times \hat{X}$ is an entourage of $\hat{X}$ (Chapter II, § 3, no. 7, Proposition 12); the hypothesis implies that, whenever $(z, z') \in U'$ and $u \in H$, we have $(\hat{u}(z), \hat{u}(z')) \in V$. Since V is closed and $\hat{u}$ is continuous, we have also $(\hat{u}(t), \hat{u}(t')) \in V$ for all $(t, t') \in \overline{U'}$ and all $u \in H$; this completes the proof.

PROPOSITION 5. *Let* G, G′ *be two topological groups endowed with their left uniformities, and let* H *be a set of homomorphisms of* G *into* G′. *Then the following conditions are equivalent:*

a) H *is equicontinuous at the identity element* e *of* G,

b) H *is equicontinuous,*

c) H *is uniformly equicontinuous.*

It is enough to show that a) implies c). Let V′ be a neighbourhood of the identity element e' of G′; then, by hypothesis, there is a neighbourhood V of e in G such that $u(\mathrm{V}) \subset \mathrm{V}'$ for all $u \in \mathrm{H}$; since the elements of H are homomorphisms, the relation $x^{-1}y \in \mathrm{V}$ implies that we have

$$(u(x))^{-1}\, u(y) = u(x^{-1}y) \in \mathrm{V}'.$$

In view of the definition of the entourages of the left uniformities of G and G′ (Chapter III, § 3, no. 1), the result follows.

3. CLOSURE OF AN EQUICONTINUOUS SET

PROPOSITION 6. *Let* X *be a topological* (resp. *uniform*) *space, let* Y *be a uniform space and let* H *be a subset of* $\mathcal{F}$ (X; Y). *Then* H *is equicontinuous at a point* $x_0 \in \mathrm{X}$ (resp. *uniformly equicontinuous*) *if and only if the closure* $\overline{\mathrm{H}}$ *of* H *in* $\mathcal{F}_s$ (X; Y) *is equicontinuous at* x_0 (resp. *uniformly equicontinuous*).

The condition is sufficient, trivially. To show that it is necessary, consider an entourage V of Y which is *closed* in Y × Y; by hypothesis, there is a neighbourhood U of x_0 in X (resp. an entourage M of X) such that the relation $x \in \mathrm{U}$ (resp. $(x', x'') \in \mathrm{M}$) implies $(h(x_0), h(x)) \in \mathrm{V}$ [resp. $(h(x'), h(x'')) \in \mathrm{V}$] for all $h \in \mathrm{H}$. Since V is closed, the mappings $h \in \mathcal{F}$ (X; Y) which satisfy the relation $(h(x_0), h(x)) \in \mathrm{V}$ for all $x \in \mathrm{U}$ [resp. the relation $(h(x'), h(x'')) \in \mathrm{V}$ for all $(x', x'') \in \mathrm{M}$] form a closed subset of $\mathcal{F}_s$ (X; Y) (§ 1, no. 2, Remark 6); since this closed subset contains H, it contains $\overline{\mathrm{H}}$. Hence the result, since the closed entourages of Y form a fundamental system of entourages (Chapter II, § 1, no. 2, Proposition 2, Corollary 2).

4. POINTWISE CONVERGENCE AND COMPACT CONVERGENCE ON EQUICONTINUOUS SETS

THEOREM 1. *Let* X *be a topological* (resp. *uniform*) *space, let* Y *be a uniform space and let* H *be an equicontinuous* (resp. *uniformly equicontinuous*) *subset of* $\mathcal{C}$ (X; Y). *Then the following uniformities on* H *are identical: the uniformity of compact* (resp. *precompact*) *convergence, the uniformity of pointwise convergence and the uniformity of pointwise convergence in a dense subset* D *of* X.

It is enough to show that the last uniformity on H is finer than the first; in other words that, given an entourage V of Y and a compact (resp. precompact) subset A of X, there exists an entourage W of Y and a finite subset F of D such that the relation

$$(2)\qquad u \in \mathrm{H},\quad v \in \mathrm{H}\quad \text{and}\quad (u(x), v(x)) \in \mathrm{W}\quad \text{for all}\quad x \in \mathrm{F}$$

implies

$$(3)\qquad (u(x),\ v(x)) \in \mathrm{V}\quad \text{for all}\quad x \in \mathrm{A}.$$

Suppose first that A is compact and H equicontinuous. Given a symmetric entourage W of Y, every point $x \in \mathrm{X}$ has a neighbourhood $\mathrm{U}(x)$ such that the relation $x' \in \mathrm{U}(x)$ implies $(u(x), u(x')) \in \mathrm{W}$ for all $u \in \mathrm{H}$. We can therefore cover the compact set A by a finite number of open sets U_i such that, for each pair of points x', x'' of the same set U_i, we have $(u(x'), u(x'')) \in \overset{2}{\mathrm{W}}$ for all $u \in \mathrm{H}$. Let a_i be a point of $\mathrm{D} \cap \mathrm{U}_i$, let F be the set of the a_i, and suppose that (2) is true; then for each $x \in \mathrm{A}$ there exists an index i such that a_i and x belong to the same set U_i, so that we have $(u(x), u(a_i)) \in \overset{2}{\mathrm{W}}$ and $(v(a'), v(x)) \in \overset{2}{\mathrm{W}}$; thus (2) implies (3) provided that W is chosen so that $\overset{5}{\mathrm{W}} \subset \mathrm{V}$.

If A is precompact and H uniformly equicontinuous, we use Proposition 4 of no. 2; it is enough to note that $\overline{i(\mathrm{A})}$ is compact in $\hat{\mathrm{X}}$, $i(\mathrm{D})$ dense in $\hat{\mathrm{X}}$ and that the entourages of Y are the inverse images of those of $\hat{\mathrm{Y}}$ under the mapping $j \times j$.

Corollary. *Under the hypotheses of Theorem* 1, *the closure* $\overline{\mathrm{H}}$ *of* H *in* $\mathcal{F}(\mathrm{X}; \mathrm{Y})$ *with respect to the topology of pointwise convergence is the same as the closure of* H *in* $\mathcal{C}(\mathrm{X}; \mathrm{Y})$ *with respect to the topology of compact* (resp. *precompact*) *convergence.*

For the set $\overline{\mathrm{H}}$ is equicontinuous (resp. uniformly equicontinuous) by Proposition 6 of no. 3, and hence is contained in $\mathcal{C}(\mathrm{X}; \mathrm{Y})$; the result follows immediately from the fact that, on $\overline{\mathrm{H}}$, the two topologies under consideration are the same, by virtue of Theorem 1.

5. COMPACT SETS OF CONTINUOUS MAPPINGS

Theorem 2 (Ascoli). *Let* X *be a topological* (resp. *uniform*) *space, let* $\mathfrak{S}$ *be a covering of* X, *let* Y *be a uniform space and* H *a set of mappings of* X *into* Y *such that, for each* $\mathrm{A} \in \mathfrak{S}$ *and each* $u \in \mathrm{H}$, *the restriction of* u *to* A *is continuous* (resp. *uniformly continuous*). *Then, for* H *to be precompact with respect to the uniformity of* $\mathfrak{S}$*-convergence, it is necessary in all cases and also sufficient* *if the sets* $A \in \mathfrak{S}$ *are compact* (resp. *precompact*) *that the following conditions should be satisfied*:

a) *For each* $A \in \mathfrak{S}$, *the set* $H|A \subset \mathcal{F}(A; Y)$ *of restrictions to* A *of functions of* H *is equicontinuous* (resp. *uniformly equicontinuous*).

b) *For each* $x \in X$, *the set* $H(x) \subset Y$ *of points* $u(x)$ $(u \in H)$ *is precompact.*

1) Let us show first that conditions a) and b) are *necessary*. We know (§ 1, no. 2, Remark 6) that the mapping $u \to u(x)$ of $\mathcal{F}_{\mathfrak{S}}(X; Y)$ into Y is uniformly continuous; hence, if H is precompact, so is $H(x)$ (Chapter II, § 4, no. 2, Proposition 2), which proves b). To prove a), consider a set $A \in \mathfrak{S}$, a point $x_0 \in A$ and an entourage V of Y; since H is precompact it can be covered by a finite number of $W(A, V)$-small sets; in other words there is a finite sequence (u_i) of elements of H such that, for each $u \in H$, we have

$$(4) \qquad (u(x),\ u_i(x)) \in V \quad \text{for all} \quad x \in A$$

for at least one index i.

Since each of the $u_i|A$ is continuous at x_0 (resp. uniformly continuous) there is a neighbourhood U_i of x_0 in A (resp. an entourage M_i of A) such that

$$(5) \qquad x \in U_i \quad \text{implies} \quad (u_i(x), u_i(x_0)) \in V,$$

(resp. such that

$$(6) \qquad (x', x'') \in M_i \quad \text{implies} \quad (u_i(x'), u_i(x'')) \in V.)$$

Let U (resp. M) be the intersection of the U_i (resp. M_i); it is a neighbourhood of x_0 in A (resp. an entourage of A). For each $u \in H$ there is an index i for which (4) holds; writing condition (4) for x_0 and for x (resp. for x' and x'') and taking account of (5) [resp. (6)], we see immediately that the relation $x \in U$ [resp. $(x', x'') \in M$] implies $(u(x),\ u(x_0)) \in \overset{3}{V}$ [resp. $(u(x'),\ u(x'')) \in \overset{3}{V}$], for each $u \in H$; and this establishes a).

2) Now let us show that the conditions a) and b) are *sufficient* if the sets $A \in \mathfrak{S}$ are compact (resp. precompact). Condition b) implies that H is precompact with respect to the uniformity of pointwise convergence (Chapter II, § 4, no. 2, Proposition 3). But it follows from condition a) and Theorem 1 of no. 4 that on $H|A$ the uniformity of pointwise convergence in A coincides with the uniformity of uniform convergence in A; hence $H|A$ is precompact in $\mathcal{F}_u(A; Y)$, which implies that H is precompact with respect to the uniformity of $\mathfrak{S}$-convergence (§ 1, no. 2).

Note that condition b) of Theorem 2 is automatically satisfied if Y is a *precompact* space.

COROLLARY 1. *Let* X *be a topological* (resp. *uniform*) *space, let* Y *be a Hausdorff uniform space and let* H *be an equicontinuous* (resp. *uniformly equicontinuous*) *subset of* $\mathcal{C}(X; Y)$. *Suppose that* $H(x)$ *is relatively compact in* Y *for each* $x \in X$. *Then* H *is relatively compact in* $\mathcal{C}(X; Y)$ *with respect to the topology of compact* (resp. *precompact*) *convergence.*

Let $\overline{H}$ be the closure of H in $\mathcal{F}_s(X; Y)$. $\overline{H}$ is equicontinuous (resp. uniformly equicontinuous) (no. 3, Proposition 6). Moreover, we have $\overline{H}(x) \subset \overline{H(x)}$ (§ 1, no. 2, Remark 6) and therefore $\overline{H}(x)$ is also relatively compact; hence Theorem 2 shows that $\overline{H}$ is precompact with respect to $\mathfrak{S}$-convergence, where $\mathfrak{S}$ denotes the set of all compact (resp. precompact) subsets of X. Moreover, since $\overline{H(x)}$ is compact, and therefore complete, $\overline{H}$ is complete with respect to the uniformity of pointwise convergence (Chapter II, § 3, no. 5, Proposition 10 and no. 4, Proposition 8) and therefore also with respect to the uniformity of $\mathfrak{S}$-convergence (§ 1, no. 5, Proposition 5, Corollary 2); $\overline{H}$ is therefore compact, since it is precompact, complete and Hausdorff (§ 1, no. 2, Proposition 1).

COROLLARY 2. *Let* X *be a topological* (resp. *uniform*) *space, let* Y *be a* complete *Hausdorff uniform space and let* H *be an equicontinuous* (resp. *uniformly equicontinuous*) *subset of* $\mathcal{C}(X; Y)$. *Suppose that* $H(x)$ *is relatively compact in* Y *for all* $x \in D$, *where* D *is a dense subset of* X. *Then* H *is relatively compact in* $\mathcal{C}(X; Y)$ *with respect to the topology of compact* (resp. *precompact*) *convergence.*

It is enough to show that $H(x)$ is relatively compact for *all* $x \in X$, for we can then apply Corollary 1. Since Y is complete it is enough to show that $H(x)$ is precompact for all $x \in X$. Now if V is any symmetric entourage of Y, there is a neighbourhood U of x such that $(u(x), u(x')) \in V$ for all $x' \in U$ and all $u \in H$. By hypothesis there exists $x' \in U \cap D$, and since $H(x')$ is relatively compact in Y, there exists a finite number of points $y_k \in Y$ such that $H(x')$ is contained in the union of the sets $V(y_k)$; hence $H(x)$ is contained in the union of the sets $\overset{2}{V}(y_k)$, and the proof is complete.

COROLLARY 3. *Let* X *be a* locally compact *space,* Y *a Hausdorff uniform space,* H *a subset of* $\mathcal{C}(X; Y)$. *Then* H *is relatively compact in* $\mathcal{C}_c(X; Y)$ *if and only if* H *is equicontinuous and* $H(x)$ *relatively compact in* Y *for all* $x \in X$.

In view of Corollary 1 it is enough to show that, if H is relatively compact in $\mathcal{C}_c(X; Y)$, then H is equicontinuous. Now each point $x \in X$ has a compact neighbourhood A, and it follows from Theorem 2 that H/A is equicontinuous; this implies that H is equicontinuous at x, and the result is proved.

Remark. Let X be a topological space, Y a uniform space and $\mathfrak{S}$ a set of subsets of X. Then on every *precompact* subset H of $\mathcal{F}_{\mathfrak{S}}$ (X; Y), the uniformity of $\mathfrak{S}$-convergence is the same as the uniformity of pointwise convergence in $B = \bigcup_{A \in \mathfrak{S}} A$. We can reduce to the case where B = X and Y is Hausdorff and complete; for if j is the canonical injection $B \to X$ and i the canonical mapping $Y \to \hat{Y}$, the uniformity of $\mathfrak{S}$-convergence on $\mathcal{F}$ (X; Y) is the inverse image of the uniformity of $\mathfrak{S}$-convergence on $\mathcal{F}(B; \hat{Y})$ under the mapping $\theta : u \to i \circ u \circ j$ (§ 1, no. 4, Proposition 4), and H is precompact if and only if $\theta(H)$ is (Chapter II, § 4, no. 2, Proposition 3). This being so, if B = X and Y is Hausdorff and complete, $\mathcal{F}_{\mathfrak{S}}$ (X; Y) is Hausdorff and complete (§ 1, no. 2, Proposition 1 and no. 5, Theorem 1); hence the closure $\overline{H}$ of H in this space is *compact*. On $\overline{H}$, the topology of pointwise convergence is Hausdorff (§ 1, no. 2, Proposition 1) and coarser than that of $\mathfrak{S}$-convergence; hence these two topologies coincide (Chapter I, § 9, no. 4, Theorem 2, Corollary 3) and consequently so do the uniformities of $\mathfrak{S}$-convergence and pointwise convergence (Chapter II, § 4, no. 1, Theorem 1).

3. SPECIAL FUNCTION SPACES

1. SPACES OF MAPPINGS INTO A METRIC SPACE

Let X be a set, Y a uniform space, $(f_\iota)_{\iota \in I}$ a family of pseudometrics defining the uniform structure of Y (Chapter IX, § 1, no. 4), and let $\mathfrak{S}$ be a set of subsets of X. For each $\iota \in I$, each set $A \in \mathfrak{S}$, and each pair (u, v) of mappings of X into Y, write

$$g_{\iota, A}(u, v) = \sup_{x \in A} f_\iota(u(x), v(x));$$

it follows immediately that $g_{\iota, A}$ is a *pseudometric* on $\mathcal{F}$ (X; Y) and that the family of pseudometrics $(g_{\iota, A})_{\iota \in I, A \in \mathfrak{S}}$ defines the uniformity of *$\mathfrak{S}$-convergence* on $\mathcal{F}$ (X; Y). In particular:

PROPOSITION 1. *If* Y *is a metrizable uniform space, the uniformity of uniform convergence on* $\mathcal{F}$ (X; Y) *is metrizable.*

For if d is a metric on Y compatible with its uniform structure, the structure of uniform convergence on $\mathcal{F}$ (X; Y) is defined by the single pseudometric

$$\delta(u, v) = \sup_{x \in X} d(u(x), v(x));$$

in general this pseudometric is not finite, but it is equivalent to a finite one (Chapter IX, § 1, no. 2), and since the uniformity of uniform convergence is Hausdorff (§ 1, no. 2, Proposition 1), it is metrizable.

COROLLARY. *Let* X *be a topological space and let* Y *be a metrizable uniform space. Suppose that there is a sequence* (K_n) *of compact subsets of* X *such that every compact subset of* X *is contained in some* K_n. *Then the uniformity of compact convergence on* $\mathscr{F}(X; Y)$ *is metrizable.*

Since the K_n cover X, $\mathscr{F}_c(X; Y)$ is isomorphic to a uniform subspace of $\prod_n \mathscr{F}_u(K_n; Y)$ (§ 1, no. 2, Remark 3), and the corollary therefore follows from Proposition 1 (Chapter IX, § 2, no. 4, Theorem 1, Corollary 2).

Note that this corollary applies in particular if X is *locally compact and σ-compact* (Chapter I, § 9, no. 9, Proposition 15, Corollary 1).

Now let Y be a metric space and let d be its metric. If X is any set and $\mathfrak{S}$ any set of subsets of X, we shall denote by $\mathscr{B}_{\mathfrak{S}}(X; Y)$ the set of all mappings $u : X \to Y$ such that $u(A)$ is *bounded* for each $A \in \mathfrak{S}$. Unless the contrary is expressly stated we shall regard $\mathscr{B}_{\mathfrak{S}}(X; Y)$ as endowed with the uniformity of $\mathfrak{S}$-convergence, which is defined by the following family of pseudometrics on $\mathscr{B}_{\mathfrak{S}}(X; Y)$:

$$d_A(u, v) = \sup_{x \in A} d(u(x), v(x)) \qquad (A \in \mathfrak{S})$$

which are *finite* by hypothesis. When $\mathfrak{S} = \{X\}$, we write $\mathscr{B}(X; Y)$ in place of $\mathscr{B}_{\mathfrak{S}}(X; Y)$. A mapping $u : X \to Y$ is said to be *bounded* if it belongs to $\mathscr{B}(X; Y)$, i.e. if $u(X)$ is a bounded subset of Y.

PROPOSITION 2. *Let* X *be a set and* Y *a metric space. The set* $\mathscr{B}(X; Y)$ *of bounded mappings is both open and closed in the space* $\mathscr{F}_u(X; Y)$.

If u is bounded, then every mapping $v : X \to Y$ such that for all $x \in X$, we have $d(u(x), v(x)) \leqslant 1$ is bounded, because

$$d(v(x), v(x_0)) \leqslant d(u(x), u(x_0)) + 2;$$

hence $\mathscr{B}(X; Y)$ is open. On the other hand, if u lies in the closure of $\mathscr{B}(X; Y)$ in $\mathscr{F}_u(X; Y)$, there is a mapping $u_0 \in \mathscr{B}(X; Y)$ such that $d(u(x), u_0(x)) \leqslant 1$ for all $x \in X$; hence u is bounded.

COROLLARY 1. *Let* X *be a set and* Y *a metric space. Then* $\mathscr{B}_{\mathfrak{S}}(X; Y)$ *is closed in* $\mathscr{F}_{\mathfrak{S}}(X; Y)$. *In particular, if* Y *is complete then* $\mathscr{B}_{\mathfrak{S}}(X; Y)$ *is complete with respect to the uniformity of* $\mathfrak{S}$*-convergence.*

For $\mathscr{B}_{\mathfrak{S}}(X; Y)$ is the inverse image of the subset $\prod_{A \in \mathfrak{S}} \mathscr{B}(A; Y)$ of the product $\prod_{A \in \mathfrak{S}} \mathscr{F}_u(X; Y)$ under the canonical mapping of $\mathscr{F}_{\mathfrak{S}}(X; Y)$ into $\prod_{A \in \mathfrak{S}} \mathscr{F}_u(A; Y)$; the first assertion therefore follows from § 1, no. 2, Remark 3, and the second follows from the first, if we take account of Theorem 1 of § 1, no. 5.

Corollary 2. *Let* X *be a topological space and* Y *a metric space. Then the space of all bounded continuous mappings of* X *into* Y *is both open and closed in* $\mathscr{C}_u(X; Y)$; *it is complete if* Y *is complete.*

The space in question is $\mathscr{B}(X; Y) \cap \mathscr{C}_u(X; Y)$; the first assertion follows from Proposition 2; the second follows from the first (§ 1, no. 6, Theorem 2, Corollary 1).

2. SPACES OF MAPPINGS INTO A NORMED SPACE

Consider, more particularly, the situation in which Y is a *normed* vector space over a non-discrete valued division ring K (Chapter IX, § 3, no. 3). Let us denote by $\|y\|$ the norm of $y \in Y$. The set $\mathscr{F}(X; Y) = Y^X$ is then canonically endowed with a K-vector space structure. A mapping $u: X \to Y$ is bounded if and only if the real-valued function $x \to \|u(x)\|$ is bounded in X. If u, v are bounded mappings of X into Y, it is clear that $u + v$ and λu $(\lambda \in K)$ are bounded; in other words, $\mathscr{B}(X; Y)$ is a *vector subspace* of $\mathscr{F}(X; Y)$. Moreover, $\|u\| = \sup_{x \in X} \|u(x)\|$ is a *norm* on $\mathscr{B}(X; Y)$; for it satisfies the triangle inequality and $\|u\| = 0$ implies $u = 0$, and for each $\lambda \in K$ we have

$$\|\lambda u\| = \sup_{x \in X} \|\lambda u(x)\| = \sup_{x \in X} |\lambda| . \|u(x)\| = |\lambda| . \sup_{x \in X} \|u(x)\| = |\lambda| . \|u\|.$$

Moreover, it is immediately verified that the uniformity on $\mathscr{B}(X; Y)$ defined by this norm is the uniformity of uniform convergence. Unless the contrary is expressly stated, whenever $\mathscr{B}(X; Y)$ is considered as a normed space, it is the norm defined above which is in question.

Proposition 3. *If the normed space* Y *is complete, then every series* (u_n) *of bounded mappings of* X *into* Y *which is absolutely convergent in the normed space* $\mathscr{B}(X; Y)$ (i.e. which is such that $\sum_{n=0}^{\infty} \|u_n\| < +\infty$; cf. Chapter IX, § 3, no. 6) *is uniformly convergent in* X.

For since $\mathscr{B}(X; Y)$ is complete (no. 1, Proposition 2, Corollary 1), the result follows from Chapter IX, § 3, no. 6, Proposition 11 and the definition of a uniformly convergent series.

Remark. 1) If $\sum_{n=0}^{\infty} \|u_n\| < +\infty$, then $\sum_{n=0}^{\infty} \|u_n(x)\| \leqslant \sum_{n=0}^{\infty} \|u_n\| < +\infty$ for each $x \in X$; in other words, for each $x \in X$ the series with general term $u_n(x)$ is absolutely convergent in the space Y. The converse is false. To avoid all confusion we shall sometimes say that the series with general term u_n is *normally convergent*, meaning that the series with general term $\|u_n\|$ is convergent. A series can be uniformly convergent in X without being normally convergent; this is the case, for example, for the series (u_n) in the space $\mathcal{B}(\mathbf{R}, \mathbf{R})$, defined as follows: $u_n(x) = (1/n) \sin x$ if $x \in [n\pi, (n+1)\pi]$, $u_n(x) = 0$ otherwise.

When Y is a *normed algebra* (Chapter IX, § 3, no. 7) over a non-discrete valued field K, then $\mathcal{B}(X; Y)$ is a K-algebra, and the norm $\|u\|$ is compatible with the algebra structure, since

$$\|uv\| = \sup_{x \in X} \|u(x)v(x)\| \leqslant \sup_{x \in X} \|u(x)\| . \|v(x)\|$$
$$\leqslant \sup_{x \in X} \|u(x)\| . \sup_{x \in X} \|v(x)\| = \|u\| . \|v\|.$$

Thus $\mathcal{B}(X; Y)$ is now a *normed algebra* over K.

PROPOSITION 4. *Let* X_i $(1 \leqslant i \leqslant n)$ *and* Y *be normed vector spaces over a non-discrete valued division ring* K, *and let* $X = \prod_{i=1}^{n} X_i$. *Then the set of all multilinear mappings of* X *into* Y *is closed in the space* $\mathcal{F}_s(X; Y)$.

This set consists of all $u \in F(X; Y)$ which satisfy all the relations

$$(1) \quad \begin{aligned} u(x_1, \ldots, x_i' + x_i'', \ldots, x_n) &= u(x_1, \ldots, x_i', \ldots, x_n) \\ &\quad + u(x_1, \ldots, x_i'', \ldots, x_n), \\ u(x_1, \ldots, \lambda x_i, \ldots, x_n) &= \lambda u(x_1, \ldots, x_i, \ldots, x_n) \end{aligned}$$

($1 \leqslant i \leqslant n$, x_i, x_i', x_i'' arbitrary elements of X_i, λ an arbitrary element of K); since both sides of the relations (1) are continuous functions of u on $\mathcal{F}_s(X; Y)$ (§ 1, no. 2, Remark 6), the result follows (Chapter I, § 8, no. 1, Proposition 2).

PROPOSITION 5. *Under the hypotheses of Proposition* 4, *the set* $\mathcal{L}(X_1, \ldots, X_n; Y)$ *of continuous multilinear mappings of* X *into* Y *is closed in* $\mathcal{F}(X; Y)$ *with respect to the topology of bounded convergence; it is complete with respect to the uniformity of bounded convergence if* Y *is complete.*

For if $\mathfrak{S}$ is the set of all bounded subsets of X, $\mathscr{L}(X_1, \ldots, X_n; Y)$ is the intersection of the set of all multilinear mappings of X into Y and the set $\mathscr{B}_{\mathfrak{S}}(X; Y)$ (Chapter IX, § 3, no. 5, Theorem 1); the result thus follows from Proposition 4 and Proposition 2, Corollary 1.

For the remainder of this subsection, K *denotes a non-discrete valued field.*

Then $\mathscr{L}(X_1, \ldots, X_n; Y)$ is a *vector subspace* of $\mathscr{F}(X; Y)$. Let B be the unit ball in X, the set of all $(x_i)_{1 \leqslant i \leqslant n}$ such that $\sup_{1 \leqslant i \leqslant n} \|x_i\| \leqslant 1$. Then the mapping $u \to u|B$ of $\mathscr{L}(X_1, \ldots, X_n; Y)$ into $\mathscr{B}(B; Y)$ is *injective*; moreover, the inverse image, under this mapping, of the uniformity of *uniform* convergence on $\mathscr{B}(B; Y)$ is the uniformity of *bounded* convergence on $\mathscr{L}(X_1, \ldots, X_n; Y)$. For every bounded subset of X is contained in a set of the form μB (for some $\mu \in K^*$), and if u is an element of $\mathscr{L}(X_1, \ldots, X_n; Y)$, to say that $\|u(z)\| \leqslant a$ for all $z \in \mu B$ is equivalent to saying that $\|u(z)\| \leqslant a/|\mu|^n$ for all $z \in B$. It is easily verified that the number

$$\|u\| = \sup_{z \neq 0} \frac{\|u(z)\|}{\|z\|}$$

is a *norm* on $\mathscr{L}(X_1, \ldots, X_n; Y)$ and defines the uniformity of bounded convergence on this set, and clearly we have

$$(2) \qquad \|u(x_1, \ldots, x_n)\| \leqslant \|u\| . \|x_1\| \ldots \|x_n\|.$$

Unless the contrary is expressly stated, whenever $\mathscr{L}(X_1, \ldots, X_n; Y)$ is considered as a normed space, it is the norm defined above which is in question.

PROPOSITION 6. *The multilinear mapping*

$$(u, x_1, \ldots, x_n) \to u(x_1, \ldots, x_n)$$

of the normed space $\mathscr{L}(X_1, \ldots, X_n; Y) \times X_1 \times \cdots \times X_n$ *into* Y *is continuous.*

This is an immediate consequence of the inequality (2) (Chapter IX, § 3, no. 5, Theorem 1).

PROPOSITION 7. *Let* X, Y, Z *be three normed spaces over* K. *The canonical mapping of the normed space* $\mathscr{L}(X, Y; Z)$ *into the space of linear mappings of* X *into* $\mathscr{L}(Y; Z)$ *which sends each* $u \in \mathscr{L}(X, Y; Z)$ *to the mapping* $x \to u(x, .)$ *is an isometry of* $\mathscr{L}(X; Y; Z)$ *onto* $\mathscr{L}(X; \mathscr{L}(Y; Z))$.

This follows immediately from the definitions and the relation

$$\sup_{\|x\| \leqslant 1} \Big(\sup_{\|y\| \leqslant 1} \|u(x, y)\| \Big) = \sup_{\|x\| \leqslant 1, \|y\| \leqslant 1} \|u(x, y)\|.$$

PROPOSITION 8. *Let* X, Y, Z *be three normed spaces over* K. *The bilinear mapping* $(u, v) \to v \circ u$ *of* $\mathcal{L}(X; Y) \times \mathcal{L}(Y; Z)$ *into* $\mathcal{L}(X; Z)$ *is continuous.*

For if $u \in \mathcal{L}(X; Y)$ and $v \in \mathcal{L}(Y; Z)$ we have

$$\|v \circ u\| \leqslant \|u\| \cdot \|v\|, \tag{3}$$

since for all $x \in X$ we have $\|v(u(x))\| \leqslant \|v\| \cdot \|u(x)\| \leqslant \|v\| \cdot \|u\| \cdot \|x\|$ by reason of (2).

In particular, on the set $\mathcal{L}(X)$ of *continuous endomorphisms* of a normed space X over K, the norm $\|u\|$ is compatible with the K-*algebra* structure of $\mathcal{L}(X)$.

Remark. 2) The set $\mathcal{L}(\mathbf{R}^m; \mathbf{R}^n)$ of linear (necessarily continuous) mappings of $\mathbf{R}^m$ into $\mathbf{R}^n$ can be identified with the set $\mathbf{M}_{n,m}(\mathbf{R})$ of matrices with n rows and m columns with coefficients in $\mathbf{R}$ and hence can be identified with $\mathbf{R}^{mn}$; on $\mathcal{L}(\mathbf{R}^m; \mathbf{R}^n)$, the uniformity of bounded convergence (with respect to the Euclidean metric on $\mathbf{R}^m$), of compact convergence, and of pointwise convergence are then identified with the *additive* uniformity on $\mathbf{R}^{mn}$. Take the norm of $x = (x_i) \in \mathbf{R}^n$ to be

$$\|x\| = \sup_i |x_i|,$$

and let (e_j) be the canonical basis of $\mathbf{R}^m$; if u and v are two linear mappings of $\mathbf{R}^m$ into $\mathbf{R}^n$ such that $\|u(e_j) - v(e_j)\| \leqslant \varepsilon$ for $1 \leqslant j \leqslant m$, then we have $|\alpha_{ij} - \beta_{ij}| \leqslant \varepsilon$ for each pair (i,j) [$U = (\alpha_{ij})$ and $V = (\beta_{ij})$ being the matrices of u, v respectively]; and conversely, if these inequalities are satisfied, we have $\|u(x) - v(x)\| \leqslant ma\varepsilon$ for every point x of a cube of centre 0 and side a in $\mathbf{R}^m$.

3. COUNTABILITY PROPERTIES OF SPACES OF CONTINUOUS FUNCTIONS

THEOREM 1. *Let* X *be a compact space.*

a) *If* X *is metrizable and if* Y *is any metrizable uniform space of countable type* (Chapter IX, § 2, no. 8), *then the metrizable space* $\mathcal{C}_u(X; Y)$ *of continuous mappings of* X *into* Y, *endowed with the topology of uniform convergence, is of countable type.*

b) *Conversely, if the metrizable space* $\mathcal{C}_u(X; \mathbf{R})$ *is of countable type, then* X *is metrizable.*

a) Let d (resp. d') be a metric compatible with the topology of X (resp. with the uniformity of Y); then $\delta(f, g) = \sup_{x \in X} d'(f(x), g(x))$ is a metric defining the uniformity of uniform convergence on the space $\mathcal{C}(X; Y)$, the functions of $\mathcal{C}(X; Y)$ being bounded because X is compact (no. 1). For each pair of integers $m > 0$, $n > 0$, let G_{mn} be the set of functions $f \in \mathcal{C}(X; Y)$ such that the relation $d(x, x') \leqslant 1/m$ implies $d(f(x), f(x')) \leqslant 1/n$. Every function $f \in \mathcal{C}(X; Y)$ is uniformly continuous (Chapter II, § 4, no. 1, Theorem 2) and therefore, for each $n > 0$, $\mathcal{C}(X; Y)$ is the union of the sets G_{mn} $(m > 0)$. Let $\{a_1, \ldots, a_{p(m)}\}$ be a finite subset of X such that the open balls with centres a_i and radii $1/m$ cover X $(1 \leqslant i \leqslant p(m))$; and let $(b_r)_{r \in \mathbf{N}}$ be a countable sequence which is dense in Y. For each mapping $\varphi : [1, p(m)] \to \mathbf{N}$, let H_φ be the set of those $f \in G_{mn}$ such that $d'(f(a_k), b_{\varphi(k)}) \leqslant 1/n$ for $1 \leqslant k \leqslant p(m)$. By the definition of the b_r, G_{mn} is the union of the sets H_φ for $\varphi \in \mathbf{N}^{p(m)}$; let C_{mn} be the set of mappings $\varphi \in \mathbf{N}^{p(m)}$ such that $H_\varphi \neq \emptyset$, and for each $\varphi \in C_{mn}$ let g_φ be an element of H_φ; finally, let L_{mn} denote the countable set of g_φ for $\varphi \in C_{mn}$. Let $f \in G_{mn}$, and let φ be an element of C_{mn} such that $f \in H_\varphi$; then it follows immediately from the definitions that we have $d'(f(x), g_\varphi(x)) \leqslant 4/n$ for all $x \in X$, i.e. $\delta(f, g_\varphi) \leqslant 4/n$. Hence the union of the sets L_{mn} is dense in $\mathcal{C}_u(X; Y)$, because for every integer $n > 0$ and every $f \in \mathcal{C}(X; Y)$ there exists m such that $f \in G_{mn}$, and we have just seen that the distance from f to L_{mn} is $\leqslant 4/n$.

b) Let $I = [0, 1]$. Since $\mathcal{C}_u(X; I)$ is a uniform subspace of $\mathcal{C}_u(X; \mathbf{R})$, it is of countable type. Let (f_n) be a sequence which is dense in $\mathcal{C}_u(X; I)$. Consider the product space $K = I^{\mathbf{N}}$ and the mapping $\psi : x \to (f_n(x))$ of X into K, which is obviously continuous. The mapping ψ is *injective*; for, by definition of the sequence (f_n), the relation $f_n(x) = f_n(x')$ for all n implies, on passing to the limit, $f(x) = f(x')$ for *every* function $f \in \mathcal{C}(X; I)$; but this is impossible if $x \neq x'$ by virtue of Axiom (O_{IV}) applied to the point x and to a neighbourhood V of x which does not contain x' (Chapter IX, § 1, no. 5, Theorem 2). It follows that the compact space X is homeomorphic to the subspace $\psi(X)$ of K (Chapter I, § 9, no. 4, Theorem 2, Corollary 2); since K is metrizable and of countable type, so is $\psi(X)$ and therefore so is X.

Q.E.D.

COROLLARY. *Let* X *be a locally compact space whose topology admits a countable base, and let* Y *be a metrizable uniform space of countable type.*

a) *The space* $\mathcal{L}$ *of continuous mappings of* X *into* Y *which have a limit at infinity, endowed with the topology of uniform convergence in* X, *is a metrizable space of countable type.*

b) *The space* $\mathcal{C}_c(X; Y)$ *of continuous mappings of* X *into* Y, *endowed with the topology of compact convergence, is a metrizable space of countable type.*

a) Let X′ be the compact space obtained by adjoining a point at infinity to X (Chapter I, § 9, no. 8, Theorem 4); by definition, every function $f \in \mathcal{L}$ can be uniquely extended to a continuous function $\bar{f} : X' \to Y$, and $f \to \bar{f}$ is therefore a bijection of L onto $\mathcal{C}(X'; Y)$; and this bijection is a homeomorphism of the space L onto $\mathcal{C}_u(X'; Y)$ by Proposition 6 of § 1, no. 6. Since X′ is metrizable (Chapter IX, § 2, no. 9, Proposition 16, Corollary) the result follows from Theorem 1, applied to X′ and Y.

b) Let (U_n) be a covering of X by relatively compact open sets, such that every compact subset of X is contained in some U_n (Chapter I, § 9, no. 9, Proposition 15, Corollary 1). If $\mathfrak{S}$ is the set of the $\overline{U}_n$, the topology of compact convergence on $\mathcal{C}(X; Y)$ is the same as the topology of $\mathfrak{S}$-convergence. Consequently (§ 1, no. 2, Remark 3) the space $\mathcal{C}_c(X; Y)$ is homeomorphic to a subspace of the product $\prod_n \mathcal{C}_u(\overline{U}_n; Y)$; since each of the compact spaces $\overline{U}_n$ has a countable base, it is metrizable (Chapter IX, § 2, no. 9, Proposition 16); each of the $\mathcal{C}_u(\overline{U}_n; Y)$ is therefore metrizable and of countable type by Theorem 1, and hence so is $\mathcal{C}_c(X; Y)$.

Note that the space of all bounded continuous real-valued functions on **R**, endowed with the topology of uniform convergence, is not of countable type (Exercise 4).

4. THE COMPACT-OPEN TOPOLOGY

THEOREM 2. *Let* X *be a topological space,* Y *a uniform space. For each compact subset* K *of* X *and each open subset* U *of* Y, *let* $T(K, U)$ *denote the set of all continuous mappings* $u : X \to Y$ *such that* $u(K) \subset U$. *Then the sets* $T(K, U)$ *generate the topology of compact convergence on* $\mathcal{C}(X; Y)$.

Let Y′ be the Hausdorff uniform space associated with Y (Chapter II, § 3, no. 8) and let $i : Y \to Y'$ be the canonical mapping of Y onto Y′. The topology of compact convergence is the coarsest topology for which the mappings $u \to (i \circ u)|K$ of $\mathcal{C}(X; Y)$ into $\mathcal{C}_u(K; Y')$ are continuous, as K runs through the set of all compact subsets of X (§ 1, no. 4, Proposition 4). Hence we obtain a subbase of the topology of $\mathcal{C}_c(X; Y)$ by taking a subbase of the topology of $\mathcal{C}_c(K; Y')$ for each compact subset K of X and then taking the union [in $\mathfrak{P}(\mathcal{C}(X; Y))$] of the inverse images of these subbases in $\mathcal{C}(X, Y)$. On the other hand, every open subset of Y is of the form $\overset{-1}{i}(U')$, where U′ is open in Y′ (Chapter II, § 3, no. 7, Proposition 12); hence, for each compact subset $K' \supset K$, $T(K, \overset{-1}{i}(U'))$ is the inverse image of $T(K, U')$ under the mapping

$$\mathcal{C}(X; Y) \to \mathcal{C}_u(K'; Y').$$

It thus remains for us to prove the theorem when X is *compact* and Y is *Hausdorff*; we shall make these assumptions from now on.

Let us first show that $\boldsymbol{T}(K, U)$ is *open* in $\mathcal{C}_c(X; Y)$. Let u_0 be a point of this set; since $u_0(K)$ is compact (Chapter I, § 9, no. 4, Theorem 2, Corollary 1) and contained in the open set U, there exists a symmetric entourage V of Y such that $V(u_0(K)) \subset U$ (Chapter II, § 4, no. 3, Proposition 4, Corollary). Let W be the neighbourhood of u_0 in $\mathcal{C}_c(X; Y)$ consisting of all continuous mappings $u : X \to Y$ such that $(u(x), u_0(x)) \in V$ for all $x \in K$. For such mappings we clearly have $u(K) \subset V(u_0(K)) \subset U$; hence $u \in \boldsymbol{T}(K, U)$ and therefore $W \subset \boldsymbol{T}(K, U)$, which proves our assertion.

Conversely, if W is a neighbourhood of a point $u_0 \in \mathcal{C}_c(X; Y)$, let us show that W contains the intersection of a finite number of neighbourhoods of the form $\boldsymbol{T}(K, U)$. We may suppose that W is the set of all $u \in \mathcal{C}(X; Y)$ such that $(u(x), u_0(x)) \in V$ for all $x \in X$, V being a given entourage of Y. Since u_0 is continuous on X, it is uniformly continuous (Chapter II, § 4, no. 1, Theorem 2). Let V_1 be a symmetric entourage of Y, open in $Y \times Y$ and such that $\overset{2}{V_1} \subset V$. X can be covered by a finite number of compact sets K_i $(1 \leqslant i \leqslant n)$ such that each $u_0(K_i)$ is V_1-small $(1 \leqslant i \leqslant n)$. Let U_i be the open set $V_1(u_0(K_i))$, and let $u : X \to Y$ be a continuous mapping contained in the intersection of the n sets $\boldsymbol{T}(K_i, U_i)$ (which are neighbourhoods of u_0). Then, for every $x \in K_i$, $u(x)$ belongs to U_i and therefore $u_0(x)$ and $u(x)$ are $\overset{2}{V_1}$-close, hence V-close. Since each $x \in X$ belongs to some K_i, we have $u \in W$ and the proof is complete.

This result leads us to make the following definition:

DEFINITION 1. *Let* X, Y *be two topological spaces, not necessarily uniformizable. For each compact subset* K *of* X *and each open subset* U *of* Y, *let* $\boldsymbol{T}(K, U)$ *be the set of all* $u \in \mathcal{C}(X; Y)$ *such that* $u(K) \subset U$. *The topology on* $\mathcal{C}(X; Y)$ *generated by the sets* $\boldsymbol{T}(K, U)$ *is called the topology of compact convergence or the compact-open topology; and we denote by* $\mathcal{C}_c(X; Y)$ *the topological space obtained by endowing* $\mathcal{C}(X; Y)$ *with this topology.*

If Y is a uniform space it follows from Theorem 2 that this definition agrees with that given in § 1, no. 3.

If H is a subset of $\mathcal{C}(X; Y)$ we shall say that the topology induced on H by that of $\mathcal{C}_c(X; Y)$ is the compact-open topology on H.

Example. Let I be the interval [0, 1] in **R**. If Y is any topological space, the space $\mathcal{C}_c(I; Y)$ is called the *space of paths* in Y. For each $y \in Y$, the subspace $\Omega_y(Y)$ of $\mathcal{C}_c(I; Y)$ consisting of paths u such that $u(0) = u(1) = y$ is called the *space of loops* (in Y) *at the point* y.

Remarks. 1) Likewise, the topology induced on $\mathcal{C}(X; Y)$ by the product topology on $Y^X = \mathcal{F}(X; Y)$ is called the *topology of pointwise convergence* (Y being not necessarily uniformizable); it is generated by sets of the form $T(\{x\}, U)$ as x runs through X and U runs through the set of all open subsets of Y, and it is therefore *coarser* than the compact-open topology. We deduce that, *if* Y *is Hausdorff, the space* $\mathcal{C}_c(X; Y)$ *is Hausdorff* (Chapter I, § 8, no. 1, Proposition 5, Corollary).

2) Let $\mathfrak{S}$ be a subbase of the topology of Y, and let $\mathfrak{K}$ be a set of compact subsets of X with the following property:

(R) If L is any compact subset of X and V is any neighbourhood of L, there exists a *finite* number of sets $K_i \in \mathfrak{K}$ such that $L \subset \bigcup_i K_i \subset V$.

Then the sets $T(K, U)$, where $K \in \mathfrak{K}$ and $U \in \mathfrak{S}$, form a *subbase* for the compact-open topology on $\mathcal{C}(X; Y)$. To prove this, we have to show that if L is any compact subset of X and V any open subset of Y, and if $u \in T(L, V)$, then there exists a finite number of pairs (K_i, U_i) such that $K_i \in \mathfrak{K}$, $U_i \in \mathfrak{S}$ and $u \in \bigcap_i T(K_i, U_i) \subset T(L, V)$. Note first that for every finite sequence (s_k) of sets of $\mathfrak{S}$ and every compact subset M of X, we have $T\left(M, \bigcap_k S_k\right) = \bigcap_k T(M, S_k)$ by definition. We may therefore first of all replace $\mathfrak{S}$ by the set of finite intersections of sets of $\mathfrak{S}$, i.e. we may suppose that $\mathfrak{S}$ is a *base* of the topology of Y. By hypothesis, $u(L)$ is quasi-compact and contained in V, hence there exists a finite number of sets $U_i \in \mathfrak{S}$ contained in V which cover $u(L)$. The sets $\overset{-1}{u}(U_i)$ are open in X and cover L. For each $x \in L$ there is therefore a compact neighbourhood N_x of x in L, contained in some one of the $\overset{-1}{u}(U_i)$. We can cover L with a finite number of these sets $N_{x_j} = L_j$; for each j, let us denote by $i(j)$ one of the indices i such that $L_j \subset \overset{-1}{u}(U_i)$. This being so, for each index j there exists [by (R)] a finite number of sets $K_{jk} \subset \overset{-1}{u}(U_{i(j)})$, belonging to $\mathfrak{K}$, which cover L_j. For each $v \in \bigcap_{j,k} T(K_{jk}, U_{i(j)})$ we have $\bigcup_k v(K_{jk}) \subset U_{i(j)}$ and therefore $v(L_j) \subset U_{i(j)}$, and $v(L) = \bigcup_j v(L_j) \subset \bigcup_j U_{i(j)} \subset V$; thus our assertion is proved.

THEOREM 3. *Let* X, Y, Z *be three topological spaces and let* f *be a mapping of* $X \times Y$ *into* Z. *If* f *is continuous then* $\tilde{f}: x \to f(x, .)$ *is a continuous mapping of* X *into* $\mathcal{C}_c(Y; Z)$. *The converse is true if* Y *is locally compact.*

Suppose that f is continuous. To show that $\tilde{f}$ is continuous we have to prove that, for each compact subset K of Y and each open subset U of Z, the inverse image V of $T(K, U)$ under $\tilde{f}$ is open in X. Let $x_0 \in V$; for each $y \in K$, we have $f(x_0, y) \in U$, and since f is continuous there is a neighbourhood V_y of x_0 in X and a neighbourhood W_y of y in Y such that $f(V_y \times W_y) \subset U$. Since K is compact, there exists a finite number of points $y_i \in K$ such that the sets W_{y_i} $(1 \leqslant i \leqslant n)$ cover K. Let V′ be the intersection of the neighbourhoods V_{y_i} of x_0, which is a neighbourhood of x_0; if $x \in V'$ and $y \in K$, we have $f(x, y) \in U$, since y is contained in some one of the W_{y_i} and x is contained in each V_{y_i}; hence $V' \subset V$, and therefore V is a neighbourhood of each of its points and consequently is open in X.

Conversely, suppose that $\tilde{f}$ is continuous and that Y is locally compact, and let us show that f is continuous. Let $x_0 \in X$, let $y_0 \in Y$ and let U be an open neighbourhood of $f(x_0, y_0)$ in Z; we shall show that there is a neighbourhood V of x_0 in X and a neighbourhood W of y_0 in Y such that $f(V \times W) \subset U$. Since $y \to f(x_0, y)$ is continuous, there is a *compact* neighbourhood W of y_0 such that $f(\{x_0\} \times W) \subset U$. On the other hand, since $\tilde{f}$ is continuous, the set V of $x \in X$ such that $f(x, .) \in T(W, U)$ [i.e. such that $f(x, y) \in U$ for all $y \in W$] is an open subset of X and therefore a neighbourhood of x_0; and we have $f(V \times W) \subset U$.

Q.E.D.

COROLLARY 1. *Let* X *be a locally compact space,* Y *a topological space,* H *a subset of* $\mathscr{C}(X; Y)$. *Then the compact-open topology on* H *is the coarsest for which the mapping* $(u, x) \to u(x)$ *of* $H \times X$ *into* Y *is continuous.*

For, by Theorem 3, this mapping is continuous if and only if the canonical injection $H \to \mathscr{C}_c(X; Y)$ is continuous.

Remark. 3) Let X be a locally compact space and Y a Hausdorff topological space. If $\mathscr{T}$ is a topology on a subset H of $\mathscr{C}(X; Y)$ such that the mapping $(u, x) \to u(x)$ is continuous on $H \times X$ and if also H is *compact* with respect to $\mathscr{T}$, then $\mathscr{T}$ is the compact-open topology. For it is finer than the latter by Corollary 1, and since the compact-open topology is Hausdorff, the two topologies are identical. Note that if in addition Y is *completely regular*, then H is *equicontinuous* with respect to every uniformity compatible with the topology of Y (§ 2, no. 5, Theorem 2, Corollary 3), and for every compact subset K of X the set

$$H(K) = \bigcup_{x \in K} H(x)$$

is compact, since it is the image of $H \times K$ under the continuous mapping $(u, x) \to u(x)$.

COROLLARY 2. *Let* X, Y, Z *be three topological spaces such that* X *is Hausdorff and* Y *is locally compact. Then the restriction to* $\mathscr{C}(X \times Y; Z)$ *of the canonical bijection* $\mathscr{F}(X \times Y; Z) \to \mathscr{F}(X; \mathscr{F}(Y; Z))$ (*Set Theory,* R, § 4, no. 14) *is a homeomorphism of* $\mathscr{C}_c(X \times Y; Z)$ *onto* $\mathscr{C}_c(X; \mathscr{C}_c(Y; Z))$.

This restriction is certainly a bijection

$$\rho : \mathcal{C}(X \times Y;\ Z) \rightarrow \mathcal{C}(X;\ \mathcal{C}_c(Y;\ Z))$$

by Theorem 3; it remains therefore to be shown that the compact-open topology on $\mathcal{C}(X \times Y; Z)$ is the inverse image under ρ of the compact-open topology on $\mathcal{C}(X; \mathcal{C}_c(Y; Z))$. Since the sets $\boldsymbol{T}(K, U)$, where K is a compact subset of Y and U is an open subset of Z, form a subbase of the topology of $\mathcal{C}_c(Y; Z)$, it follows from Remark 2 that the topology of $\mathcal{C}_c(X; \mathcal{C}_c(Y; Z))$ is generated by the sets of the form $\boldsymbol{T}(J, \boldsymbol{T}(K, U))$, where K and U are as above and J is a compact subset of X. Now the image of $\boldsymbol{T}(J, \boldsymbol{T}(K, U))$ under $\overset{-1}{\rho}$ is precisely $\boldsymbol{T}(J \times K, U)$, and is therefore an open set; so we have shown that ρ is continuous. To prove that ρ is a homeomorphism, we note first that the sets of the form $J \times K$ in $X \times Y$ (where J is a compact subset of X, and K is a compact subset of Y) satisfy the condition (R) of Remark 2: for if L is a compact subset of $X \times Y$ and V is a neighbourhood of L in $X \times Y$, the projections $M = \mathrm{pr}_1(L)$, $N = \mathrm{pr}_2(L)$ are compact, since X and Y are Hausdorff and $V \cap (M \times N)$ is a neighbourhood of L in the compact space $M \times N$, so that every point of L has a neighbourhood in $M \times N$ of the form $J \times K \subset V$, where $J \subset M$ and $K \subset N$ are compact; since L can be covered by a finite number of these neighbourhoods, the assertion is proved. Therefore sets of the form $\boldsymbol{T}(J \times K; U)$, where J is a compact subset of X, K a compact subset of Y and U an open subset of Z, generate the topology of $\mathcal{C}_c(X \times Y; Z)$. But we have already seen that the image of $\boldsymbol{T}(J \times K, U)$ under ρ is the open set $\boldsymbol{T}(J, \boldsymbol{T}(K, U))$ in $\mathcal{C}_c(X; \mathcal{C}_c(Y; Z))$; hence ρ is a homeomorphism.

Note that if in addition Z is assumed to be uniformizable, Corollary 2 is a trivial consequence of § 1, no. 4, Proposition 2.

Proposition 9. *Let* X, Y, Z *be three topological spaces,* Y *being locally compact. Then the mapping* $(u, v) \rightarrow v \circ u$ *of* $\mathcal{C}_c(X; Y) \times \mathcal{C}_c(Y; Z)$ *into* $\mathcal{C}_c(X; Z)$ *is continuous.*

We have to show that, for every compact subset K of X and every open subset U of Z, the set R of pairs (u, v) such that $v(u(K)) \subset U$ is open in $\mathcal{C}_c(X; Y) \times \mathcal{C}_c(Y; Z)$. Let $(u_0, v_0) \in R$; then $u_0(K)$ is a compact subset of the locally compact space Y, contained in the open set $\overset{-1}{v_0}(U)$, and hence there is a compact neighbourhood L of $u_0(K)$t contained in $\overset{-1}{v_0}(U)$ (Chapter I, § 9, no. 7, Proposition 10). The set V of all $u \in \mathcal{C}_c(X; Y)$ such that $u(K) \subset \mathring{L}$ is a neighbourhood of u_0, and the set W of all $v \in \mathcal{C}_c(Y; Z)$ such that $v(L) \subset U$ is a neighbourhood of v_0; furthermore, the relation $(u, v) \in V \times W$ implies $v(u(K)) \subset U$. Hence the result.

5. TOPOLOGIES ON GROUPS OF HOMEOMORPHISMS

PROPOSITION 10. *Let* X *be a uniform space and let* H *be an equicontinuous set of homeomorphisms of* X *onto itself. If* H *and* H^{-1} *are endowed with the topology of pointwise convergence in* X, *then the mapping* $u \to u^{-1}$ *of* H^{-1} *onto* H *is continuous.*

It is enough to show that, for each $x_0 \in X$, the mapping $u \to u^{-1}(x_0)$ of H^{-1} into X is continuous at every point $u_0 \in H^{-1}$. Let V be a symmetric entourage of X, and let $y_0 = u_0^{-1}(x_0)$. By hypothesis there is a symmetric entourage U of X such that the relation $(x, x_0) \in U$ implies $(u^{-1}(x), u^{-1}(x_0)) \in V$ for all $u \in H^{-1}$. Take an element $u \in H^{-1}$ which is $W(\{y_0\}, U)$-close to u_0; then we have $(u(y_0), u_0(y_0)) \in U$, i.e. $(u(y_0), x_0) \in U$. It follows that $(y_0, u^{-1}(x_0)) \in V$, i.e. $(u_0^{-1}(x_0), u^{-1}(x_0)) \in V$; this completes the proof.

COROLLARY. *Let* X *be a uniform space and let* H *be an equicontinuous group of homeomorphisms of* X *onto itself. Then the topology of pointwise convergence in* X *is compatible with the group structure of* H.

This is a consequence of Proposition 10, together with § 2, no. 1, Proposition 1, Corollary 5.

PROPOSITION 11. *Let* X *be a compact space and let* Γ *be the group of all homeomorphisms of* X *onto itself. Then the topology of uniform convergence in* X *is compatible with the group structure of* Γ.

We know already (no. 4, Proposition 9) that the mapping $(u, v) \to v \circ u$ of $\Gamma \times \Gamma$ into Γ is continuous with respect to this topology; thus we have to show that $u \to u^{-1}$ is continuous at every point u_0 of Γ. Since u_0^{-1} is uniformly continuous on X, given any symmetric entourage V of X there exists an entourage W of X such that the relation $(x, x') \in W$ implies $(u_0^{-1}(x), u_0^{-1}(x')) \in V$. Hence, if $u \in \Gamma$ is such that $(u_0(x), u(x)) \in W$ for all $x \in X$, it follows that $(x, u_0^{-1}(u(x))) \in V$ for all $x \in X$, and therefore (as u is bijective) $(u^{-1}(x), u_0^{-1}(x)) \in V$ for all $x \in X$. This completes the proof.

Now let X be a locally compact space and let Γ be the group of all homeomorphisms of X onto itself. The topology of *compact convergence* in X *is not necessarily compatible* with the group structure of Γ (Exercise 17). Let X′ denote the compact space obtained by adjoining a point at infinity ω to X. Every homeomorphism u of X onto itself extends uniquely to a homeomorphism u' of X′ onto itself such that $u'(\omega) = \omega$ (Chapter I, § 10, no. 3, Corollary to Proposition 7), so that Γ can be identified with *the subgroup of the group* Γ' *of all homeomorphisms of* X′ *onto itself, consisting of all homeomorphisms which leave* ω *fixed.* The topology induced on Γ by that of $\mathcal{C}_u(X'; X')$ is therefore *compatible* with the group structure of Γ (Proposition 11), and Γ is *closed* in Γ' [with respect to the topology induced by that of $\mathcal{C}_u(X'; X')$] because it is defined by the equation $u(\omega) = \omega$ (§ 1, no. 2, Remark 6). We denote by $\mathcal{T}_\beta$ the group topology thus defined on Γ; it is *finer* than the topology of compact convergence and can also (by virtue of § 1, no. 6, Proposition 6) be defined as the topology of uniform convergence on X, when X is endowed with the uniformity induced by the unique uniformity of X'.

The topology $\mathcal{T}_\beta$ can be characterized as follows:

PROPOSITION 12. *On the group Γ of all homeomorphisms of a locally compact space X, the topology $\mathcal{T}_\beta$ is the coarsest for which the mappings $u \to u$ and $u \to u^{-1}$ of Γ into $\mathcal{C}_c(X; X)$ are continuous.*

Let us denote the latter topology for the moment by $\mathcal{T}'$. Since $u \to u^{-1}$ is continuous with respect to $\mathcal{T}_\beta$ and since $\mathcal{T}_\beta$ is finer than the topology of compact convergence, it is clear that $\mathcal{T}_\beta$ is finer than $\mathcal{T}'$. To prove the converse, endow X' with its unique uniformity; let $u_0 \in \Gamma$ and let V be an entourage of X'; then we have to prove that there is a compact subset K of X and a symmetric entourage W of X' such that the relations

$$u \in \Gamma,\ (u_0(x), u(x)) \in W \text{ and } (u_0^{-1}(x), u^{-1}(x)) \in W \text{ for all } x \in K$$

imply

$$(u_0(x), u(x)) \in V \text{ for all } x \in X.$$

Let V_1 be a symmetric open entourage of X' such that $\overset{2}{V_1} \subset V$; then $K_1 = X' - V_1(\omega)$ is a compact subset of X. Choose a symmetric open entourage W of X' such that $W \subset V$ and $W(\omega) \cap W(u_0^{-1}(K_1)) = \varnothing$; this is possible by Proposition 4 of Chapter II, § 4, no. 3. Let $K_2 = X' - W(\omega)$, which is a compact subset of X. We shall see that W and the compact set $K = K_1 \cup K_2$ do what is required. Since $W \subset V$, it is enough to show that the relation

$$(u_0^{-1}(x), u^{-1}(x)) \in W \quad \text{for all} \quad x \in K_1 \quad (u \in \Gamma)$$

implies that

$$(u(y), \omega) \in V_1 \quad \text{for all} \quad y \in W(\omega);$$

for we shall then also have $(u_0(y), \omega) \in V_1$ and thence

$$(u_0(y), u(y)) \in \overset{2}{V_1} \subset V$$

for all $y \in W(\omega) = X' - K_2$. Now if we had $y \in W(\omega)$ and

$$u(y) \in X' - V_1(\omega) = K_1,$$

it would follow that $y \in u^{-1}(K_1) \subset W(u_0^{-1}(K_1))$, contrary to the choice of W; the proof is therefore complete.

In general the group Γ, endowed with $\mathcal{C}_\beta$, is not locally compact; but we have the following criterion:

THEOREM 4. *Let* G *be a subgroup of the group* Γ *of all homeomorphisms of a locally compact space* X. *Suppose that, in the space* $\mathcal{C}_c(X; X)$, *there is a neighbourhood* V *of the identity mapping* e *such that* $V \cap G = H$ *is symmetric in* G *and relatively compact in* $\mathcal{C}_c(X; X)$. *Then the closure* $\overline{G}$ *of* G *in* Γ *with respect to the topology* $\mathcal{C}_\beta$ *is a locally compact group with respect to the topology induced by* $\mathcal{C}_\beta$; *this topology induced on* $\overline{G}$ *is the same as the topology of compact convergence, and the closure* $\overline{H}$ *of* H *in* $\mathcal{C}_c(X; X)$ *is a neighbourhood of* e *in* $\overline{G}$ *with respect to this topology.*

Let us show first that $\overline{H}$ is *contained in* Γ and that the topology induced on $\overline{H}$ by $\mathcal{C}_\beta$ is *the same as the topology of compact convergence.* Let $u_0 \in \overline{H}$; u_0 is therefore the limit, in $\mathcal{C}_c(X; X)$, of an ultrafilter Φ on H. Since Φ^{-1} (the image of Φ under $u \to u^{-1}$) is an ultrafilter base on $H \subset \overline{H}$, it converges in the compact subspace $\overline{H}$ of $\mathcal{C}_c(X; X)$ to an element v_0. The mapping $(u, v) \to uv$ converges to u_0v_0 with respect to $\Phi \times \Phi^{-1}$ (no. 4, Proposition 9); *a fortiori*, $u \to uu^{-1} = e$ converges to u_0v_0 with respect to Φ, hence $u_0v_0 = e$ since $\mathcal{C}_c(X; X)$ is Hausdorff. Similarly $v_0u_0 = e$; hence u_0 is a homeomorphism of X, i.e., $u_0 \in \Gamma$. Thus $\overline{H}$ is contained in Γ. Moreover, this argument shows that $\overline{H}^{-1} = \overline{H}$ and that, for every ultrafilter Φ on $\overline{H}$ which converges to u_0, Φ^{-1} converges in $\mathcal{C}_c(X; X)$ to u_0^{-1}; hence the mapping $u \to u^{-1}$ of $\overline{H}$ into $\mathcal{C}_c(X; X)$ is continuous when $\overline{H}$ carries the topology of compact convergence (Chapter I, § 7, no. 4, Proposition 9, Corollary 1). Proposition 12 then shows that, on $\overline{H}$, the topology of compact convergence is the same as the topology induced by $\mathcal{C}_\beta$.

Furthermore, since the topology $\mathcal{C}_\beta$ on Γ is finer than the topology of compact convergence, $\overline{H}$ is also the closure of H with respect to $\mathcal{C}_\beta$. But H is a neighbourhood of e in G with respect to the topology of compact convergence, and *a fortiori* with respect to the topology induced by $\mathcal{C}_\beta$; it follows (Chapter I, § 3, no. 1, Proposition 2) that $\overline{H}$ is a neighbourhood of e in $\overline{G}$ with respect to the topology induced by $\mathcal{C}_\beta$, and hence $\overline{G}$ is locally compact in this topology. If W is the interior of V with respect to the topology of compact convergence, then $W \cap \Gamma$ is open in $\mathcal{C}_\beta$, hence $W \cap \overline{G}$ is contained in the closure of $H = V \cap G$ with respect to $\mathcal{C}_\beta$ (Chapter I, § 1, no. 6, Proposition 5); this shows that $\overline{H}$ is also a neighbourhood of e in $\overline{G}$ with respect to the topology of compact convergence. Finally, for each $u_0 \in \Gamma$, the bijections $v \to u_0 \circ v$ and $v \to u_0^{-1} \circ v$ of $\mathcal{C}_c(X; X)$ onto itself are continuous (no. 4, Proposition 9), and hence, if $u_0 \in \overline{G}$, $u_0\overline{H}$ is a neighbourhood of u_0 in $\overline{G}$ with respect to the topology of compact convergence. This completes the proof.

COROLLARY. *Let* G *be a group of homeomorphisms of a locally compact space* X. *If the closure* $\overline{G}$ *of* G *in* $\mathcal{C}_c(X; X)$ *is compact, then* $\overline{G}$ *is a group of homeomorphisms of* X, *and the topology of compact convergence is compatible with the group structure of* $\overline{G}$, *which is therefore a compact topological group.*

A group of homeomorphisms of a locally compact space X which is locally compact but not compact with respect to the topology of compact convergence is *locally closed* in $\mathcal{C}_c(X; X)$ by virtue of Chapter I, § 9, no. 7, Proposition 12, *but is not necessarily closed.*

For example, in the ring $\mathcal{L}(\mathbf{R}^n)$ of endomorphisms of $\mathbf{R}^n$, identified with the ring $\mathbf{M}_n(\mathbf{R})$ of square $n \times n$ matrices over $\mathbf{R}$ and endowed with the topology of compact convergence, the group $\mathbf{GL}(n, \mathbf{R})$, identified with the group of non-singular matrices, is locally compact but dense (Chapter VI, § 1, no. 6, Proposition 6).

4. APPROXIMATION OF CONTINUOUS REAL-VALUED FUNCTIONS

1. APPROXIMATION OF CONTINUOUS FUNCTIONS BY FUNCTIONS BELONGING TO A LATTICE

In this section we shall study the set $\mathcal{C} = \mathcal{C}(X; \mathbf{R})$ of continuous real-valued functions (*) defined on a *compact* space X, and we shall always suppose that $\mathcal{C}$ is endowed with the topology of *uniform convergence.* From § 3, no. 2 we know that this topology is defined by the norm

$$\|f\| = \sup_{x \in X} |f(x)|$$

and that this norm is compatible with the **R**-algebra structure of $\mathcal{C}$. With this norm and this algebra structure, $\mathcal{C}$ is a *complete normed algebra* over **R** (§ 1, no. 6, Theorem 2, Corollary 1).

(*) The real-valued functions under consideration in this section are assumed always to be *finite.*

If H is a subset of $\mathcal{C}$, we shall say that a continuous real-valued function f on X *can be uniformly approximated* by functions of H if f lies in the *closure* of H in the space $\mathcal{C}$, i.e. if, for each $\varepsilon > 0$, there exists a function $g \in H$ such that $|f(x) - g(x)| \leqslant \varepsilon$ for *all* $x \in X$. To say that *every* continuous real-valued function on X can be uniformly approximated by functions of H therefore means that H is *dense* in $\mathcal{C}$.

On the set $\mathcal{C}$, the relation $f \leqslant g$ [which means that $f(x) \leqslant g(x)$ for all $x \in X$)] is an order relation, with respect to which $\mathcal{C}$ is a *lattice*. Clearly we have $\| |u| - |v| \| \leqslant \|u - v\|$, and therefore $u \to |u|$ is a *uniformly continuous* mapping of $\mathcal{C}$ into itself. It follows that

$$(u, v) \to \sup (u, v) = \tfrac{1}{2}(u + v + |u - v|)$$

and

$$(u, v) \to \inf (u, v) = \tfrac{1}{2}(u + v - |u - v|)$$

are uniformly continuous on $\mathcal{C} \times \mathcal{C}$.

PROPOSITION 1. *Let* X *be a compact space and let* H *be a set of continuous real-valued functions defined on* X. *Let* f *be a continuous real-valued function on* X *such that for each* $x \in X$ *there exists a function* $u_x \in H$ *such that* $u_x(x) > f(x)$ [resp. $u_x(x) < f(x)$]. *Then there exists a finite number of functions* $u_{x_i} = f_i \in H$ $(1 \leqslant i \leqslant n)$ *such that, if* $v = \sup(f_1, f_2, \ldots, f_n)$ [resp. $w = \inf(f_1, f_2, \ldots, f_n)$], *we have* $v(x) > f(x)$ [resp. $w(x) < f(x)$] *for all* $x \in X$.

For each $x \in X$, let G_x be the open set consisting of all $z \in X$ such that $u_x(z) > f(z)$ [resp. $u_x(z) < f(z)$]. Since $x \in G_x$ by hypothesis, X is the union of the sets G_x as x runs through X. Since X is compact there exists a finite number of points x_i $(1 \leqslant i \leqslant n)$ such that the G_{x_i} cover X, and it is clear that the functions $f_i = u_{x_i}$ satisfy the conditions of the proposition.

THEOREM 1 (Dini). *Let* X *be a compact space, and let* H *be a set of continuous real-valued functions on* X *which is directed with respect to the relation* $\leqslant$ (resp. $\geqslant$). *If the upper* (resp. *lower*) *envelope* f *of* H *is finite and continuous on* X, *then* f *can be uniformly approximated by functions belonging to* H (or, equivalently, *the section filter of* H *converges uniformly to* f *in* X).

Given any $\varepsilon > 0$, for each $x \in X$ there exists a function $u_x \in H$ such that $u_x(x) > f(x) - \varepsilon$. By Proposition 1 and the fact that H is directed with respect to the relation $\leqslant$, there exists $g \in H$ such that $g(x) > f(x) - \varepsilon$ for all $x \in X$; on the other hand, we have $g(x) \leqslant f(x)$ by definition, and therefore the theorem is proved.

COROLLARY. *Let* (u_n) *be an increasing* (resp. *decreasing*) *sequence of continuous real-valued functions on* X. *If the upper* (resp. *lower*) *envelope* f *of the sequence* (u_n) *is finite and continuous on* X, *then the sequence* (u_n) *converges uniformly to* f *in* X.

It is clear that the conclusion of Theorem 1 is no longer necessarily valid if X is no longer assumed to be compact, as is shown by the example of the decreasing sequence of functions $x/(n+x)$ in $\mathbf{R}_+$.

PROPOSITION 2. *Let* X *be a compact space, and let* H *be a set of continuous real-valued functions on* X *such that, given any two functions* $u \in \mathrm{H}$, $v \in \mathrm{H}$, *the functions* $\sup(u, v)$ *and* $\inf(u, v)$ *are in* H. *Then a continuous real-valued function* f *on* X *can be uniformly approximated by functions belonging to* H *if and only if, for each real number* $\varepsilon > 0$ *and each pair* x, y *of points of* X, *there is a function* $u_{x,y} \in \mathrm{H}$ *such that* $|f(x) - u_{x,y}(x)| < \varepsilon$ *and* $|f(y) - u_{x,y}(y)| < \varepsilon$.

The condition is clearly necessary; let us show that it is sufficient. For each $\varepsilon > 0$, we shall show that there is a function $g \in \mathrm{H}$ such that $|f(z) - g(z)| < \varepsilon$ for all $z \in \mathrm{X}$. Let x be any point of X, and let H_x be the set of all functions $u \in \mathrm{H}$ such that $u(x) < f(x) + \varepsilon$. By hypothesis, for each $y \in \mathrm{X}$, the function $u_{x,y}$ belongs to H_x and we have $u_{x,y}(y) > f(y) - \varepsilon$. Hence, by Proposition 1, there is a finite number of functions of H_x whose upper envelope v_x is such that $v_x(z) > f(z) - \varepsilon$ for all $z \in \mathrm{X}$; on the other hand, we have $v_x(x) < f(x) + \varepsilon$ by the definition of H_x; finally, $v_x \in \mathrm{H}$ by hypothesis. Proposition 1 therefore shows that there exists a finite number of functions v_{x_i} whose lower envelope g is such that $g(z) < f(z) + \varepsilon$ for all $z \in \mathrm{X}$; but since we have $v_{x_i}(z) > f(z) - \varepsilon$ for all $z \in \mathrm{X}$ and for every index i, we have also $g(z) > f(z) - \varepsilon$ for all $z \in \mathrm{X}$. Since $g \in \mathrm{H}$ by hypothesis, the proof is complete.

Remark. When the set H satisfies the conditions of Proposition 2, it is a *lattice* with respect to the ordering $f \leqslant g$. But it should be remarked that a subset H of $\mathcal{C}$ can be a lattice with respect to this ordering without it being necessarily the case that the least upper bound (resp. greatest lower bound) *in* H of two functions u, v of H is the same as their least upper bound (resp. greatest lower bound) *in* $\mathcal{C}$. * An example is provided by the *convex* mappings of a compact interval of $\mathbf{R}$ into $\mathbf{R}$. *

COROLLARY. *Suppose that* H *is such that, whenever* $u \in \mathrm{H}$ *and* $v \in \mathrm{H}$, *we have* $\sup(u, v) \in \mathrm{H}$ *and* $\inf(u, v) \in \mathrm{H}$; *and is such that, given any two distinct points* x, y *of* X *and any two real numbers* α, β, *there is a function* $g \in \mathrm{H}$ *such that* $g(x) = \alpha$ *and* $g(y) = \beta$. *Then every continuous real-valued function on* X *can be uniformly approximated by functions belonging to* H.

DEFINITION 1. *If* X *is any set, a set* H *of mappings of* X *into a set* Y *is said to separate the elements of a subset* A *of* X (or to be a *separating set* for the elements of A) *if, given any two distinct elements* x, y *of* A, *there is a function* $f \in$ H *such that* $f(x) \neq f(y)$.

For example, if X is a *completely regular* space (Chapter IX, § 1, no. 5) then the set of all continuous mappings of X into [0, 1] separates the points of X.

THEOREM 2 (Stone). *Let* X *be a compact space, and let* H *be a vector subspace of* $\mathcal{C}$(X; **R**) *such that* 1) *the constant functions belong to* H; 2) *if* $u \in$ H, *then* $|u| \in$ H; 3) H *separates the points of* X. *Then every continuous real-valued function on* X *can be uniformly approximated by functions of* H.

It is enough to show that H satisfies the conditions of the Corollary to Proposition 2. By hypothesis, if $u \in$ H and $v \in$ H, we have

$$\sup(u, v) = \tfrac{1}{2}(u + v + |u - v|) \in \mathrm{H}$$

and

$$\inf(u, v) = \tfrac{1}{2}(u + v - |u - v|) \in \mathrm{H}.$$

On the other hand, let x and y be any two distinct points of X, and let α, β be any two real numbers. By hypothesis, there is a function $h \in$ H such that $h(x) \neq h(y)$: say $h(x) = \gamma$ and $h(y) = \delta$. Since the constants belong to H, the function

$$g(z) = \alpha + (\beta - \alpha)\frac{h(z) - \gamma}{\delta - \gamma}$$

belongs to H and is such that $g(x) = \alpha$ and $g(y) = \beta$.

2. APPROXIMATION OF CONTINUOUS FUNCTIONS BY POLYNOMIALS

Given a set H of real-valued functions defined on a set X, we say that a real-valued function defined on X is a *polynomial* (resp. a *polynomial with no constant term*) *with real coefficients, in the functions* of H, if it is of the form $x \to g(f_1(x), f_2(x), \ldots, f_n(x))$ where g is a polynomial (resp. a polynomial with no constant term) in n indeterminates (n arbitrary) with real coefficients, and the f_i $(1 \leqslant i \leqslant n)$ belong to H.

THEOREM 3 (Weierstrass-Stone). *Let* X *be a compact space and let* H *be a set of continuous real-valued functions on* X *which separates the points of* X. *Then every continuous real-valued function on* X *can be uniformly approximated by polynomials (with real coefficients) in the functions of* H.

An equivalent statement of the theorem is that *any subalgebra of* $\mathcal{C}$(X; **R**) *which contains the constant functions and separates the points of* X *is dense in* $\mathcal{C}$(X; **R**).

Let H_0 be the set of all polynomials in the functions of H, and let $\overline{\mathrm{H}}_0$ be the closure of H_0 in $\mathcal{C}$. If g is any polynomial in n variables with real coefficients, then $(u_1, u_2, \ldots, u_n) \to g(u_1, u_2, \ldots, u_n)$ is a continuous mapping of $\mathcal{C}^n$ into $\mathcal{C}$, which maps H_0^n into H_0, and therefore maps $\overline{H}_0^n$ into $\overline{H}_0$ (Chapter I, § 2, no. 1, Theorem 1). In particular, $\overline{H}_0$ is a vector subspace of $\mathcal{C}$ and evidently satisfies the first and third conditions of Theorem 2; we shall show that it also satisfies the second condition, and this will prove that $\overline{H}_0 = \mathcal{C}$.

Since every function $u \in \overline{H}_0$ is bounded in X, it is enough to prove the following lemma:

LEMMA 1. *For each real number* $\varepsilon > 0$ *and each compact interval* $I \subset \mathbf{R}$ *there exists a polynomial* $p(t)$ *with no constant term such that* $|p(t) - |t|| \leqslant \varepsilon$ *for all* $t \in I$.

It is enough to prove the lemma for an interval of the form $I = [-a, +a]$ and hence, replacing t by at, for the interval $I = [-1, +1]$. Since $|t| = \sqrt{t^2}$, Lemma 1 is a consequence of the following result:

LEMMA 2. *Let* (P_n) *be the sequence of polynomials without constant terms defined by*

$$p_0(t) = 0, \quad p_{n+1}(t) = p_n(t) + \tfrac{1}{2}\left(t - (p_n(t))^2\right), \qquad n \geqslant 0. \tag{1}$$

In the interval $[0, 1]$ *the sequence* (p_n) *is increasing and converges uniformly to* $\sqrt{t}$.

To prove Lemma 2 it is enough to show that, for all $t \in [0, 1]$, we have

$$0 \leqslant \sqrt{t} - p_n(t) \leqslant \frac{2\sqrt{t}}{2 + n\sqrt{t}}, \tag{2}$$

for (2) implies that $0 \leqslant \sqrt{t} - p_n(t) \leqslant 2/n$.

We prove (2) by induction on n. It is true for $n = 0$. If $n \geqslant 0$ it follows from the inductive hypothesis (2) that $0 \leqslant \sqrt{t} - p_n(t) \leqslant \sqrt{t}$, hence $0 \leqslant p_n(t) \leqslant \sqrt{t}$, and therefore from (1) we have

$$\sqrt{t} - p_{n+1}(t) = \left(\sqrt{t} - p_n(t)\right)\left(1 - \tfrac{1}{2}\left(\sqrt{t} + p_n(t)\right)\right),$$

so that $\sqrt{t} - p_{n+1}(t) \geqslant 0$, and from (2)

$$\sqrt{t} - p_{n+1}(t) \leqslant \frac{2\sqrt{t}}{2 + n\sqrt{t}}\left(1 - \frac{\sqrt{t}}{2}\right) \leqslant \frac{2\sqrt{t}}{2 + n\sqrt{t}}\left(1 - \frac{\sqrt{t}}{2 + (n+1)\sqrt{t}}\right)$$
$$= \frac{2\sqrt{t}}{2 + (n+1)\sqrt{t}}.$$

Q.E.D.

If X is not compact, the conclusion of Theorem 3 is not necessarily valid. For example, a continuous real-valued function on **R** which is bounded and not constant cannot be uniformly approximated in **R** by polynomials (cf. Exercise 6).

PROPOSITION 3. *Let* $(K_\iota)_{\iota \in I}$ *be a family of compact intervals of* **R**, $K = \prod_{\iota \in I} K_\iota$ *their product, and let* X *be a compact subspace of* K. *Then every continuous real-valued function on* X *can be uniformly approximated by polynomials in the coordinates* $x_\iota = \mathrm{pr}_\iota x$.

For if $x = (x_\iota)$ and $y = (y_\iota)$ are two distinct points of X, there is at least one index ι such that $x_\iota \neq y_\iota$; hence the family of continuous functions pr_ι satisfies the conditions of Theorem 3.

PROPOSITION 4. *Let* X *be a compact space, let* A *be a closed subspace of* X, *and let* H *be a set of continuous real-valued functions on* X *which separates the points of* $\complement A$ *and is such that* A *is the intersection of the sets* $\overset{-1}{u}(0)$ *as* u *runs through* H. *Then every continuous real-valued function on* X *which is zero on* A *can be uniformly approximated by polynomials without constant terms in the functions of* H.

Consider first the particular case in which A consists of a single point x_0. The hypotheses then imply that H separates the points of X; for, if $x \neq x_0$, then by hypothesis there is a function $u \in H$ such that $u(x) \neq 0 = u(x_0)$. Hence, for each $\varepsilon > 0$ and each continuous real-valued function f on X such that $f(x_0) = 0$, there exists (Theorem 3) a polynomial g in the functions of H such that $|f(x) - g(x)| \leqslant \varepsilon$ for all $x \in X$. In particular $|g(x_0)| \leqslant \varepsilon$, so that

$$|f(x) - (g(x) - g(x_0))| \leqslant 2\varepsilon$$

for all $x \in X$; and since $g(x) - g(x_0)$ is a polynomial in the functions of H with no constant term, the result is established in this case.

In the general case, consider the equivalence relation R on X whose classes are the set A and the sets $\{x\}$ for $x \notin A$. The quotient space X/R is Hausdorff (Chapter I, § 8, no. 6, Proposition 15) and therefore compact. Let $\varphi : X \to X/R$ be the canonical mapping. Every continuous real-valued function f on X which vanishes on A can be written in the form $f = f_1 \circ \varphi$, where f_1 is a continuous real-valued function on X/R which vanishes at the point $x'_0 = \varphi(A)$. Applying the result already proved to the space X/R and the point x'_0, we obtain the final result.

3. APPLICATION: APPROXIMATION OF CONTINUOUS REAL-VALUED FUNCTIONS DEFINED ON A PRODUCT OF COMPACT SPACES

THEOREM 4. *Let* $(X_\iota)_{\iota \in I}$ *be a family of compact spaces, and let*

$$X = \prod_{\iota \in I} X_\iota.$$

Then every continuous real-valued function on X *can be uniformly approximated by sums of a finite number of functions of the form*

$$(x_\iota) \to \prod_{\alpha \in J} u_\alpha(x_\alpha),$$

where J *is an (arbitrary) finite subset of* I *and* u_α *is a continuous real-valued function on* X_α *for each* $\alpha \in J$.

Consider the set H of "functions of one variable" $(x_\iota) \to u_\alpha(x_\alpha)$ (any $\alpha \in I$) which are continuous on X. This set separates the points of X, for if $x = (x_\iota)$ and $y = (y_\iota)$ are any two distinct points of X, there exists $\alpha \in I$ such that $x_\alpha \neq y_\alpha$ and there exists a continuous real-valued function h_α on X_α such that $h_\alpha(x_\alpha) \neq h_\alpha(y_\alpha)$. The function $x \to h_\alpha(\mathrm{pr}_\alpha x)$ then belongs to H and takes distinct values at x and y. Since every polynomial in the functions of H is of the form stated in the theorem, the result follows from Theorem 3.

If not all the X_ι are *compact*, the conclusion of Theorem 4 is not necessarily valid (cf. Exercise 9).

4. APPROXIMATION OF CONTINUOUS MAPPINGS OF A COMPACT SPACE INTO A NORMED SPACE

Let X be a compact space and let Y be a normed vector space over the field **R** (Chapter IX, § 3); the space $\mathcal{C}(X; Y)$ will always be assumed to carry the topology of uniform convergence defined by the norm $\|u\| = \sup_{x \in X} \|u(x)\|$ (§ 3, no. 2).

Given a set H of *continuous real-valued functions* defined on X, a finite family $(u_i)_{1 \leqslant i \leqslant n}$ of functions belonging to H, and a finite family $(\mathbf{a}_i)_{1 \leqslant i \leqslant n}$ of points of Y: the mapping $x \to \sum_{i=1}^{n} \mathbf{a}_i u_i(x)$ of X into Y is then *continuous*; we denote it by $\sum_{i=1}^{n} \mathbf{a}_i u_i$, and we say that it is a *linear combination* of functions of H with coefficients in Y. We say that a continuous mapping $f : X \to Y$ can be *uniformly approximated* by linear combinations of functions of H (with coefficients in Y), if f lies in the *closure* of the vector subspace of $\mathcal{C}(X; Y)$ formed by these linear combinations.

PROPOSITION 5. *Let* X *be a compact space,* Y *a normed space over* **R** *and* H *a subset of* $\mathcal{C}$ (X; **R**). *If every continuous real-valued function on* X *can be uniformly approximated by functions of* H, *then every continuous mapping* $\boldsymbol{f}$ *of* X *into* Y *can be uniformly approximated by linear combinations of functions of* H *with coefficients in* Y.

Given any real number $\varepsilon > 0$, for each $x \in X$ there exists an open neighbourhood of x in which the oscillation of $\boldsymbol{f}$ is $\leqslant \varepsilon$. Hence there is a finite open covering $(A_i)_{1 \leqslant i \leqslant n}$ of X such that the oscillation of $\boldsymbol{f}$ in each A_i is $\leqslant \varepsilon$. Let $\boldsymbol{a}_i$ be a value of $\boldsymbol{f}$ in A_i $(1 \leqslant i \leqslant n)$, and let $(u_i)_{1 \leqslant i \leqslant n}$ be a continuous partition of unity subordinate to the covering (A_i) (Chapter IX, § 4, no. 4, Corollary to Proposition 4). Let x be any point of X. For each index i such that $x \notin A_i$, we have $u_i(x) = 0$, and for each index i such that $x \in A_i$ we have $\|\boldsymbol{f}(x) - \boldsymbol{a}_i\| \leqslant \varepsilon$; it follows that

$$\left\| \boldsymbol{f}(x) - \sum_{i=1}^{n} \boldsymbol{a}_i u_i(x) \right\| = \left\| \sum_{i=1}^{n} (\boldsymbol{f}(x) - \boldsymbol{a}_i) u_i(x) \right\| \leqslant \varepsilon \sum_{i=1}^{n} u_i(x) = \varepsilon.$$

On the other hand, by hypothesis there is a function $v_i \in H$ such that

$$|u_i(x) - v_i(x)| \leqslant \frac{\varepsilon}{\sum_{j=1}^{n} \|\boldsymbol{a}_j\|}$$

for all $x \in X$ $(1 \leqslant i \leqslant n)$; hence we have

$$\left\| \boldsymbol{f}(x) - \sum_{i=1}^{n} \boldsymbol{a}_i v_i(x) \right\| \leqslant 2\varepsilon \quad \text{for all } x \in X,$$

and the proof is complete.

From Proposition 5 it follows that, to each of the propositions in which we have proved that a certain subset H of $\mathcal{C}$ (X; **R**) is dense, there corresponds an analogous proposition for continuous mappings of X into an arbitrary normed space Y. We shall write down explicitly only the proposition which corresponds in this way to Theorem 3. Given a set H of real-valued functions on X, a *polynomial in the functions of* H, *with coefficients in* Y, is defined to be any linear combination, with coefficients in Y, of products of a finite (possibly empty) family of functions belonging to H. Then:

PROPOSITION 6. *Let* X *be a compact space and let* H *be a set of continuous real-valued functions on* X *which separates the points of* X. *Then every continuous mapping of* X *into a normed space* Y *over* **R** *can be uniformly approximated by polynomials in the functions of* H *with coefficients in* Y.

From this we deduce:

PROPOSITION 7. *Let* X *be a compact space and let* H *be a set of continuous complex-valued functions on* X *which separates the points of* X. *Then every continuous mapping of* X *into a normed space* Y *over* **C** *can be uniformly approximated by polynomials in the functions* $f \in H$ *and their conjugates* $\bar{f}$ *with coefficients in* Y.

We have only to note that Y is also a normed space over **R** and to apply Proposition 6 to the set of real parts and imaginary parts of the functions $f \in H$, using the formulas

$$\mathfrak{R} f = \frac{1}{2}(f + \bar{f}), \qquad \mathfrak{I} f = \frac{1}{2i}(f - \bar{f}).$$

COROLLARY 1. *If* X *is a compact subset of the space* $\mathbf{C}^n$, *then every continuous mapping* $(z_1, z_2, \ldots, z_n) \to f(z_1, z_2, \ldots, z_n)$ *of* X *into a normed space* Y *over the field* **C** *can be uniformly approximated by polynomials in the* z_k *and* $\bar{z}_k$ *with coefficients in* Y.

We shall see later that in general it is not possible to approximate f uniformly by polynomials (with coefficients in Y) *in the variables* z_k *alone*, even if $Y = \mathbf{C}$.

COROLLARY 2. *Let* X *be a locally compact space and let* $\mathcal{C}_0(X)$ *be the normed* **C**-*algebra of continuous mappings of* X *into* **C** *which tend to* 0 *at infinity. Let* A *be a subalgebra of* $\mathcal{C}_0(X)$ *which separates the points of* X *and is such that* (i) $\bar{f} \in A$ *whenever* $f \in A$, (ii) *for each* $x \in X$, *there is an* $f \in A$ *such that* $f(x) \neq 0$. *Then* A *is dense in* $\mathcal{C}_0(X)$.

If X′ is the compact space obtained by adjoining a point at infinity ω to X, then $\mathcal{C}_0(X)$ can be identified with the subspace of $\mathcal{C}(X; \mathbf{C})$ consisting of continuous mappings which vanish at ω, the norm on $\mathcal{C}_0(X)$ being defined by

$$\|f\| = \sup_{x \in X} |f(x)| = \sup_{x \in X'} |f(x)|.$$

By virtue of Proposition 7, every $f \in \mathcal{C}_0(X)$ can be uniformly approximated by polynomials with complex coefficients in the functions belonging to A; moreover, since $f(\omega) = 0$, the argument of no. 2, Proposition 4 shows that we may suppose these polynomials to have no constant term, and then they belong to A.

Another application of Proposition 7 is the following :

PROPOSITION 8. *Let* P *be the set of all periodic continuous mappings of* $\mathbf{R}^m$ *into* $\mathbf{C}$ *whose group of periods contains* $\mathbf{Z}^m$. *Then every function belonging to* P *can be uniformly approximated in* $\mathbf{R}^m$ *by linear combinations, with complex coefficients, of functions of the form*

$$(x_1, x_2, \ldots, x_m) \to e(h_1x_1 + h_2x_2 + \cdots + h_mx_m),$$

where the h_i *are rational integers* (such linear combinations are called *trigonometric polynomials in* m *variables*).

We have only to observe that P (endowed with the topology of uniform convergence) is canonically isomorphic to the space of all continuous mappings of the compact space $\mathbf{T}^m$ into $\mathbf{C}$ (Chapter VII, § 1, no. 6), and apply Proposition 7 to the set of mappings of $\mathbf{T}^m$ into $\mathbf{C}$ which correspond to the m mappings $(x_1, x_2, \ldots, x_m) \to e(x_i)$ $(1 \leqslant i \leqslant m)$ of $\mathbf{R}^m$ into $\mathbf{C}$.

EXERCISES

§ 1

1) Let X be a set, let Y be a uniform space containing more than one point, and let $\mathfrak{S}$ be a non-empty set of non-empty subsets of X. Let $Y' \subset \mathcal{F}(X; Y)$ be the set of all constant mappings of X into Y. For each $y \in Y$, let c_y be the constant mapping of X into Y whose value is y.

a) Show that $y \to c_y$ is an isomorphism of Y onto the uniform subspace Y′ of $\mathcal{F}_{\mathfrak{S}}(X; Y)$.

b) $\mathcal{F}_{\mathfrak{S}}(X; Y)$ is Hausdorff if and only if Y is Hausdorff and $\mathfrak{S}$ is a covering of X.

c) Y′ is closed in $\mathcal{F}_{\mathfrak{S}}(X; Y)$ if and only if $\mathcal{F}_{\mathfrak{S}}(X; Y)$ is Hausdorff.

2) Let X be a set, let Y be a Hausdorff uniform space containing more than one point, and let $\mathfrak{S}_1$, $\mathfrak{S}_2$ be two sets of subsets of X which satisfy conditions (F'_I), (F'_{II}) of no. 2. Show that if $\mathfrak{S}_1 \subset \mathfrak{S}_2$ and $\mathfrak{S}_1 \neq \mathfrak{S}_2$, then the uniformity of $\mathfrak{S}_1$-convergence is strictly coarser than the uniformity of $\mathfrak{S}_2$-convergence. In particular:

(i) If X is a non-compact Hausdorff topological space, the uniformity of compact convergence is strictly coarser than the uniformity of uniform convergence.

(ii) If X is a Hausdorff topological space which has infinite compact subsets (cf. Chapter I, § 9, Exercise 4), then the uniformity of pointwise convergence is strictly coarser than the uniformity of compact convergence.

3) Let X be a set, let $\mathfrak{S}$ be a covering of X, let Y be a non-Hausdorff uniform space and let Y_0 be the Hausdorff uniform space associated with Y (Chapter II, § 3, no. 8). Show that the Hausdorff uniform space associated with $\mathcal{F}_{\mathfrak{S}}(X; Y)$ is isomorphic to $\mathcal{F}_{\mathfrak{S}}(X; Y_0)$.

4) Show that, on the set $\mathcal{C}(\mathbf{R}; \mathbf{R})$ of finite continuous real-valued functions defined on $\mathbf{R}$, the topology of uniform convergence with respect to the

additive uniformity of $\mathbf{R}$ is not the same as the topology of uniform convergence with respect to the uniformity induced on $\mathbf{R}$ by the (unique) uniformity of the extended line $\overline{\mathbf{R}}$ (although the topologies induced by these two uniformities on $\mathbf{R}$ are the same).

5) Let X be a topological space. A set $\mathfrak{S}$ of subsets of X is said to be *saturated* if it satisfies conditions (F'_I) and (F'_{II}) of no. 2 and if the closure of every set of $\mathfrak{S}$ belongs to $\mathfrak{S}$.

a) Show that if X is normal and if $\mathfrak{S}$ is saturated and covers X, then the set $\mathcal{C}(X; \mathbf{R})$ is dense in the space $\tilde{\mathcal{C}}_{\mathfrak{S}}(X; \mathbf{R})$ (no. 6, Theorem 2, Corollary 2) with respect to the topology of $\mathfrak{S}$-convergence. In particular, $\mathcal{C}(X; \mathbf{R})$ is dense in $\mathcal{F}_s(X; \mathbf{R})$. $\mathcal{C}(X; R)$ is closed in $\mathcal{F}_s(X; \mathbf{R})$ only if X is discrete.

b) Let X be a completely regular space (Chapter IX, § 1, no. 5) and let $\mathfrak{S}_1, \mathfrak{S}_2$ be two saturated sets of subsets of X, such that $\mathfrak{S}_1 \subset \mathfrak{S}_2$ and $\mathfrak{S}_1 \neq \mathfrak{S}_2$. Show that the topology of $\mathfrak{S}_1$-convergence on $\mathcal{C}(X; \mathbf{R})$ is strictly coarser than the topology of $\mathfrak{S}_2$-convergence.

6) *a*) Let X be a topological space, Y a uniform space, and let Φ be a filter on the set $\mathcal{C}(X; Y)$ which converges pointwise to a function u_0. Then u_0 is continuous at a point $x_0 \in X$ if and only if, for each entourage V of Y and each set $M \in \Phi$, there is a neighbourhood U of x_0 and a mapping $u \in M$ such that $(u_0(x), u(x)) \in V$ for all $x \in U$.

b) Let X be a quasi-compact space, Y a uniform space and Φ a filter on $\mathcal{C}(X; Y)$ which converges pointwise to a function u_0. Then u_0 is continuous on X if and only if, for each entourage V of Y and each set $M \in \Phi$, there exists a finite number of functions $u_i \in M\ (1 \leqslant i \leqslant n)$ such that, for each $x \in X$, we have $(u_0(x), u_i(x)) \in V$ for at least one index i.

¶ 7) Let X, Y be two Hausdorff uniform spaces. For each continuous mapping $f: X \to Y$, let $G(f) \subset X \times Y$ be the graph of f; $G(f)$ is closed in $X \times Y$ (Chapter I, § 8, no. 1, Proposition 2, Corollary 2); hence $f \to G(f)$ is an injective mapping of $\mathcal{C}(X; Y)$ into the set $\mathfrak{F}(X \times Y)$ of non-empty closed subsets of $X \times Y$.

a) Show that the mapping $f \to G(f)$ of $\mathcal{C}_u(X; Y)$ into $\mathfrak{F}(X \times Y)$ is uniformly continuous when $\mathfrak{F}(X \times Y)$ carries the uniformity defined in Chapter II, § 1, Exercise 5.

b) Let Γ be the image of $\mathcal{C}(X; Y)$ in $\mathfrak{F}(X \times Y)$ under the mapping $f \to G(f)$, and let $\varphi : \Gamma \to \mathcal{C}_u(X; Y)$ be the inverse mapping. Show that if X is compact, then φ is continuous on Γ (argue by contradiction).

c) Take both X and Y to be the compact interval $[0, 1]$ of $\mathbf{R}$. Show that φ is not uniformly continuous on Γ.

¶ 8) Let X, Y be two metric spaces and let (f_n) be a sequence of Borel mappings of class α of X into Y (Chapter IX, § 6, Exercise 16).

a) Suppose that the sequence (f_n) converges pointwise to a mapping f: $X \to Y$. Show that f is of class $\alpha + 1$. [If U is an open subset of Y, note that $\overset{-1}{f}(U) = \bigcup_{n \geqslant 1} \left(\bigcap_{k \geqslant 0} \overset{-1}{f}_{n+k}(U) \right)$, and use Chapter IX, § 6, Exercise 4.]

b) Suppose that the sequence (f_n) converges uniformly to f. Show that f is of class α. [Let F be a closed subset of Y, and for each integer $n > 0$, let V_n be the set of points of Y whose distance from F is $\leqslant 1/n$. Show that there is an increasing sequence $n \to m(n)$ of integers such that $\overset{-1}{f}(F) = \bigcap_n \overset{-1}{f}_{m(n)}(V_n)$.]

c) Suppose that Y is of countable type. Show that, if $f: X \to Y$ is a Borel function of class $\alpha > 0$, there exists a sequence of Borel functions $g_n: X \to Y$, of class $< \alpha$, such that (g_n) converges pointwise to f. [Show that we may take the g_n to be functions which take only a finite number of values; use Exercises 4 *h*) and 16 *c*) of Chapter IX, § 6.]

¶ 9) Let X be a Baire space, Y a metric space and (f_n) a sequence of continuous mappings of X into Y which converges pointwise to a function f. A point $x \in X$ is said to be a *point of uniform convergence* of the sequence (f_n) if, for each $\varepsilon > 0$, there exists a neighbourhood V of x and an integer p such that $d(f_m(y), f(g_n)) \leqslant \varepsilon$ for all $y \in V$ and all integers $m \geqslant p$, $n \geqslant p$, d being the metric on Y. Show that the complement S of the set of points of uniform convergence of the sequence (f_n) is meagre in X [cf. Chapter IX, § 5, Exercise 22 *a*).] Give an example where S is dense in X. [Take $X = Y = \mathbf{R}$, let $n \to r_n$ be a bijection of $\mathbf{N}$ onto $\mathbf{Q}$, and let (g_n) be a sequence of continuous mappings of $\mathbf{R}$ into $[0, 1]$ which converges to the zero function for $x \neq 0$ and to 1 at $x = 0$, the convergence being uniform in every open set of $\mathbf{R}$ which does not contain 0. Then consider the sequence of functions $f_n(x) = \sum_{p=0}^{\infty} \alpha_p g_n(x - r_p)$, where the sequence (α_p) tends to 0 in a suitable way.]

10) Show that, on the group Γ of homeomorphisms of the real line $\mathbf{R}$ onto itself, the topology of pointwise convergence is the same as the topology of compact convergence (cf. § 3, Exercise 14).

11) Let X be a topological space and G a topological group. The set $\mathcal{C}(X; G)$ of continuous mappings of X into G is a subgroup of the group G^X. Let $\mathfrak{S}$ be a set of subsets of X.

a) Suppose that for each $A \in \mathfrak{S}$, each neighbourhood V of e in G and each $u \in \mathcal{C}(X; G)$, there exists a neighbourhood W of e in G such that $sWs^{-1} \subset W$ for all $s \in u(A)$. Show that the topology of $\mathfrak{S}$-convergence is then compatible with the group structure of $\mathcal{C}(X; G)$, and that the right (resp. left) uniformity of the topological group $\mathcal{C}_{\mathfrak{S}}(X; G)$ so defined is the same as the uniformity of $\mathfrak{S}$-convergence. Consider the case of pointwise convergence, the case of compact convergence when G is locally compact, and the case where G is abelian.

b) If G is the group $\mathbf{SL}(2, \mathbf{R})$ endowed with the topology induced by that of $\mathbf{R}^4$, show that the topology of uniform convergence is not compatible with the group structure of $\mathcal{C}(\mathbf{R}; G)$.

12) Let X be a topological space and let A be a topological ring (Chapter III, § 6, no. 3). The set $\mathcal{C}(X; A)$ of continuous mappings of X into A is a subring of the ring A^X. Let $\mathfrak{S}$ be a set of subsets of X.

a) Suppose that, for each $M \in \mathfrak{S}$ and each $u \in \mathcal{C}(X; A)$, the set $u(M)$ is bounded (Chapter III, § 6, Exercise 12). Show that the topology of $\mathfrak{S}$-convergence is compatible with the ring structure of $\mathcal{C}(X; A)$. Consider the case of pointwise convergence and the case of compact convergence.

b) Let X be a locally compact σ-compact space which is not compact (Chapter I, § 9, no. 9). Show that the topology of uniform convergence is not compatible with the ring structure of $\mathcal{C}(X; \mathbf{R})$.

§ 2

1) Let f be the real-valued function on $\mathbf{R}$ which is equal to 0 for $x \leqslant 0$, to x for $0 \leqslant x \leqslant 1$, and to 1 for $x \geqslant 1$. Show that the sequence of real-valued functions $f_n(x) = f(nx - n^2)$ $(n \in \mathbf{N})$ is equicontinuous but not uniformly equicontinuous on $\mathbf{R}$, although it is formed of uniformly continuous functions.

2) Let X be a Hausdorff topological space in which every point admits a countable fundamental system of neighbourhoods, and let Y be a uniform space. Let H be a subset of $\mathcal{C}(X; Y)$ with the property that, for each compact subset K of X, the set $H|K$ of restrictions to K of mappings $u \in H$ is an equicontinuous subset of $\mathcal{C}(K; Y)$. Show that H is equicontinuous (argue by contradiction).

3) Let X, Y, Z be three metric spaces and let H be a subset of $\mathcal{C}(X \times Y; Z)$. Suppose that, for each $x_0 \in X$, the set of mappings $u(x_0, .)$, as u runs through H, is an equicontinuous subset of $\mathcal{C}(Y; Z)$, and that, for each $y_0 \in Y$, the set of mappings $u(., y_0)$, as u runs through

H, is an equicontinuous subset of $\mathcal{C}(X; Z)$. Show that if X is *complete*, then for each $b \in Y$ there is a set S_b in X whose complement is meagre in X and which is such that, for each $a \in S_b$, the set H is equicontinuous at (a, b). (Apply Chapter IX, § 5, Exercise 23, using Chapter X, § 2, Proposition 1, Corollary 1.)

4) Let X, Y, Z be three uniform spaces.

a) Let H be a uniformly equicontinuous subset of $\mathcal{C}(Y; Z)$. If H, $\mathcal{C}(X; Y)$ and $\mathcal{C}(X; Z)$ are endowed with the uniformity of uniform convergence, show that the mapping $(u, v) \to u \circ v$ of $H \times \mathcal{C}(X; Y)$ into $\mathcal{C}(X; Z)$ is uniformly continuous.

b) Let K be a uniformly equicontinuous subset of $\mathcal{C}(X; Y)$, and let L be a uniformly equicontinuous subset of $\mathcal{C}(Y; Z)$. Show that the set of mappings $v \circ u$, where u runs through K and v runs through L, is a uniformly equicontinuous subset of $\mathcal{C}(X; Z)$.

5) Let X be a topological space and let Y be a normed space over a non-discrete valued division ring K. Let H be a subset of $\mathcal{F}(X; Y)$, equicontinuous at a point $x_0 \in X$, and let $k > 0$ be a real number. Then the set H_k of linear combinations $\sum_i c_i u_i$ of functions $u_i \in H$, such that $\sum_i |c_i| \leqslant k$, is equicontinuous at the point x_0.

6) Let X be a topological space, Y a uniform space, H an equicontinuous subset of $\mathcal{C}(X; Y)$, and Φ a filter on H. Show that the set of all $x \in X$ such that $\Phi(x)$ is a Cauchy filter base on Y is closed in X.

7) Let X be a topological space, Y a complete Hausdorff uniform space, and let φ be a uniformly continuous homeomorphism of Y onto an open subset $\varphi(Y)$ of a Hausdorff uniform space Y'. Let H be an equicontinuous subset of $\mathcal{C}(X; Y)$, and let H' be the set of mappings $\varphi \circ u$, where $u \in H$. If v lies in the closure of H' in $\mathcal{F}_s(X; Y')$, show that $\overset{-1}{v}(\varphi(Y))$ is both open and closed in X. In particular, if X is connected, $v(X)$ is contained in $\varphi(Y)$ or in a connected component of $\complement\varphi(Y)$. (Observe that v is continuous, and use Exercise 6.)

8) Let H be an equicontinuous set of mappings of a topological space X into $\mathbf{R}$.

a) Show that the set of upper (resp. lower) envelopes of finite subsets of H is equicontinuous.

b) Let v be a mapping of X into $\overline{\mathbf{R}}$ which lies in the closure of H with respect to the topology of pointwise convergence. Show that v

is continuous on X and that the sets $\overset{-1}{v}(+\infty)$ and $\overset{-1}{v}(-\infty)$ are both open and closed in X (use Exercise 7).

c) Deduce from *a*) and *b*) that the upper (resp. lower) envelope w (resp. v) of H is continuous on X and that $\overset{-1}{w}(+\infty)$ [resp. $\overset{-1}{v}(-\infty)$] is both open and closed in X.

d) Suppose that X is *connected* and that there is a point $x_0 \in X$ such that $H(x_0)$ is bounded above in **R**. Show that, for every compact subset K of X, the set of restrictions to K of functions of H is uniformly bounded above in K [use *c*)].

9) Let X be a set, Y a uniform space and $\mathfrak{S}$ a covering of X. A subset H of the uniform space $\mathcal{F}_{\mathfrak{S}}(X; Y)$ is precompact if and only if (i) for each $x \in X$, $H(x)$ is precompact in Y; (ii) the uniformity of $\mathfrak{S}$-convergence and that of pointwise convergence coincide on H. (Apply Theorem 2 by considering X as a discrete space and reducing to the case where Y is Hausdorff and complete.)

10) *a*) Let X be a compact space, let (f_n) be a sequence of continuous mappings $X \to X$, which converges pointwise but not uniformly in X to a continuous function (§ 1, no. 6, Remark 2). Show that the set of mappings f_n is relatively compact in $\mathcal{C}_s(X; X)$ but not in

$$\mathcal{C}_u(X;\ X) = \mathcal{C}_c(X;\ X).$$

b) Let I be the interval $[-1, +1]$ of **R** and for each integer $n > 0$ let $u(x_n) = \sin\sqrt{x + 4n^2\pi^2}$ for all $x \geqslant 0$. Show that the set H of functions u_n is an equicontinuous subset of $\mathcal{C}(\mathbf{R}_+; I)$ and is relatively compact in $\mathcal{C}_c(\mathbf{R}_+; I)$ but not in $\mathcal{C}_u(\mathbf{R}_+; I)$. (Note that the sequence (u_n) converges pointwise to 0.)

11) Let X be a completely regular space and let $\mathfrak{S}$ be a set of subsets of X whose interiors cover X. Show that the space $\mathcal{C}_s(X; \mathbf{R})$ is not locally compact.

12) Let X be a topological space, Y a uniform space, H an equicontinuous subset of $\mathcal{C}(X; Y)$ and V a symmetric entourage of Y. Given a point $x \in X$, show that the set of points $x' \in X$ for which there is an integer n (depending on x') such that $H(x') \subset \overset{n}{V}(H(x))$ is both open and closed in X. Deduce that, if K is any compact connected subset of X and if x_0 is any point of K, there is an integer $n > 0$ such that $H(K) \subset \overset{n}{V}(H(x_0))$.

¶ 13) Let X be a topological space, let Y be a locally compact uniform space and let H be an equicontinuous subset of $\mathcal{C}(X; Y)$.

a) Let A be the set of all $x \in X$ such that $H(x)$ is relatively compact in Y. Show that A is open in X (cf. Chapter I, § 9, no. 7, Proposition 10 and Chapter II, § 4, no. 3, Proposition 4).

b) Suppose in addition that Y is complete with respect to its uniformity; then A is also closed in X [observe that if $x_0 \in \overline{A}$, then $H(x_0)$ is relatively compact in Y by considering an ultrafilter on $H(x_0)$ as the image of an ultrafilter on H]. In this case, if X is connected, then H is relatively compact in $\mathcal{C}_c(X; Y)$ if and only if, for *one* point $x_0 \in X$, the set $H(x_0)$ is relatively compact in Y.

c) Take X to be the compact interval [0, 1] of **R**, and take Y to be the interval]0, 1[endowed with the uniformity induced by that of **R**. Give an example of an equicontinuous subset H of $\mathcal{C}(X; Y)$ such that the set A defined in *a*) is the interval]0, 1].

d) Suppose that $X = Y$ and that H consists of homeomorphisms of X onto itself. Show that if H is uniformly equicontinuous, then the set A defined in *a*) is both open and closed in X.

14) Let Γ be an equicontinuous group of homeomorphisms of **R**. Show that if a homeomorphism $u \in \Gamma$ leaves at least one point of **R** fixed, and if u is increasing, then u is the identity mapping (show that otherwise the group generated by u is not equicontinuous at a suitably chosen fixed point of u). If u is decreasing, then u^2 is the identity mapping.

¶ 15) Let X be a compact metrizable space, let Γ be the group of all homeomorphisms of X, and let G be a subgroup of Γ which operates *transitively* on X. If H is the centralizer of G in Γ, show that H is equicontinuous. [Argue by *reductio ad absurdum*, supposing that H is not equicontinuous at some point $a \in X$; deduce the existence of a sequence of points $x_n \in X$ and a sequence of elements $u_n \in H$ such that

$$\lim_{n\to\infty} x_n = a, \qquad \lim_{n\to\infty} u_n(a) = b, \qquad \lim_{n\to\infty} u_n(x_n) = c,$$

where $b \neq c$. Hence show that the sequence (u_n) converges pointwise in X, but that no point of X is a point of uniform convergence of this sequence; this contradicts Exercise 9 of § 1.]

16) Let X be a Hausdorff uniform space and let Γ be an equicontinuous group of homeomorphisms of X. The equivalence relation R defined by Γ on X is open (Chapter I, § 5, no. 2).

a) Show that if every orbit of Γ is closed in X, then the orbit space X/Γ is Hausdorff.

b) Show that if every orbit of Γ is compact, then the relation R is closed (use Chapter I, § 5, no. 4, Proposition 10).

c) Give an example where X is compact but none of the orbits of Γ is closed in X (cf. Chapter III, § 2, Exercise 29).

¶ 17) Let X be a compact space and Γ a *countable* group of homeomorphisms of X. Suppose that the orbit space X/Γ is *Hausdorff* (and therefore *compact*).

a) Show that, for each $x \in X$, the orbit $\Gamma(x)$ of x is *finite* (observe that if it were infinite it would have no isolated point, and use Baire's theorem (Chapter IX, § 5, no. 3, Theorem 1)).

b) For each $x_0 \in X$, let $\Delta(x_0)$ be the normal subgroup of Γ consisting of homeomorphisms which leave fixed all the points of the orbit $\Gamma(x_0)$. Show that if $\Delta(x_0)$ is finitely generated, then Γ is equicontinuous at x_0. [Let f_i $(1 \leqslant i \leqslant m)$ be the generators of $\Delta(x_0)$, and let g_k $(1 \leqslant k \leqslant n)$ be representatives of each of the cosets of $\Delta(x_0)$ in Γ other than $\Delta(x_0)$ itself. Take a neighbourhood V of x_0 such that none of the sets $f_i(V)$, $f_i^{-1}(V)$ meets any of the sets $g_k(V)$, and then use the fact that the equivalence relation defined by Γ is closed (Chapter I, § 10, no. 4, Proposition 8).]

c) Without making any hypothesis on $\Delta(x_0)$, suppose that x_0 has a countable fundamental system of *connected* neighbourhoods in X. Show that Γ is equicontinuous at x_0 [same method as in *b*)]. Deduce that if X is locally connected, then Γ is equicontinuous.

d) Take X to be the compact subspace of **R** consisting of the points $0, 1, 1/n$ and $1 + 1/n$ (n an integer $\geqslant 2$). Give an example of a countable group Γ of homeomorphisms of X, such that X/Γ is Hausdorff but Γ is not equicontinuous.

¶ 18) Let G be a Hausdorff topological group operating continuously on a Hausdorff topological space X. G is said to be *proper at a point* $x_0 \in X$ if the orbit $G.x_0$ is *closed* in X and if G operates properly on $G.x_0$ (Chapter III, § 4, no. 1).

a) Let G be a locally compact topological group operating continuously on a Hausdorff uniform space X. Suppose that the set of homeomorphisms $x \to s.x$ of X, as s runs through G, is *equicontinuous*. Show that if G is proper at a point $x_0 \in X$, then there is a neighbourhood V of x_0 such that the set of elements $s \in G$ such that $s.V \cap V \neq \emptyset$ is relatively compact in G (cf. Chapter III, § 4, no. 4, Proposition 7). Deduce that the set D of points of X at which G is proper is open in X and that G operates properly on D. If in addition X is locally compact, show that D is both open and closed in X (use no. 1, Proposition 1, Corollary 2).

b) In the real plane $\mathbf{R}^2$, let E be the set consisting of the origin (0, 0) and the points $(0, 2^{-n})$ as n runs through the integers $\geqslant 0$. For each $n \in \mathbf{Z}$ such that $n \neq 0$ let u_n be the restriction to E of an affine linear mapping of $\mathbf{R}^2$ into itself, such that $u_n(0, 0) = (n, 0)$, and $u_n(0, 2^{-|n|}) = (0, \theta^n)$ where θ is an irrational number such that $0 < \theta < 1$. Let X be the (locally compact) subspace of $\mathbf{R}^2$ which is the union of E and the sets $u_n(\mathrm{E})$ for $n \in \mathbf{Z}, n \neq 0$. Furthermore, let u_0 denote the identity mapping of E, and extend u_n $(n \in \mathbf{Z})$ to the whole of X by putting $u_n(u_m(x)) = u_{n+m}(x)$ for all $x \in \mathrm{E}$ and all $m \in \mathbf{Z}$. The set G of mappings u_n is a group of homeomorphisms of X which is given a discrete topology so that G operates continuously on X. Show that G is proper at every point of X, but is not equicontinuous and does not operate properly on X.

c) Let X be the (not locally compact) subspace of $\mathbf{R}^2$ (identified with $\mathbf{C}$) which is the union of the half-plane $y > 0$ and the origin. Define a homeomorphism u of X onto itself by putting $u(0, 0) = (0, 0)$ and $u(re^{i\omega}) = re^{i\omega'}$ where $\omega' = \frac{1}{2}\omega$ if $0 < \omega \leqslant \frac{1}{2}\pi$, $\omega' = \omega - \frac{1}{4}\pi$ for $\frac{1}{2}\pi \leqslant \omega \leqslant \frac{3}{4}\pi$, $\omega' = 2\omega - \pi$ for $\frac{3}{4}\pi \leqslant \omega < \pi$ $(r < 0, 0 < \omega < \pi)$. Let G be the group of homeomorphisms of X generated by u, and endow G with the discrete topology. Show that G is equicontinuous on X and that the set of points of X at which G is proper is the half-plane $y > 0$.

19) Let X be a Hausdorff uniform space and let G be a group of homeomorphisms of X onto itself.

a) Show that if G is equicontinuous and discrete with respect to the topology of pointwise convergence, then G is closed in $\mathcal{F}_s(\mathrm{X}; \mathrm{X})$.

b) Show that if G, endowed with the discrete topology, is proper at one point of X at least (Exercise 18), then G is closed in $\mathcal{F}_s(\mathrm{X}; \mathrm{X})$ and the topology of pointwise convergence induces the discrete topology on G.

c) Let X be the topological sum of two spaces X_1, X_2, each homeomorphic to $\mathbf{R}$, so that X can be identified with the product space $\mathbf{R} \times \{1, 2\}$. Endow X with the product uniformity. For each pair (m, n) of rational integers let u_{mn} denote the homeomorphism of X defined by $u_{mn}(x_1) = x_1 + m\alpha + n\beta$, $u_{mn}(x_2) = x_2 + m\gamma + n\delta$, where $x_1 \in \mathrm{X}_1$ and $x_2 \in \mathrm{X}_2$, and $\alpha, \beta, \gamma, \delta$ are four non-zero real numbers such that α/β and γ/δ are irrational and distinct. Show that the u_{mn} form a group of homeomorphisms of X which is equicontinuous and discrete with respect to the topology of pointwise convergence, but that G (with the discrete topology) is not proper at any point of X.

20) Let G be a group of homeomorphisms of a Hausdorff uniform space X. Suppose that G is equicontinuous and that G, endowed with the discrete topology, operates properly on X. Suppose, furthermore, that the set of points $x \in X$, which are not fixed points of any homeomorphism $u \in G$ other than the identity, is dense in X. Under these conditions, show that there is an open set F in X such that $F \cap U(F) = \emptyset$ for all $u \in G$ other than the identity, and such that the canonical image of F in the orbit space X/G is a dense open subset of X/G, homeomorphic to F (use Exercise 16 and Zorn's lemma).

§ 3

1) Let X be a topological space and Y a metric space. Show that the set of all mappings of X into Y whose oscillation (Chapter IX, § 2, no. 3) at every point of X is $\leqslant \alpha$ (where α is a given real number > 0) is closed in the space $\mathcal{F}_u(X; Y)$.

2) Let X be a set; show that the mapping $u \to \sup_{x \in X} u(x)$ of $\mathcal{B}(X; \mathbf{R})$ into $\mathbf{R}$ is continuous.

3) Let X be a metric space, d the metric on X. For each $x \in X$ let d_x be the real-valued function $y \to d(x, y)$ which is continuous on X. Show that the mapping $x \to d_x$ is an *isometry* of X onto a subspace of $\mathcal{C}_u(X; \mathbf{R})$ [endowed with the pseudometric $\delta(u, v) = \sup_{x \in X} |u(x) - v(x)|$].

4) A completely regular space X is such that the metrizable space $\mathcal{C}_u(X; \mathbf{R})$ is of countable type if (and only if) X is compact and metrizable. [By considering the Stone-Čech compactification of X (Chapter IX, § 1, Exercise 7) show that X must be metrizable, and then observe that the space $\mathcal{C}_u(\mathbf{Z}; \mathbf{R})$ is not of countable type.]

5) Let X be a completely regular space.

a) If every point of the space $\mathcal{C}_c(X; \mathbf{R})$ has a countable fundamental system of neighbourhoods, then there exists an increasing sequence (K_n) of compact subsets of X such that every compact subset of X is contained in some K_n; and $\mathcal{C}_c(X; Y)$ is then metrizable for any metrizable space Y.

b) $\mathcal{C}_c(X; \mathbf{R})$ is metrizable and of countable type if and only if all the compact subspaces K_n are metrizable (use Exercise 4). For every metrizable space Y of countable type, $\mathcal{C}_c(X; Y)$ is then metrizable and of countable type.

¶ 6) *a*) Let X be a topological space, Y a metric space, d the metric on Y, n an integer > 0 and S a subset of the space $X^n \times Y^n \times \mathbf{R}_+^*$. For each point $z = ((x_i), (y_i), r) \in S$, let U_z be the open subset of $\mathcal{C}_s(X; Y)$ consisting of continuous mappings f of X into Y such that

$d(f(x_i), y_i) < r$ for $1 \leqslant i \leqslant n$. Show that, if D is a dense subset of S, we have $\bigcup_{z \in S} U_z = \bigcup_{z \in D} U_z$.

b) Deduce from *a*) that if the topologies of X and Y have countable bases, then every subspace of $\mathcal{C}_s(X; Y)$ is a Lindelöf space (Chapter I, § 9, Exercise 15) and is therefore paracompact (Chapter IX, § 4, Exercise 23).

7) Let X be a space which has a countable dense subset, let Y be a Hausdorff space in which every point has a countable fundamental system of neighbourhoods (resp. a metrizable space) and let H be a subset of $\mathcal{C}(X; Y)$. Show that if $\mathcal{T}$ is a topology on H which is finer than the topology of pointwise convergence in X and with respect to which H is compact, then every point of H has a countable fundamental system of neighbourhoods in the topology $\mathcal{T}$ (resp. $\mathcal{T}$ is metrizable). (Note that if D is any countable dense subset of X, then the topology $\mathcal{T}$ is finer than the topology of pointwise convergence in D and that the latter topology is Hausdorff.)

8) *a*) Let f be the continuous mapping $(x, y) \to xy$ of $\mathbf{R} \times \mathbf{R}$ into $\mathbf{R}$. Show that the mapping $x \to f(x, .)$ of $\mathbf{R}$ into $\mathcal{C}_u(\mathbf{R}; \mathbf{R})$ is not continuous.

b) Let X, Y be two topological spaces, let Z be a uniform space and let $f: X \times Y \to Z$ be a mapping. Show that if $f(x, .)$ is continuous on Y for all $x \in X$ and if the mapping $x \to f(x, .)$ of X into $\mathcal{C}_u(Y; Z)$ is continuous, then f is continuous on $X \times Y$.

¶ 9) *a*) Let X, Y, Z be three topological spaces. Suppose that X and Y are Hausdorff and that every point of each of these spaces has a countable fundamental system of neighbourhoods. Let f be a mapping of $X \times Y$ into Z such that $f(x, .)$ is continuous on Y for each $x \in X$ and such that the mapping $x \to f(x, .)$ of X into $\mathcal{C}_c(Y; Z)$ is continuous. Show that f is continuous on $X \times Y$ (cf. § 1, no. 6, Theorem 2, Corollary 3).

b) Show that the conclusion of *a*) holds good when the hypothesis on X is replaced by the hypothesis that X is locally compact and Z is a uniform space (use § 2, Exercise 2, and Ascoli's theorem) [cf. Exercise 11 *b*)].

¶ 10) Let X, Y be two topological spaces. For each subset A of X and each subset B of Y let $T(A, B)$ denote the set of all $u \in \mathcal{C}(X; Y)$ such that $u(A) \subset B$.

Let $\mathfrak{U} = (U_\alpha)$ be an open covering of X. Let $\mathcal{T}_{\mathfrak{U}}$ denote the topology on $\mathcal{C}(X; Y)$ generated by the sets $T(F, V)$, where V runs through the set of all open sets of Y and F runs through the set of closed subsets of X contained in at least one U_α.

a) Show that $\mathfrak{T}_{\mathfrak{U}}$ is finer than the compact-open topology (use Theorem 3 of Chapter IX, § 4, no. 3). If X is regular, the mapping $(u, x) \to u(x)$ of $\mathcal{C}(X; Y) \times X$ into Y is continuous when $\mathcal{C}(X, Y)$ carries the topology $\mathfrak{T}_{\mathfrak{U}}$.

b) Let $\mathfrak{T}$ be a topology on $\mathcal{C}(X; Y)$ with respect to which the mapping $(u, x) \to u(x)$ of $\mathcal{C}(X; Y) \times X$ into Y is continuous. Let u_0 be a continuous mapping of X into Y, let x_0 be a point of X and let V be an open set in Y which contains $u_0(x_0)$. Show that there exists an open neighbourhood U of x_0 in X such that $T(U, V)$ is a neighbourhood of u_0 with respect to $\mathfrak{T}$.

c) Suppose that X is completely regular. Show that if, among the topologies $\mathfrak{T}$ on $\mathcal{C}(X; \mathbf{R})$ for which $(u, x) \to u(x)$ is continuous, there is a topology $\mathfrak{T}_0$ coarser than all the others, then $\mathfrak{T}_0$ must be the compact-open topology. [Let $T(U, V)$ be a neighbourhood of 0 in $\mathcal{C}(X; \mathbf{R})$ with respect to $\mathfrak{T}_0$. Let (W_α) be any open covering of $\overline{U}$; let $\mathfrak{W}$ be the covering of X formed by the sets W_α and $\complement\overline{U}$. Using *a*) and arguing by contradiction, show that a finite number of the W_α cover $\overline{U}$.]

d) Show that, if X is completely regular but not locally compact, the mapping $(u, x) \to u(x)$ of $\mathcal{C}_c(X; \mathbf{R}) \times X$ into $\mathbf{R}$ is not continuous at every point. [Consider a point x_0 which has no compact neighbourhood and argue by contradiction, using *b*).]

¶ 11) Let X, Y be two non-empty topological spaces whose product $X \times Y$ is normal. Let $\mathfrak{T}$ be a topology on the set $\mathcal{C}(X; \mathbf{R})$ such that, for every continuous real-valued function f on $X \times Y$, the mapping $y \to f(.,y)$ of Y into $\mathcal{C}(X; \mathbf{R})$ is continuous.

a) Let y_0 be a limit of a sequence (z_n) of points of Y all distinct from y_0, let A be a countably infinite closed subset of X, all of whose points are isolated, and let I be a bounded open interval in $\mathbf{R}$. Show that, with respect to the topology $\mathfrak{T}$, the set $T(A, I)$ (Exercise 10) has no interior point. [If $u_0 \in T(A, I)$, construct a continuous mapping $f: X \times Y \to \mathbf{R}$ such that $f(.,y_0) = u_0$ and that $f(., z_n) \notin T(A, I)$ for all $z_n \neq y_0$.]

b) Hence show that if, in addition, the mapping $(u, x) \to u(x)$ of $\mathcal{C}(X; \mathbf{R}) \times X$ into $\mathbf{R}$ is continuous with respect to the topology $\mathfrak{T}$, if X is locally paracompact (Chapter IX, § 4, Exercise 27), and if there exists in X a sequence of points (x_n) which converges to a point distinct from all the x_n, then, X must be locally compact [use Exercise 10 *b*) and Chapter IX, § 4, Exercise 25 *c*)].

c) Deduce from *b*) that if X is metrizable but not locally compact, then there are points of $\mathcal{C}_c(X; \mathbf{R})$ which have no countable fundamental system of neighbourhoods [use Exercise 9 *a*)]. Give a direct proof, using Exercise 5 *a*).

12) Let X be a topological space, Y and Z two uniform spaces, $\mathfrak{S}$ a set of subsets of X and $\mathfrak{T}$ a set of subsets of Y.

a) Let $u_0 \in \mathcal{C}(X; Y)$. Show that if, for each $A \in \mathfrak{S}$, $u_0(A)$ is contained in some set of $\mathfrak{T}$, then the mapping $v \to v \circ u_0$ of $\mathcal{C}_{\mathfrak{T}}(Y; Z)$ into $\mathcal{C}_{\mathfrak{S}}(X; Z)$ is continuous.

b) Let H be a subspace of $\mathcal{C}_{\mathfrak{S}}(X; Y)$ and let $v_0 \in \mathcal{C}(Y; Z)$. Show that if v_0 is uniformly continuous on H(A) for all $A \in \mathfrak{S}$, then the mapping $u \to v_0 \circ u$ of H into $\mathcal{C}_{\mathfrak{S}}(X; Z)$ is continuous.

c) Let $u_0 \in \mathcal{C}(X; Y)$ and let $v_0 \in \mathcal{C}(Y; Z)$. Suppose that, for each $A \in \mathfrak{S}$, there exists $B \in \mathfrak{T}$ and an entourage V of Y satisfying the conditions: (i) $V(u_0(A)) \subset B$, (ii) v_0 is uniformly continuous on B. Then the mapping $(u, v) \to v \circ u$ of $\mathcal{C}_{\mathfrak{S}}(X; Y) \times \mathcal{C}_{\mathfrak{T}}(Y; Z)$ into $\mathcal{C}_{\mathfrak{S}}(X; Z)$ is continuous at (u_0, v_0). In particular, the mapping $(u, v) \to v \circ u$ of $\mathcal{C}_u(X; Y) \times \mathcal{C}_u(Y; Z)$ into $\mathcal{C}_u(X; Z)$ is continuous at every point (u_0, v_0) such that v_0 is uniformly continuous on Y.

d) Let v_0 be the homeomorphism $x \to x^3$ of **R** onto itself. Show that the mapping $u \to v_0 \circ u$ of $\mathcal{C}_u(\mathbf{R}; \mathbf{R})$ into itself is not continuous at every point.

¶ 13) Let X, Y, Z be three Hausdorff topological spaces.

a) Show that for all $u_0 \in \mathcal{C}(X; Y)$ and all $v_0 \in \mathcal{C}(Y; Z)$, the mappings $v \to v \circ u_0$ of $\mathcal{C}_c(Y; Z)$ into $\mathcal{C}_c(X; Z)$ and $u \to v_0 \circ u$ of $\mathcal{C}_c(X; Y)$ into $\mathcal{C}_c(X; Z)$ are continuous.

b) Suppose that every point of Y (resp. Z) has a countable fundamental system of neighbourhoods and that X has a countable dense subset. Let H be a compact subset of $\mathcal{C}_c(X; Y)$. Show that the mapping $(u, v) \to v \circ u$ of $H \times \mathcal{C}_c(Y; Z)$ into $\mathcal{C}_c(X; Z)$ is continuous. [Start by showing that if K is any compact subset of X, then H(K) is a compact subset of Y, by using Exercises 7 and 9 *a*).]

c) Show that the mapping $(v, v) \to v \circ u$ of $\mathcal{C}_c(\mathbf{Q}; \mathbf{Q}) \times \mathcal{C}_c(\mathbf{Q}; \mathbf{Q})$ into $\mathcal{C}_c(\mathbf{Q}; \mathbf{Q})$ is not continuous at any point.

14) Let Γ be the group of all homeomorphisms of the real plane $\mathbf{R}^2$, endowed with the topology of *pointwise* convergence. Show that the mapping $(u, v) \to v \circ u$ of $\Gamma \times \Gamma$ into Γ is not continuous. [Consider a sequence of homeomorphisms (v_n) such that v_n leaves fixed every point (x, y) for which $y \notin [1/(n+1), 1/n]$, and such that the restriction

of v_n to the line $y = 2/(2n + 1)$ is the translation $(x, y) \to (x + 1, y)$; on the other hand, consider the sequence (u_n) of translations

$$(x, y) \to (x, y + 2/(2n + 1)).]$$

15) *a*) Let X be a uniform space, let $\mathfrak{S}$ be a set of subsets of X, let Γ be the group of all homeomorphisms of X, and let $u_0 \in \Gamma$. Suppose that for each $A \in \mathfrak{S}$ there exist $B \in \mathfrak{S}$ and an entourage V of X such that: α) u_0^{-1} is uniformly continuous on V(A); β) the relations $u \in \Gamma$ and $(u(x_0), u(x)) \in V$ for all $x \in B$ together imply $A \subset u(B)$. Under these conditions, show that $u \to u^{-1}$ (defined on Γ) is continuous at u_0 with respect to the topology of $\mathfrak{S}$-convergence on Γ.

b) Show that the condition β) may be replaced by the following: β') there exists a connected set $C \subset X$ such that $V(A) \subset C$ and $V(C)$ is contained in the interior of $u_0(B)$. [Show that $\beta') \Longrightarrow \beta)$ by *reductio ad absurdum.*]

¶ 16) Let X be a uniform space and let Γ be the group of all *automorphisms* of the uniform structure of X.

a) Show that the topology of uniform convergence on Γ is compatible with its group structure (use Exercises 12 and 15). Moreover, the uniformity induced on Γ by the uniformity of uniform convergence in X coincides with the *right* uniformity of the topological group Γ.

b) If X is the compact interval [0, 1] of **R**, show that the topological group Γ defined in *a*) has no completion [construct a sequence of homeomorphisms (u_n) of X which is uniformly convergent but such that the sequence (u_n^{-1}) is not uniformly convergent].

c) Show that if X is a complete Hausdorff uniform space, then the topological group Γ is complete with respect to its *two-sided* uniformity (Chapter III, § 3, Exercise 6). [Observe that if Φ is a Cauchy filter on Γ with respect to this uniformity, then $\Phi(x)$ and $\Phi^{-1}(x)$ converge in X for all $x \in X$.]

¶ 17) *a*) Show that if X is a locally compact, locally connected space, then the topology induced on the group Γ of all homeomorphisms of X by the compact-open topology coincides with the topology $\mathfrak{T}_\beta$ defined in no. 5 [use Exercise 15 *b*) and Proposition 12 of no. 5].

b) Let X be the locally compact subspace of **R** consisting of the points 0 and 2^n ($n \in \mathbf{Z}$). If Γ is the group of all homeomorphisms of X, show that the topology induced on Γ by the compact-open topology is not compatible with the group structure of Γ.

18) Let X be a locally compact space endowed with a uniformity compatible with its topology, and let Γ be the group of homeomorphisms

of X. For each entourage V of X and each compact subset K of X, let $G(K, V)$ denote the set of pairs (u, v) of homeomorphisms of X such that $(u(x), v(x)) \in V$ and $(u^{-1}(x), v^{-1}(x)) \in V$ for all $x \in K$.

a) Show that the sets $G(K, V)$ form a fundamental system of entourages of a uniformity $\mathfrak{U}$ on Γ and that the topology induced by this uniformity is the topology $\mathfrak{T}_\beta$ defined in no. 5.

b) Show that if X is complete with respect to its uniformity, then Γ is complete with respect to the uniformity $\mathfrak{U}$.

c) Show that Γ is complete with respect to the two-sided uniformity defined by the topology $\mathfrak{T}_\beta$ [use Exercise 16 *c*)].

d) Take X to be the locally compact subspace of **R** consisting of points of the form $n + 2^{-m}$ ($n \in \mathbf{Z}$, m an integer $\geqslant 1$). Show that on the group Γ neither the uniformity $\mathfrak{U}$ nor the uniformity of compact convergence is comparable with any of the three uniformities (left, right and two-sided) defined by the topology $\mathfrak{T}_\beta$.

19) Let H be an equicontinuous group of homeomorphisms of a uniform space X; endowed with the topology of pointwise convergence, H is a topological group (no. 5, Corollary to Proposition 10).

a) Show that the left uniformity on H is finer than the uniformity of pointwise convergence and that these two uniformities coincide if H is uniformly equicontinuous.

b) Let u be the homeomorphism of the real line **R** defined by $u(x) = x + 1$ for $x \leqslant 0$, $u(x) = x + 1/(x + 1)$ for $x \geqslant 0$. Let H be the subgroup generated by u in the group of all homeomorphisms of **R**. Show that H is equicontinuous and is discrete with respect to the topology of pointwise convergence, but that the uniformity of pointwise convergence on H is not the discrete uniformity (observe that the limit $u^{n+1}(x) - u^n(x)$ is 0 as n tends to $+\infty$).

c) Take X to be the discrete space **N** of natural integers, endowed with the metric d such that $d(m, n) = 1$ whenever $m \neq n$, and take H to be the group of isometries of **N**, which is uniformly equicontinuous. Show that the topological group obtained by endowing H with the topology of pointwise convergence has no completion [same method as in Exercise 16 *b*)].

d) Suppose that X is Hausdorff and complete and that H is uniformly equicontinuous. Show that the Cauchy filters on H with respect to the two-sided uniformity of the topological group H converge in the space $\mathcal{F}_s(X; X)$; that the set H′ of their limit points is a uniformly equicontinuous group of homeomorphisms of X, which (endowed with the topology of pointwise convergence) is complete with respect to its two-sided uniformity; and that H is dense in H′.

20) Let X be the locally compact subspace of $\mathbf{R}^2$ consisting of the lines $y = 0, y = 1/n$ $(n \geqslant 1)$, and let G be the group of all homeomorphisms u of X whose restriction to each of the lines $y = 0, y = 1/n$ is a translation of the form $(x, y) \to (x + a_y, y)$. Show that G satisfies the conditions of Theorem 4 of no. 5, but that the topologies of compact and pointwise convergence on G are distinct.

¶ 21) Let X be a locally compact space, and let T be a subset of $\mathscr{C}(X; X)$ consisting of *surjective* mappings. Let $\mathscr{G}$ be a topology on T which is finer than the topology of pointwise convergence and with respect to which T is *locally compact*, and consider the following property of the pair $(T, \mathscr{G})$:

(A) For all $u \in T$ and $v \in T$, we have $u \circ v \in T$; and for each $u \in T$ the mappings $v \to u \circ v$ and $v \to v \circ u$ of T into itself are *continuous* with respect to $\mathscr{G}$.

a) Suppose that X is *compact* and *metrizable*, that the pair $(T, \mathscr{G})$ satisfies (A) and there exists a group G of homeomorphisms of X which is *dense* in T with respect to the topology $\mathscr{G}$. Show that the set H of *bijective* mappings in T is the complement of a *meagre* set in T. [Note first, using Exercise 7, that every point of T has a compact metrizable neighbourhood (with respect to $\mathscr{G}$). Let (V_n) be a fundamental system of neighbourhoods in T of the identity element e of G, and let H_n be the set of all $v \in T$ for which there exists $u \in T$ such that $v \circ u \in V_n$; observe that H contains the intersection of the sets H_n.]

b) Show that, under the hypotheses of *a*), the restriction to $G \times X$ of the mapping $\pi : (u, x) \to u(x)$ of $T \times X$ into X is *continuous*. [Show first that it is enough to prove that, for each $x_0 \in X$, there exists $u_0 \in H$ such that π is continuous at (u_0, x_0), by establishing that this implies the continuity of π at (e, x_0). If V is a compact metrizable neighbourhood of e in T, show next, by using *a*) and Chapter IX, § 5, Exercise 23, that there exists $u_0 \in H \cap V$ such that π is continuous at (u_0, x_0).]

¶ 22) Let X be a *compact* space, and let T be a *subgroup* of the group of all homeomorphisms of X, endowed with a locally compact topology $\mathscr{G}$ for which condition (A) of Exercise 21 is satisfied. Let G be a *countable* subgroup of T, and let f be a continuous real-valued function on X. Consider the continuous mapping $x \to \varphi(x)$ of X into I^G, where $I = f(X) \subset \mathbf{R}$ and $\varphi(x) = (f(u(x)))_{u \in G}$. Let Y be the image of X under φ; Y is a compact metrizable subspace of I^G.

a) Show that the set of all $v \in T$ such that $\varphi \circ v = \varphi$ is a subgroup K of T whose normalizer in T contains G. Let R be the equivalence relation $u \circ v^{-1} \in K$ on T, and let T′ be the quotient space T/R and $p : T \to T/R$ the canonical mapping. Show that T′ is locally compact, and that if we set $p(u)v = p(u \circ v)$, then T operates on the right on T′ in such a way that $t' \to t'v$ is continuous on T′ for all $v \in T$.

b) Let $G' = p(G)$, let H′ be the closure of G′ in T′, and let $H = \overset{-1}{p}(H')$. Show that, if $u \in H$, the relation $p(v) = p(v')$ implies $p(u \circ v) = p(u \circ v')$ so that H operates on the left on T′; moreover, if we set $up(v) = p(u \circ v)$ for $u \in H, v \in T$, then the mapping $t' \to ut'$ is continuous on T′. Finally, if u_1, u_2 are elements of H such that $p(u_1) = p(u_2)$, we have $u_1t' = u_2t'$; hence, by passing to the quotient, we define a mapping $(u', t') \to u't'$ of $H' \times T'$ into T′, such that $t' \to u't'$ is continuous on T′ for all $u' \in H'$. Show that if $u' \in H'$ and $v' \in H'$, then $u'v' \in H'$ and that the law of composition so defined on H′ induces a group structure on G′ (with respect to which G′ is isomorphic to GK/K).

c) If $u \in H$, the relation $\varphi(x) = \varphi(y)$ implies $\varphi(u(x)) = \varphi(u(y))$ and consequently there is a unique mapping $\tilde{u} : Y \to Y$ such that $\varphi \circ u = \tilde{u} \circ \varphi$. Show that $\tilde{u}$ is continuous and surjective and that $\tilde{u}_1 = \tilde{u}_2$ if and only if $p(u_1) = p(u_2)$; this allows us to write $\tilde{u} = \tilde{u}'$, where $u' = p(u)$, and defines an injective mapping $\psi : u' \to \tilde{u}'$ of H′ into $\mathcal{C}_s(Y; Y)$. Show that ψ is continuous and that $\psi(u'v') = \psi(u') \circ \psi(v')$; using ψ to identify H′ with a subset of $\mathcal{C}_s(Y; Y)$ and denoting by $\mathfrak{T}'$ the topology on H′ induced by that of T′, show that $(H', \mathfrak{T}')$ satisfies property (A) of Exercise 21.

d) Show that, if G is endowed with the topology induced by $\mathfrak{T}$, the mapping $(u, x) \to u(x)$ of $G \times X$ into X is continuous. [Show that, for every continuous real-valued function f on X, the mapping $(u, x) \to f(u(x))$ is continuous on $G \times X$, by using *c*) and Exercise 21; then apply Chapter IX, § 1, no. 5, Proposition 4.]

¶ 23) Let X be a compact space and let T be a subset of $\mathcal{C}(X; X)$ which is stable under the law of composition $(u, v) \to u \circ v$. Suppose that T is endowed with a topology $\mathfrak{T}$ finer than that of pointwise convergence and with respect to which T is locally compact; also suppose that, for every *countable* stable subset S of T, the restriction to $S \times X$ of the mapping $\pi : (u, x) \to u(x)$ of $T \times X$ into X is continuous. Show that, under these conditions, π is continuous on $T \times X$. [Let V be a compact subset of T (with respect to $\mathfrak{T}$); show that, for each countable subset D of V, the closure of D with respect to $\mathfrak{T}$ is contained in the closure of D with respect to the topology of uniform convergence, by using Corollary 1 of Theorem 3 (no. 4). Deduce that V is relatively compact in $\mathcal{C}_u(X; X)$ (Chapter II, § 4, Exercise 6), and hence equicontinuous.]

24) Let X be a locally compact space, T a group of homeomorphisms of X, $\mathcal{T}$ a locally compact topology on T, which is finer than the topology of pointwise convergence and for which condition (A) of Exercise 21 is satisfied. Show that the mapping $(u, x) \to u(x)$ of $T \times X$ into X is continuous with respect to $\mathcal{T}$. [Extend the elements of T to homeomorphisms of the one-point compactification X′ of X; then, by using Exercise 22 *d*), show that the result of Exercise 23 is applicable.]

¶ 25) Let G be a group endowed with a topology $\mathcal{T}$ with respect to which G is locally compact and the translations $t \to st$ and $t \to ts$ are continuous on G for all $s \in G$. Show that $\mathcal{T}$ is compatible with the group structure of G (*Theorem of R. Ellis*). [Identify G with the group of left translations of G, endowed with the topology of pointwise convergence, so that G is identified with a group of homeomorphisms of the one-point compactification X of G, endowed with the topology of pointwise convergence. Using Exercise 24, deduce that on G the topology of pointwise convergence in X is the same as the topology of uniform convergence in X (no. 4, Theorem 3, Corollary 1), and use Proposition 11 of no. 5.]

¶ 26) *a*) Let X be a locally compact uniform space, G a group of homeomorphisms of X, and T the closure of G in $\mathcal{C}_s(X; X)$. Show that T is compact with respect to the topology of pointwise convergence and is a group of homeomorphisms if and only if G is relatively compact in $\mathcal{C}_c(X; X)$. [Use Exercises 12 and 24 and the Corollary to Theorem 4 (no. 5).]

b) Let G be a locally compact, non-compact topological group and let X be its one-point compactification; G can be identified with a group of homeomorphisms of X. Show that G is not equicontinuous, although its closure in $\mathcal{C}_s(X; X)$ is compact.

27) Let X be a Hausdorff uniform space and let G be an equicontinuous group of homeomorphisms of X.

a) A point $x_0 \in X$ is said to be *almost periodic* with respect to G if the orbit of x_0 with respect to G is relatively compact in X. Let Y be the (compact) closure of this orbit in X; then $u(Y) \subset Y$ for all $u \in G$. Let $\tilde{u}$ be the restriction of u to Y considered as an element of $\mathcal{C}(Y; Y)$; show that $\tilde{u}$ is a homeomorphism of Y onto itself and that the closure Γ in $\mathcal{C}_u(Y; Y)$ of the image of G under the mapping $u \to \tilde{u}$ is a compact group of homeomorphisms (no. 6, Theorem 4, Corollary) which is transitive on Y.

b) Suppose that X is complete. Then x_0 is almost periodic with respect to G if and only if, for each neighbourhood V of x_0 in X, there is a finite number of elements $u_i \in$ G such that, for all $u \in$ G, we have $u_i^{-1}(u(x_0)) \in$ V for at least one index i.

c) Suppose that x_0 is almost periodic and that G is endowed with a topology compatible with its group structure. With the notation of *a*), the mapping $u \rightarrow \tilde{u}$ of G into Γ (endowed with the topology of uniform convergence) is continuous if and only if, for each pair of distinct points x, y of Y, there is a neighbourhood U of e in G and a neighbourhood V of y such that the relation $u \in$ U implies $u(x) \notin$ V [use the fact that the topologies induced on Γ by those of $\mathcal{C}_u$ (Y; Y) and $\mathcal{C}_s$ (Y; Y) are the same].

28) Let G be a topological group and let X be the Banach space of bounded continuous mappings of G into **C** endowed with the norm

$$||f|| = \sup_{s \in G} |f(s)|.$$

For each $f \in$ X and each $s \in$ G, let $U_s f$ denote the function $t \rightarrow f(s^{-1}t)$, which belongs to X, so that the U_s, as s runs through G, form a group of isometries G′ of X. An element $f \in$ X is said to be an *almost periodic function* (on the left) on G if f is an almost periodic element of X with respect to the group G′. For this to be so it is necessary and sufficient that, for each $\varepsilon > 0$, there should exist a finite number of elements $s_i \in$ G with the property that, for each $s \in$ G, there is at least one index i such that $|f(s^{-1}t) - f(s_i^{-1}t)| \leqslant \varepsilon$ for all $t \in$ G.

a) Suppose that f is almost periodic. Let Y be the closure in X of the orbit of f under G′, and let Γ be the closure in $\mathcal{C}_u$ (Y; Y) of the set of restrictions V_s of the U_s to Y ($s \in$ G). For each $\sigma \in$ Γ, let $f_\sigma \in$ Y be the image of f under σ. For each $s \in$ G, we have

$$f_\sigma(t) = U_s^{-1} f_\sigma(s^{-1}t).$$

Deduce that there is a continuous function $\bar{f}$ on Γ such that $f(s) = \bar{f}(V_s)$ [use Exercise 27 *c*)].

b) Show by using *a*) that a function $f \in$ X is almost periodic on G if and only if it is of the form $g \circ \varphi$, where φ is the canonical mapping of G into the *compact group* G^c *associated with* G (*Set Theory*, Chapter IV, § 3, no. 3, Example 8) and g is a continuous mapping of G^c into **C**.

c) Take G to be the additive group **R**. Show that if f is almost periodic on **R**, then for each $\varepsilon > 0$ there exists a real number $T > 0$ such that every interval of **R** of length T contains a number s such that

$$|f(x + s) - f(x)| \leqslant \varepsilon \quad \text{for all} \quad x \in \mathbf{R}$$

(s is called an "ε-period" of f). [Use *b*) and Exercise 2 of Chapter V, § 1.]

29) Let X be a compact space and let G be a topological group operating continuously on X, such that every orbit of G is dense in X. Show that if, for every continuous real-valued function f on X, there exists $x_0 \in X$ such that $s \to f(s.x_0)$ is almost periodic on G, then G is equicontinuous (use the fact that X is homeomorphic to a closed subspace of a cube). Consider the converse, when G is abelian.

§ 4

1) Let X be a Lindelöf space (Chapter I, § 9, Exercise 15), let H be a directed set (with respect to the relation $\geqslant$) of continuous real-valued functions on X such that the lower envelope g of the functions of H is continuous. Show that there exists a decreasing sequence of functions (f_n) belonging to H which converges pointwise to g.

2) Let I be a compact interval in $\mathbf{R}$ and let (f_n) be a sequence of *monotone* real-valued functions defined on I which converges pointwise in I to a *continuous* function g. Show that g is monotone and that the sequence (f_n) converges uniformly to g in I.

¶ 3) Let Γ be a simply transitive group of homeomorphisms of $\mathbf{R}$ endowed with the topology of pointwise convergence. For each $x \in \mathbf{R}$ let s_x denote the element of Γ such that $s_x(0) = x$. Show that the mapping $x \to s_x$ is a homeomorphism of $\mathbf{R}$ onto Γ [use the fact that if $s \in \Gamma$ is such that $x < s(x)$ for *some* $x \in \mathbf{R}$, then $y < s(y)$ for all $y \in \mathbf{R}$]; hence show that Γ is isomorphic to $\mathbf{R}$ as a topological group [use Exercise 17 *a*) of § 3 and Theorem 1 of Chapter V, § 3].

4) Let X be a compact space and let H be a vector subspace of $\mathcal{C}(X; \mathbf{R})$ such that $|u| \in H$ whenever $u \in H$. Then a continuous real-valued function f on X can be uniformly approximated by functions belonging to H if and only if, for each pair x, y of points of X, the function f satisfies every linear relation $\alpha f(x) = \beta f(y)$ where (α, β) are such that $\alpha\beta \geqslant 0$ and $\alpha g(x) = \beta g(y)$ for every function $g \in H$. [Apply Proposition 2 of no. 1, observing that the image of H under the mapping $u \to (u(x)\ u(y))$ is the whole plane $\mathbf{R}^2$, or a line of equation $\alpha X = \beta Y$ with $\alpha\beta \geqslant 0$, or else consists of the single point $(0, 0)$.]

5) Let X be a compact space and let H be a set of continuous real-valued functions on X. Let R be the equivalence relation "$u(x) = u(y)$ for all $u \in H$". A continuous real-valued function f on X can then be uniformly approximated by polynomials (resp. polynomials with no

constant term) in the functions of H if and only if f is constant on each equivalence class mod R [resp. if and only if f is constant on each equivalence class mod R and vanishes at all points x such that $u(x) = 0$ for all $u \in \mathrm{H}$].

¶ 6) Let X be a completely regular space and let $\mathscr{C}^\infty(\mathrm{X}; \mathbf{R})$ denote the normed subalgebra of $\mathscr{B}(\mathrm{X}; \mathbf{R})$ consisting of all bounded continuous real-valued functions on X. In order that every subalgebra of $\mathscr{C}^\infty(\mathrm{X}; \mathbf{R})$, which separates the points of X and contains the constant functions, should be dense in $\mathscr{C}^\infty(\mathrm{X}; \mathbf{R})$, it is necessary and sufficient that X should be compact. [Let $\beta\mathrm{X}$ be the Stone-Čech compactification of X (Chapter IX, § 1, Exercise 7). Show that if $\beta\mathrm{X} - \mathrm{X}$ is not empty then there are non-dense subalgebras of $\mathscr{C}^\infty(\mathrm{X}; \mathbf{R})$ which separate the points of X and contain the constant functions, by observing that $\mathscr{C}^\infty(\mathrm{X}; \mathbf{R})$ may be identified with $\mathscr{C}(\beta\mathrm{X}; \mathbf{R})$.]

¶ 7) Let X be a non-compact, completely regular space.

a) Show that the following properties are equivalent:

α) The subalgebra of $\mathscr{C}^\infty(\mathrm{X}; \mathbf{R})$ consisting of functions of the form $c + f$, where c is a constant and f has compact support (Chapter IX, § 4, no. 3), is dense in $\mathscr{C}^\infty(\mathrm{X}; \mathbf{R})$.

β) If $\beta\mathrm{X}$ is the Stone-Čech compactification of X, then $\beta\mathrm{X} - \mathrm{X}$ consists of a single point.

γ) There is only one uniformity on X, compatible with its topology, with respect to which X is precompact.

δ) If A, B are two completely separated closed subsets of X (Chapter IX, § 1, Exercise 11) then one of A, B is compact.

ζ) If $\mathscr{U}$ is any uniformity on X compatible with its topology and if V is any entourage of $\mathscr{U}$, there is a subset L of X whose complement is compact, such that $\mathrm{L} \times \mathrm{L} \subset \mathrm{V}$.

θ) There is only one uniformity on X compatible with its topology. [Note that α) implies that X is locally compact and therefore open in $\beta\mathrm{X}$, and hence show that $\alpha) \Longrightarrow \beta)$; to prove that $\beta) \Longrightarrow \alpha)$ use the Weierstrass-Stone theorem. To show that $\beta) \Longrightarrow \gamma)$, use the fact that the uniformity induced on X by that of $\beta\mathrm{X}$ is the finest uniformity compatible with the topology of X, with respect to which X is precompact. To show that $\gamma) \Longrightarrow \delta)$, use Chapter IX, § 1, Exercise 11. To show that $\delta) \Longrightarrow \zeta)$, show first that if V is any entourage of $\mathscr{U}$, there exists $z \in \mathrm{X}$ such that $\mathrm{V}(z)$ is not compact; otherwise X would be complete with respect to $\mathscr{U}$ and paracompact (Chapter II, § 4, Exercise 9), therefore normal; show then with the help of δ) that X would be countably compact (Chapter IX, § 2, Exercise 14), and finally use Chapter

II, § 4, Exercise 6. Then apply this result to an entourage V defined by $f(x, x') \leqslant 1$, where f is a continuous pseudometric on X, and deduce from δ) that there is a subset L of X, whose complement is compact, such that $L \times L \subset V$. Finally, to establish that $\zeta) \Longrightarrow \theta)$, observe that every ultrafilter on X either converges or is finer than the filter of complements of relatively compact subsets of X.]

b) Show that a space X which satisfies the equivalent conditions of *a*) is pseudo-compact (Chapter IX, § 1, Exercise 22). [Use Exercise 12 of Chapter IX, § 1, or else observe that if X is not pseudo-compact, there is a sequence (x_n) of points of X which has no cluster point and a continuous mapping $f: X \to [0, 1]$ such that $f(x_{2n}) = 0$ and $f(x_{2n+1}) = 1$.]

c) Let Y be a non-compact, completely regular space. Show that the subspace X of βY which is the complement of a point of $\beta Y - Y$ satisfies the conditions of *a*).

8) Let $(X_i)_{1 \leqslant i \leqslant n}$ be a finite family of compact spaces, let $X = \prod_{i=1}^{n} X_i$ be their product, and for each index i let F_i be a closed subset of X_i. Show that every continuous real-valued function on X which vanishes on the union of the sets $F_i \times \prod_{j \neq i} X_j$ $(1 \leqslant i \leqslant n)$ can be uniformly approximated by finite sums of functions of the form

$$(x_1, \ldots, x_n) \to u_1(x_1) \cdots u_n(x_n),$$

where u_i is a continuous real-valued function on X_i which vanishes on F_i $(1 \leqslant i \leqslant n)$.

* 9) Let $(r_n)_{n \geqslant 1}$ be the sequence of distinct rational numbers belonging to the interval $I = [0, 1]$ of **R**, arranged in some order. Define by induction a sequence of closed intervals $I_n \subset I$, as follows: I_n has as its midpoint the point r_{k_n} with the smallest index not contained in $\bigcup_{p<n} I_p$; its length is $\leqslant 1/4^n$ and it meets no interval I_p for which $p < n$; (I_n) is thus a sequence of mutually disjoint closed intervals. On the product space $I \times \mathbf{R}$ we define a continuous real-valued function u as follows: for each integer $n \geqslant 1$, the function $x \to u(x, n)$ is equal to 1 at an interior point of I_n, is equal to 0 on the exterior of I_n and takes its values in $[0, 1]$; on the other hand, for each $x \in I$, the function $y \to u(x, y)$ is affine-linear in each of the intervals $[n, n+1]$. Show that u cannot be uniformly approximated in $I \times \mathbf{R}$ by linear combinations of functions of the form $v(x)w(y)$, where v is continuous on I and w is continuous and bounded on **R**. [In the space $\mathcal{C}^{\infty}(\mathbf{R}; \mathbf{R})$ of continuous bounded real-valued functions on **R**, consider the set of

partial functions $y \to u(x, y)$ as x runs through I; show that there exists an infinite sequence (u_n) of these functions such that $\|u_n\| = 1$ and $\|u_m - u_n\| = 1$ whenever $m \neq n$. Deduce that there cannot exist any finite-dimensional vector subspace E of $\mathcal{C}^\infty(\mathbf{R}; \mathbf{R})$ such that the distance of each u_n from E is $\leqslant 1/4$; for otherwise there would exist a sequence (v_n) of points of E such that $\|v_n\| = 2$ and $\|v_n - v_m\| \geqslant 1/2$ whenever $m \neq n$, contrary to the fact that every finite-dimensional subspace of $\mathcal{C}^\infty(\mathbf{R}; \mathbf{R})$ is locally compact.] *

10) Use the Weierstrass-Stone theorem to give a new proof of Urysohn's theorem (Chapter IX, § 4, no. 2, Theorem 2) for closed subspaces of a compact space X. [If F is a closed subspace of X, consider the set H of restrictions to F of continuous mappings $X \to \mathbf{R}$, and observe that if $f \in H$ and $|f(x)| \leqslant a$ for all $x \in F$, then there exists a function $g \in H$ which coincides with f on F and is such that $|g(x)| \leqslant a$ in X.]

¶ 11) *a*) Let X be a completely regular space and let Y be a closed subspace of X; the mapping $\varphi : u \to u|Y$ of $\mathcal{C}^\infty(X; \mathbf{R})$ into $\mathcal{C}^\infty(Y; \mathbf{R})$ is a continuous linear mapping of norm $\leqslant 1$. If X is normal, φ is a *surjective strict morphism*; and if X_Y denotes the quotient space of X obtained by identifying all the points of Y (X_Y is normal by Exercise 15 of Chapter IX, § 4), then the kernel of φ can be identified (as a normed space) with the subspace of $\mathcal{C}^\infty(X_Y; \mathbf{R})$ consisting of functions which vanish at ω, the canonical image of Y in X_Y.

b) Suppose that X is metrizable and that the frontier of Y in X is of countable type. Show that $\varphi : u \to u|Y$ is a strict morphism which has a *right inverse* (Chapter III, § 6, no. 2, Proposition 3). [Let d be a metric compatible with the topology of X; let (a_n) be a sequence of points of Fr (Y), dense in Fr (Y), and let $V_{n,m}$ be the set of points $x \in X$ such that $d(x, a_n) < 1/m$ and $d(x, Y) > 1/2m$; construct a family of continuous mappings $f_{n,m} : X \to [0, 1]$ such that the support of $f_{n,m}$ is contained in $V_{n,m}$ and such that $\sum_{n,m} f_{n,m}(x) = 1$ for all $x \in \complement Y$, the family $(f_{n,m})$ being uniformly summable in some neighbourhood of each point of $\complement Y$. Show then that, for every $v \in \mathcal{C}^\infty(Y; \mathbf{R})$, the function u which coincides with v on Y and is equal to $\sum_{n,m} v(a_n) f_{n,m}(x)$ for all $x \in \complement Y$ belongs to $\mathcal{C}^\infty(X; \mathbf{R})$ and is such that $\varphi(u) = v$.] (*)

12) *a*) Let X be a compact space, and let $f : X \to Y$ be a continuous *surjection* of X onto a compact space Y. Show that the mapping

(*) This result does not extend to the case where X is the non-metrizable compact space $\beta\mathbf{N}$ and Y is the closed set $\beta\mathbf{N} - \mathbf{N}$.

${}^a f : u \to u \circ f$ of $\mathcal{C}(Y; \mathbf{R})$ into $\mathcal{C}(X; \mathbf{R})$ is a (norm-preserving) isomorphism of $\mathcal{C}(Y; \mathbf{R})$ onto a closed subalgebra of $\mathcal{C}(X; \mathbf{R})$ containing the identity element.

b) Conversely, let A be a closed subalgebra of $\mathcal{C}(X; \mathbf{R})$ containing the identity element. Show that there exists a continuous surjection f of X onto a compact space Y such that ${}^a f$ is an isomorphism of $\mathcal{C}(Y; \mathbf{R})$ onto A. [If $(u_\lambda)_{\lambda \in L}$ is a family which is dense in A, consider the continuous mapping $x \to (u_\lambda(x))$ of X into $\mathbf{R}^L$ and use the Weierstrass-Stone theorem.]

c) Deduce from *b*) that, for each sequence (u_n) of elements of $\mathcal{C}(X; \mathbf{R})$, there exists a *metrizable* compact space Y and a continuous surjective mapping $f : X \to Y$ such that the u_n belong to the image of $\mathcal{C}(Y; \mathbf{R})$ under ${}^a f$.

¶ 13) Let X be a compact space and let A denote the normed algebra $\mathcal{C}(X; \mathbf{R})$. For each subset M of A, let V(M) denote the set of all $x \in X$ such that $u(x) = 0$ for all $u \in M$. For each subset Y of X, let $\mathfrak{I}(Y)$ denote the set of all $u \in A$ such that $u(x) = 0$ for all $x \in Y$.

a) V(M) is a closed subspace of X and $\mathfrak{I}(Y)$ is a closed ideal of A. If $\mathfrak{a}$ is the ideal of A generated by a subset M of A, then $V(M) = V(\mathfrak{a})$; V and I are order-reversing mappings (with respect to the order relation of inclusion), and we have $V\{0\} = X$; $V(A) = \varnothing$; $V(\{u\}) = \varnothing$ for every unit u of A;

$$V\Big(\bigcup_{\lambda \in L} M_\lambda\Big) = V\Big(\sum_{\lambda \in L} M_\lambda\Big) = \bigcap_{\lambda \in L} V(M_\lambda)$$

for every family $(M_\lambda)_{\lambda \in L}$ of subsets of A; $V(M.M') = V(M) \cup V(M')$; $\mathfrak{I}(X) = \{0\}$; $\mathfrak{I}(\varnothing) = A$; $\mathfrak{I}\Big(\bigcup_{\lambda \in L} Y_\lambda\Big) = \bigcap_{\lambda \in L} \mathfrak{I}(Y_\lambda)$ for every family $(Y_\lambda)_{\lambda \in L}$ of subsets of X.

b) Show that if $\mathfrak{a}$ is any ideal of A, then $\mathfrak{I}(V(\mathfrak{a})) = \bar{\mathfrak{a}}$, and that if Y is any subset of X, then $V(\mathfrak{I})(Y) = \overline{Y}$. (For the first of these relations, observe that $V(\bar{\mathfrak{a}}) = V(\mathfrak{a})$ and that we may therefore suppose $\mathfrak{a}$ closed; note that if x, y are two distinct points of $X - V(\mathfrak{a})$, there exists $u \in \mathfrak{a}$ such that $u(x) \neq u(y)$, and use Proposition 4 of no. 2. To prove the second relation, use Urysohn's theorem.)

c) Deduce from *b*) that $x \to \mathfrak{I}(\{x\})$ is a bijection of X onto the set of maximal ideals of A, which are therefore closed. An ideal of A is closed if and only if it is an intersection of maximal ideals. The ring A has no radical.

d) Show that the mapping $e \to V(\{e\})$ is a bijection of the set of idempotents of A onto the set of subsets of X which are both open and closed.

A point $x \in X$ is isolated if and only if the maximal ideal $\mathfrak{I}(\{x\})$ is principal.

14) Let X be a topological space. For each function $u \in \mathcal{C}(X; \mathbf{R})$ such that $u(x) > 0$ for all $x \in X$, let V_u be the set of all $f \in \mathcal{C}(X; \mathbf{R})$ such that $|f| \leqslant u$.

a) Show that the sets V_u are neighbourhoods of 0 in $A = \mathcal{C}(X; \mathbf{R})$ with respect to a Hausdorff topology $\mathcal{T}$ compatible with the ring structure of A.

b) The topology induced by $\mathcal{T}$ on $\mathcal{C}^\infty(X; \mathbf{R})$ is finer than the topology of uniform convergence in X. These two topologies coincide if and only if X is pseudo-compact (Chapter IX, § 1, Exercise 21). If X is not pseudo-compact, the topology induced by $\mathcal{T}$ on the set of constant functions is the discrete topology, so that $\mathcal{T}$ is not compatible with the **R**-vector space structure of A.

c) Show that A is a Gelfand ring with respect to the topology $\mathcal{T}$ (Chapter III, § 6, Exercise 11).

¶ 15) Let X be a completely regular space and let βX be its Stone-Čech compactification. Endow the ring $A = \mathcal{C}(X; \mathbf{R})$ with the topology $\mathcal{T}$ defined in Exercise 14. For each $f \in A$, let $V(f)$ be the closure in βX of the subset $\overset{-1}{f}(0)$ of X. For each subset M of A, let V(M) denote $\bigcap_{f \in M} V(f)$. For each subset Y of βX, let $\mathfrak{I}(Y)$ denote the set of all $f \in A$ such that $Y \subset V(f)$, and write $\mathfrak{I}(x)$ in place of $\mathfrak{I}(\{x\})$. We have $V(f) = \varnothing$ if and only if f is a unit in A.

a) If f, g are two elements of A such that $\overset{-1}{f}(0) \cap \overset{-1}{g}(0) = \varnothing$, show that $V(f) \cap V(g) = \varnothing$. [Consider the function $|f|/(|f| + |g|)$. Deduce that, for each subset M of A, we have $V(M) = V(\mathfrak{a})$, where $\mathfrak{a}$ is the ideal generated by M in A′.

b) Show that, for each subset Y of βX, $\mathfrak{I}(Y)$ is an ideal of A [use *a*)] and that $V(\mathfrak{I}(Y))$ is the closure $\overline{Y}$ of Y in βX.

c) Let $u \in A$ be such that $u(x) > 0$ for all $x \in X$, and let $g \in A$. Show that there exists $f \in A$ such that $|f - g| \leqslant u$ and such that $V(f)$ is a neighbourhood of $V(g)$. [Define f to be equal to $g + u$ if $g(x) + u(x) < 0$, to $g - u$ if $g(x) - u(x) > 0$, and to 0 otherwise; consider the function $f' = \inf((u + g)^+, (u - g)^-)$ and use *a*).]

d) Let $\mathfrak{a}$ be an ideal of A. Show that if $f \in A$ is such that $V(f)$ is a neighbourhood of $V(\mathfrak{a})$, then $f \in \mathfrak{a}$. [Note first that there exists $g \in \mathfrak{a}$ such that $V(g)$ is contained in the interior of $V(f)$, and then consider the function h defined on X which is equal to $f(x)/g(x)$ if $f(x) \neq 0$, and equal to 0 if $f(x) = 0$.]

e) Deduce from c) and d) that $\mathfrak{I}(V(\mathfrak{a})) = \bar{\mathfrak{a}}$ for every ideal $\mathfrak{a}$ of A. [Show first that $V(M) = V(\bar{M})$ for all subsets M of A, arguing by contradiction.] Deduce that, for each subset Y of βX, $\mathfrak{I}(Y)$ is a closed ideal of A. [Note that $\mathfrak{I}(Y) = \mathfrak{I}(V(\mathfrak{I}(Y)))$.]

f) Use e) to show that $x \to \mathfrak{I}(x)$ is a bijection of βX onto the set of maximal ideals (necessarily closed) of A. An ideal of A is closed if and only if it is an intersection of maximal ideals. The ring A has no radical.

¶ 16) We retain the hypotheses and notation of Exercise 15.

a) Let $\mathfrak{a}$ be a closed ideal in the ring A. Show that in the quotient ring $A/\mathfrak{a}$ the set P of canonical images of functions $f \geqslant 0$ of A is the set of elements $\geqslant 0$ with respect to an ordering which makes $A/\mathfrak{a}$ a *lattice-ordered ring* [observe, using Exercise 15 e), that if f and g belong to $\mathfrak{a}$, then so do $|f| + |g|$ and every function $h \in A$ such that $|h| \leqslant |f|$]. The canonical image in $A/\mathfrak{a}$ of the set of constant functions is isomorphic to $\mathbf{R}$ as an ordered field.

b) In particular, for each $x \in \beta X$, the field $A/\mathfrak{I}(x)$ is canonically endowed with an ordered field structure. The maximal ideal $\mathfrak{I}(x)$ is said to be *real* if $A/\mathfrak{I}(x)$ is isomorphic to $\mathbf{R}$, *hyper-real* otherwise. For $\mathfrak{I}(x)$ to be hyper-real it is necessary and sufficient that $x \in \beta X - X$ and that there is a unit $f \in A$ such that $\lim_{y \to x, y \in X} f(y) = 0$. Deduce that all the maximal ideals of A are real if and only if X is pseudo-compact (Chapter IX, § 1, Exercise 21).

c) For each $x \in \beta X$, let φ_x be the canonical homomorphism $A \to A/\mathfrak{I}(x)$. For $\varphi_x(f)$ to be $\geqslant 0$ in $A/\mathfrak{I}(x)$, it is necessary and sufficient that there should exist $g \in \mathfrak{I}(x)$ such that $f(y) \geqslant 0$ in $\overset{-1}{g}(0)$ [observe that the relation $\varphi_x(f) \geqslant 0$ is equivalent to $f \equiv |f| \pmod{\mathfrak{I}(x)}$]. For $\varphi_x(f)$ to be > 0, it is necessary and sufficient that there should exist $g \in \mathfrak{I}(x)$ such that $f(y) > 0$ in $\overset{-1}{g}(0)$. [Note that, if $f \notin \mathfrak{I}(x)$, there exists $g \in \mathfrak{I}(x)$ such that $\overset{-1}{f}(0) \cap \overset{-1}{g}(0) = \varnothing$, by using Exercise 15.]

d) Show that if $\mathfrak{I}(x)$ is hyper-real, the transcendence degree of the field $A/\mathfrak{I}(x)$ over $\mathbf{R}$ is at least equal to $\mathfrak{c} = \operatorname{Card}(\mathbf{R})$. [Let $f > 0$ be a unit of A such that $\varphi_x(f) = u$ is infinitely large with respect to $\mathbf{R}$. Show that as r runs through the elements of a base of $\mathbf{R}$ over $\mathbf{Q}$ (consisting of numbers > 0), the elements $\varphi_x(f^r)$ of $A/\mathfrak{I}(x)$ are algebraically independent.]

* *e*) For each point $a = (a_1, \ldots, a_n) \in \mathbf{R}^n$, let $f_a(X)$ denote the polynomial $X^n + a_1X^{n-1} + \cdots + a_n$. For each real number ξ, let $\nu(\xi)$ be the sum of the multiplicities of the zeros z of f_a in $\mathbf{C}$ such that $\mathrm{R}(z) = \xi$; and for each integer k such that $1 \leqslant k \leqslant n$, let $\rho_k(a)$ be the smallest real number ξ such that $\sum_{\eta \leqslant \xi} \nu(\eta) \geqslant k$. Show that each of the functions ρ_k is continuous on $\mathbf{R}^n$ (use Rouché's theorem).

f) Show that, for each $x \in \beta X$, the ordered field $A/\mathfrak{I}(x)$ is *real-closed*. [Let $F(X) = X^n + f_1X^{n-1} + \cdots + f_n$ be a polynomial of odd degree with coefficients in A; the functions $g_k(y) = \rho_k(f_1(y), \ldots, f_n(y))$, where the ρ_k are the functions defined in *e*), are continuous on X, and for each $y \in X$ there is an index k such that $F(g_k(y)) = 0$; deduce that the product $F(g_1) \cdots F(g_n)$ belongs to $\mathfrak{I}(x)$.] *

g) Let X be an infinite discrete space and let $y \to F_y$ be a bijection of X onto the set of finite subsets of X; for each $z \in X$ let M_z be the set of all $y \in X$ such that $z \in F_y$; these sets form a base of a filter $\mathfrak{F}$ on X. Let x be a point of βX such that the corresponding ultrafilter on X (Chapter I, § 9, Exercise 27) is finer than $\mathfrak{F}$. Let B be a subset of A such that Card (B) $\leqslant$ Card (X), and let $y \to g_y$ be a surjective mapping $X \to B$. For each $z \in X$ put $f(z) = 1 + \sup_{y \in F_z} g_y(z)$. Show that $f(z) > g_y(z)$ for all $z \in M_y$, and deduce that $\varphi_x(f) > \varphi_x(g_y)$ for all $y \in Y$ [use *c*)]. Conclude that Card $(A/\mathfrak{I}(x)) >$ Card X.

¶ 17) We retain the hypotheses and notation of Exercise 15.

a) A maximal ideal $\mathfrak{I}(x)$ is real if and only if, for every infinite sequence (g_n) of elements of $\mathfrak{I}(x)$, the intersection of the sets $\overset{-1}{g_n}(0)$ is not empty. [Use Exercise 16 *b*); to show that the condition is necessary show that, if $\bigcap_n \overset{-1}{g_n}(0) = \emptyset$, the function $f(y) = \sum_{n=0}^{\infty} \inf(g_n(y), 2^{-n})$ is continuous, is a unit in A and tends to 0 as y tends to x while remaining in X; note that $f(y) \leqslant 2^{-m}$ for $y \in \bigcap_{k \leqslant m} \overset{-1}{g_k}(0)$, and use Exercise 15 *a*) and the fact that the $V(g)$, where $g \in \mathfrak{I}(x)$, form a fundamental system of neighbourhoods of x in βX.]

b) X is said to be *real-compact* if the only real maximal ideals $\mathfrak{I}(x)$ are those for which $x \in X$. Show that every completely regular Lindelöf space (Chapter I, § 9, Exercise 15) is real-compact [use *a*)]. A completely regular pseudo-compact space is not real-compact unless it is compact.

c) Let υX be the set of all $x \in \beta X$ such that $\mathfrak{I}(x)$ is real. Show that every continuous real-valued function $f \in A$ can be extended by continuity to υX, so that $\mathcal{C}(X; \mathbf{R})$ and $\mathcal{C}(\upsilon X; \mathbf{R})$ can be canonically identified. Show that the subspace υX of βX is real-compact [note that if f is a unit in $\mathcal{C}(X; \mathbf{R})$, then its continuous extension to υX is a unit in $\mathcal{C}(\upsilon X; \mathbf{R})$ and that $\beta(\upsilon X) = \beta X$]. υX is called the *real-compactification* of X.

d) Show that υX is the completion of X with respect to the coarsest uniformity on X for which all the functions $f \in A$ are uniformly continuous. (Note that a Cauchy filter $\mathfrak{F}$ on X with respect to this uniformity converges to a point $x \in \beta X$; if $x \notin \upsilon X$, there would be a function $f \in A$ which was unbounded in a set of F.]

e) Deduce from *d*) that if $u : X \to Y$ is any continuous mapping of a completely regular space X into a real-compact space Y, then u can be extended by continuity to a mapping $\upsilon X \to Y$. The space υX is therefore the solution of the universal mapping problem in which Σ is the structure of a real-compact space, and the α-mappings and morphisms are continuous mappings (*Set Theory*, Chapter IV, § 3, no. 1).

f) Show that every closed subspace of a real-compact space is real-compact, that every product of real-compact spaces is real-compact, and that in a Hausdorff space any intersection of real-compact subspaces is real-compact. [For example, to show that a closed subspace X of a real-compact space Y is real-compact, consider the extension $\bar{j}$ to υX of the canonical injection $j : X \to Y$, and note that $\bar{j}^{-1}(X) = X$; deduce that X is closed in υX. The proofs are similar in the other two cases.]

g) Deduce from *b*), *d*) and *f*) that a completely regular space is real-compact if and only if it is homeomorphic to a closed subspace of a product $\mathbf{R}^I$.

18) Let X be a real-compact space.

a) Show that, for each non-zero homomorphism $u : \mathcal{C}(X; \mathbf{R}) \to \mathbf{R}$, there exists a unique point $x \in X$ such that $u(f) = f(x)$ for all f in $\mathcal{C}(X; \mathbf{R})$.

b) Let Y be another real-compact space. Show that, for every **R**-algebra homomorphism $v : \mathcal{C}(Y; \mathbf{R}) \to \mathcal{C}(X; \mathbf{R})$ which maps identity element to identity element, there exists a unique continuous mapping $w : X \to Y$ such that $v(g) = g \circ w$ for all $g \in \mathcal{C}(Y; \mathbf{R})$. In particular, $\mathcal{C}(X; \mathbf{R})$ and $\mathcal{C}(Y; \mathbf{R})$ are isomorphic if and only if X and Y are homeomorphic.

¶ 19) Let X be a compact space, and let A be the normed **C**-algebra $\mathcal{C}(X; \mathbf{C})$. For each closed ideal $\mathfrak{a}$ of A, show that the relation $f \in \mathfrak{a}$ implies $\bar{f} \in \mathfrak{a}$ (note that $\bar{f}$ is the uniform limit of functions of the form gf). Extend the results of Exercise 13 to the algebra A.

¶ 20) *a*) Let X be a compact space, let A be an **R**-algebra of finite rank with an identity element, and endow A with the topology defined in Chapter VI, § 1, no. 5. Let B be a subalgebra of the **R**-algebra $\mathcal{C}$ (X; A), and suppose that, for each $\varepsilon > 0$, each pair of distinct points x, y of X and each pair of elements u, v of A, there exists $f \in B$ such that $\|f(x) - u\| \leqslant \varepsilon$ and $\|f(y) - v\| \leqslant \varepsilon$ ($\| \ \|$ being any norm defining the topology of A); and suppose, further, that if u and v belong to **R**, there is a function $f \in B$ satisfying these conditions and taking its values in **R**. Show that, under these hypotheses, B is dense in $\mathcal{C}$ (X; A) with respect to the topology of uniform convergence. [Show first that every function of $\mathcal{C}$ (X; **R**) can be uniformly approximated by functions of B and then that the same is true of every constant $a \in A$, by using the hypotheses and a partition of unity.]

b) Take A = **H**, the division ring of quaternions over **R**. Let B be a subalgebra of the **R**-algebra $\mathcal{C}$ (X; **H**) such that: (i) for each $f \in B$, the conjugate $\bar{f}$ of f [defined by $\bar{f}(x) = \overline{f(x)}$ for all $x \in X$] belongs to B; (ii) for each $\varepsilon > 0$, each pair of distinct points x, y of X and each pair of elements u, v of **H**, there exists $f \in B$ such that $\|f(x) - u\| \leqslant \varepsilon$ and $\|f(y) - v\| \leqslant \varepsilon$. Show that B is dense in $\mathcal{C}$ (X; **H**) [use *a*)].

c) Extend the results of Exercise 13 to the algebra $\mathcal{C}$ (X; **H**), and show in particular that every closed ideal of this algebra is two-sided [argue as in Exercise 19, using *b*)].

¶ 21) *a*) Let X be a totally disconnected compact space, and let A be a topological ring. Endow the ring $\mathcal{C}$ (X; A) with the topology of uniform convergence, which is compatible with the ring structure. Let $\mathfrak{a}$ be a left ideal of $\mathcal{C}$ (X; A) and, for each $x \in X$, let $\mathfrak{I}(x)$ be the closure in A of the set of $f(x)$ for $f \in \mathfrak{a}$; $\mathfrak{I}(x)$ is a closed left ideal of A. Show that $\bar{\mathfrak{a}}$ is identical with the set of all $f \in \mathcal{C}$ (X; A) such that $f(x) \in \mathfrak{I}(x)$ for all $x \in X$ (note that, for each $x \in X$, the neighbourhoods of x which are both open and closed form a fundamental system of neighbourhoods of x).

b) Suppose in addition that A has a fundamental system of neighbourhoods of 0 which are two-sided ideals. Let B be a subring of $\mathcal{C}$ (X; A) which contains the constants and separates the points of X. Show that B is dense in $\mathcal{C}$ (X; A). [For each closed subset F of X, each $x \notin F$ and each neighbourhood U of 0 in A, show that there exists $f \in B$ such that $f(x) = 1$ and $f(y) \in U$ for all $y \in F$.]

HISTORICAL NOTE

(Numbers in brackets refer to the bibliography at the end of this note.)

The notion of an arbitrary function was almost unknown at the beginning of the nineteenth century. *A fortiori*, the idea of studying sets of functions in general and of endowing them with a topological structure did not appear before Riemann's time (see the Historical Note to Chapter I), and it was only towards the end of the nineteenth century that this idea came into systematic use.

Nevertheless, the idea of convergence of a *sequence* of real-valued functions had been used, more or less consciously, since the beginnings of the infinitesimal calculus. Of course, by convergence we mean here *pointwise* convergence; other types of convergence could not have been described until the notions of a convergent series and a continuous function had been precisely defined by Bolzano and Cauchy. The latter at first did not recognize the distinction between pointwise convergence and uniform convergence, and believed that he had proved that the sum of any convergent series of continuous functions is continuous ([1], (2), vol. 3, p. 120). The error was pointed out almost immediately by Abel, who at the same time showed that every power series is continuous inside its interval of convergence, by a classical argument which, in this particular case, used essentially the notion of uniform convergence ([2], pp. 223-224). It remained only to formulate this notion generally, and this was done independently by Stokes and Seidel in 1847-1848, and by Cauchy himself in 1853 ([1], (1), vol. 12, p. 30) (*).

Under the influence of Weierstrass and Riemann, the systematic study of the notion of uniform convergence and related questions was developed in the last third of the nineteenth century by the German school (Hankel, du Bois-Raymond) and above all by the Italians; Dini and Arzelà made precise the necessary conditions for the limit of a sequence of continuous functions to be continuous, while Ascoli introduced the fundamental notion of equicontinuity and proved the theorem which characterizes compact sets of continuous functions [3] (a theorem popularized later by Montel in his theory of "normal families", which are relatively compact sets of analytic functions).

(*) In a work dated 1841 but first published in 1894 ([4a], p. 67), Weierstrass uses with perfect clarity the notion of uniform convergence (which he calls by this name for the first time) for power series in one or several complex variables.

On the other hand, Weierstrass himself discovered [4*b*] the possibility of uniform approximation of a continuous real-valued function in one or more real variables on a bounded set by polynomials. This result immediately aroused lively interest and led to many "quantitative" studies (*).

The modern contribution to these questions has been above all to endow them with the full generality of which they are capable, by considering functions whose domain and range are no longer restricted to **R** or finite-dimensional spaces, and thus placing them in their natural context with the help of general topological concepts. In particular, the theorem of Weierstrass, which had already shown itself to be a powerful weapon in classical analysis, has been extended in recent years by M. H. Stone to much more general situations; developing an idea introduced by H. Lebesgue (in a proof of Weierstrass's theorem) he has shown clearly the important part played in the theory of approximation of continuous real-valued functions by lattices of functions (approximation by "lattice polynomials", cf. § 4, Proposition 2 and Theorem 2), and has shown also how the generalized Weierstrass theorem has as immediate consequences a whole series of analogous theorems of approximation, which can thus be grouped together in a much more coherent fashion. We have more or less followed his exposition [5].

BIBLIOGRAPHY

[1] A.-L. Cauchy, *Œuvres*, Paris (Gauthier-Villars), 1882-1932.

[2] N. H. Abel, *Œuvres*, vol. 1, p. 223-224, edited by Sylow and Lie (Christiania) 1881.

[3] G. Ascoli, Sulle curve limiti di una varietà data di curve, *Mem. Accad. Lincei* (3), vol. 18 (1883), pp. 521-586.

[4] K. Weierstrass, *Mathematische Werke*, Berlin (Mayer und Müller), 1894-1903:

a) Zur Theorie der Potenzreihen, Bd. 1, pp. 67-74;

b) Über die analytische Darstellbarkeit sogenannter willkürlicher Functionen reeller Argumente, Bd. 3, pp. 1-37.

[5] M. H. Stone, The generalized Weierstrass approximation theorem, *Mathematics Magazine*, **21** (1948), pp. 167-183 and 237-254.

[6] C. De La Vallée-Poussin, *Leçons sur l'approximation des fonctions d'une variable réelle*, Paris, (Gauthier-Villars), 1919.

(*) See, e.g., C. De La Vallée Poussin [6].

INDEX OF NOTATION

The reference numbers indicate the chapter, section and subsection or exercise, in that order.

INDEX OF TERMINOLOGY

The reference numbers indicate the chapter, section and subsection or Exercise, in that order.

Printing: Mercedesdruck, Berlin
Binding: Buchbinderei Lüderitz & Bauer, Berlin